Lecture Notes in Computer Science 1787

Edited by G. Goos, J. Hartmanis and J. van Leeuwen

Springer
Berlin
Heidelberg
New York
Barcelona
Hong Kong
London
Milan
Paris
Singapore
Tokyo

JooSeok Song (Ed.)

Information Security and Cryptology – ICISC'99

Second International Conference
Seoul, Korea, December 9-10, 1999
Proceedings

Springer

Series Editors

Gerhard Goos, Karlsruhe University, Germany
Juris Hartmanis, Cornell University, NY, USA
Jan van Leeuwen, Utrecht University, The Netherlands

Volume Editor

JooSeok Song
Yonsei University
Department of Computer Science
Seoul, Korea
E-mail: jssong@emerald.yonsei.ac.kr

Cataloging-in-Publication Data applied for

Die Deutsche Bibliothek - CIP-Einheitsaufnahme

Information security and cryptology : second international conference ;
proceddings / ICISC '99, Seoul, Korea, December 9 - 10, 1999.
JooSeok Song (ed.). - Berlin ; Heidelberg ; New York ; Barcelona ;
Hong Kong ; London ; Milan ; Paris ; Singapore ; Tokyo : Springer,
2000
  (Lecture notes in computer science ; Vol. 1787)
  ISBN 3-540-67380-6

CR Subject Classification (1991): E.3, G.2.1, D.4.6, K.6.5, F.2.1-2, C.2, J.1

ISSN 0302-9743
ISBN 3-540-67380-6 Springer-Verlag Berlin Heidelberg New York

Springer-Verlag is a company in the BertelsmannSpringer publishing group.
© Springer-Verlag Berlin Heidelberg 2000

Typesetting: Camera-ready by author, data conversion by Christian Grosche
Printed on acid-free paper   SPIN 10719994   06/3142   5 4 3 2 1 0

Preface

The 2nd International Conference on Information Security and Cryptology (ICISC) was sponsored by the Korea Institute of Information Security and Cryptology (KIISC). It took place at Korea University, Seoul, Korea, December 9-10, 1999. Jong In Lee of Korea University was responsible for the organization.

The call for papers brought 61 papers from 10 countries on four continents. As in the last year the review process was totally blind. The information about authors or their affiliation was not given to Technical Program Committee (TPC) members. Each TPC member was random-coded and did not even know who was reviewing which paper. The 23 TPC members finally selected 20 top-quality papers for presentation at ICISC 1999 together with one invited talk. Serge Vaudenay gave an invited talk on "Provable Security for Conventional Cryptography".

Many people contributed to ICISC'99. First of all I would like to thank all the authors who submitted papers. I am grateful to the TPC members for their hard work reviewing the papers and the Organization Committee members for all the supporting activities which made ICISC'99 a success. I would like to thank the Ministry of Information and Communication of Korea (MIC) which financially sponsored ICISC'99. Special thanks go to Pil Joong Lee and Heung Youl Youm who helped me during the whole process of preparation for the conference. Last, but not least, I thank my students, KyuMan Ko, Sungkyu Chie, and Chan Yoon Jung.

December 1999   Jooseok Song

ICISC'99

December 9-10, 1999, Korea University, Seoul, Korea

The 2nd International Conference on
Information Security and Cryptology

Sponsored by
Korea Institute of Information Security and Cryptology
(KIISC)

In cooperation with
Korea Information Security Agency
(KISA)

Under the patronage of the
Ministry of Information and Communication (MIC), Korea

General Chair

Kil-Hyun Nam (President of KIISC, Korea)

Technical Program Committee

Zongduo Dai (Academica Sinica, P.R.C.)
Ed Dawson (Queensland University of Technology, Australia)
Tzonelih Hwang (National Cheng-Kung University, Taiwan, R.O.C.)
Chul Kim (Kwangwoon University, Korea)
Kwangjo Kim (Information and Communication University, Korea)
Kaoru Kurosawa (Tokyo Institute of Technology, Japan)
Kwok-Yan Lam (National University of Singapore)
Koung Goo Lee (KISA, Korea)
Pil Joong Lee (Pohang University of Science & Technology, Korea)
Chae Hoon Lim (Future Systems Incorporation, Korea)
Jong In Lim (Korea University, Korea)
Chris Mitchell (University of London, U.K.)
Sang Jae Moon (Kyungpook National University, Korea)
Kaisa Nyberg (Nokia Research Center, Finland)
Eiji Okamoto (JAIST, Japan)
Tatsuaki Okamoto (NTT, Japan)
Choon Sik Park (ETRI, Korea)
Sung Jun Park (KISA, Korea)
Bart Preneel (Katholieke Universiteit Leuven, Belgium)
Dong Ho Won (Sungkyunkwang University, Korea)
Heung Youl Youm (Soonchunhyan University, Korea)
Moti Yung (CertCo, U.S.A.)
Yuliang Zheng (Monash University, Australia)

Organizing Committee

Jong In Lim (Korea University)
Sang Kyu Park (HanYang University)
Ha Bong Chung (HongIk University)
Dong Hoon Lee (Korea University)
Sang Jin Lee (Korea University)
Howang Bin Ryou (KwangWoon University)
Seok Woo Kim (HanSei University)
Yong Rak Choi (Taejon University)
Jae Moung Kim (ETRI)
Hong Sub Lee (KISA)
Seung Joo Han (ChoSun University)
Min Surp Rhee (DanKook University)
Seog Pal Cho (SeongGyul University)
Kyung Seok Lee (KIET)
Jong Seon No (Seoul National University)

Table of Contents

Cryptographic Protocol and Authentication Design

Digital Signature and Secret Sharing Scheme

Electronic Cash, Application, Implementation

On Provable Security for Conventional Cryptography

Serge Vaudenay

Swiss Federal Institute of Technologies (EPFL)
Serge.Vaudenay@epfl.ch

Abstract Many previous results on the provable security of conventional cryptography have been published so far. We provide here handy tools based on Decorrelation Theory for dealing with them and we show how to make their proof easier. As an illustration we survey a few of these results and we (im)prove some by our technique.

This paper covers results on pseudorandomness of some block cipher constructions and on message authentication code constructions.

Decorrelation theory was introduced in [18]–[25]. Its first aim was to address provable security in the area of block ciphers in order to prove their security against differential and linear cryptanalysis. As a matter of fact, these techniques can also be used for other areas of conventional cryptography as shown in this paper.

In [25] was noticed that decorrelation distances of some integral order d was linked to the advantage of the best attacks which is limited to d samples in several classes of attacks. Namely, non-adaptive attacks was characterized by decorrelation distance with the $|||.|||_\infty$ norm, chosen input attacks was characterized by decorrelation distance with the $||.||_a$ norm, and chosen input and output attacks was characterized by decorrelation distance with the $||.||_s$ norm. This can be used to address provable security of, say, MAC construction schemes. Due to nice properties of decorrelation distances, some previous results turn out to get simpler and more systematic.

A similar systematic approach was recently addressed by Maurer [11] and it would be interesting to compare both approaches.

1 Definitions and Properties

This section recalls basic facts in decorrelation theory.

1.1 Definitions and Notations

First of all, for any random function F from a set $\mathcal{M}_1$ to a set $\mathcal{M}_2$ and any integer d we associate the "d-wise distribution matrix" which is denoted $[F]^d$, defined in the matrix set $\mathbf{R}^{\mathcal{M}_1^d \times \mathcal{M}_2^d}$ by

$$[F]^d_{(x_1,\ldots,x_d),(y_1,\ldots,y_d)} = \Pr[F(x_1) = y_1, \ldots, F(x_d) = y_d].$$

JooSeok Song (Ed.): ICISC'99, LNCS 1787, pp. 1–16, 2000.
© Springer-Verlag Berlin Heidelberg 2000

Given a metric structure D in $\mathbf{R}^{\mathcal{M}_1^d \times \mathcal{M}_2^d}$ we can define the distance between the matrices associated to two random functions F and G. This is the "d-wise decorrelation distance". If G is a random function uniformly distributed in the set of all functions from $\mathcal{M}_1$ to $\mathcal{M}_2$ (we let F^* denote such a function), this distance is called the "d-wise decorrelation bias of function F" and denoted $\mathrm{DecF}_D^d(F)$. When F is a permutation (which will usually be denoted C as for "Cipher") and G is a uniformly distributed permutation (denoted C^*) it is called the "d-wise decorrelation bias of permutation F" and denoted $\mathrm{DecP}_D^d(F)$. In previous results we used the metric structures defined by the norms denoted $\|.\|_2$ (see [20]), $\||.\||_\infty$, $\|.\|_a$, $\|.\|_s$ (see [25]). These four norms are matrix norms, which means that they are norms on $\mathbf{R}^{\mathcal{M}_1^d \times \mathcal{M}_2^d}$ with the property that

$$\|A \times B\| \leq \|A\|.\|B\|.$$

This property leads to non-trivial inequalities which can shorten many treatments on the security of conventional cryptography.

Given two random functions F and G from $\mathcal{M}_1$ to $\mathcal{M}_2$ we call "distinguisher between F and G" any oracle Turing machine $\mathcal{A}^O$ which can send $\mathcal{M}_1$-element queries to the oracle O and receive $\mathcal{M}_2$-element responses, and which finally output 0 or 1. In particular the Turing machine can be probabilistic. In the following, the number of queries to the oracle will be limited to d. The distributions on F and G induces a distribution on $\mathcal{A}^F$ and $\mathcal{A}^G$, thus we can compute the probability that these probabilistic Turing machines output 1. The advantage for distinguishing F from G is

$$\mathrm{Adv}_\mathcal{A}(F, G) = \Pr\left[\mathcal{A}^F = 1\right] - \Pr\left[\mathcal{A}^G = 1\right].$$

We consider the class Cl_{na}^d of non-adaptive distinguishers limited to d queries, which are distinguishers who must commit to d queries before receiving the responses. Similarly, we consider its extension Cl_a^d of all distinguishers limited to d queries. For instance, these distinguishers can choose the second query with the hint of the first response. Finally, when F and G are permutations, we also consider the extension Cl_s^d of distinguishers limited to d queries but who can query either the function F/G or its inverse F^{-1}/G^{-1}. For any class of distinguishers Cl we will denote

$$\mathrm{BestAdv}_{\mathrm{Cl}}(F, G) = \max_{\mathcal{A} \in \mathrm{Cl}} \mathrm{Adv}_\mathcal{A}(F, G).$$

We notice that if $\mathcal{A}$ is a distinguisher, we can always define a complementary distinguisher $\bar{\mathcal{A}} = 1 - \mathcal{A}$ which gives the opposite output. There is no need for investigating the minimum advantage when the class is closed under the complement (which is the case of the above classes) since

$$\mathrm{Adv}_{\bar{\mathcal{A}}}(F, G) = -\mathrm{Adv}_\mathcal{A}(F, G).$$

1.2 Properties

The d-wise distribution matrices have the property that if F and G are independent random functions, F from $\mathcal{M}_2$ to $\mathcal{M}_3$ and G from $\mathcal{M}_1$ to $\mathcal{M}_2$, then

$$[F \circ G]^d = [G]^d \times [F]^d.$$

As an illustrative consequence, if F^* is a uniformly distributed random function, we obtain that

$$[F]^d \times [F^*]^d = [F^*]^d \times [F]^d = [F^*]^d.$$

Thus, if we are using a matrix norm $\|.\|$, we obtain

$$\text{DecF}^d_{\|.\|}(F \circ G) \leq \text{DecF}^d_{\|.\|}(F).\text{DecF}^d_{\|.\|}(G).$$

and the same for permutations. In the sequel we will refer to the multiplicative property of decorrelation biases.

The $\||.\||_\infty$, $\|.\|_a$, $\|.\|_s$ have the quite interesting property that they characterize the best advantage of a distinguisher in Cl^d_{na}, Cl^d_a or Cl^d_s.

Lemma 1 ([18,25]). *For any random functions F and G we have*

$$\||[F]^d - [G]^d\||_\infty = 2.\operatorname*{BestAdv}_{\text{Cl}^d_{na}}(F, G)$$

$$\|[F]^d - [G]^d\|_a = 2.\operatorname*{BestAdv}_{\text{Cl}^d_a}(F, G)$$

and when F and G are permutations we also have

$$\|[F]^d - [G]^d\|_s = 2.\operatorname*{BestAdv}_{\text{Cl}^d_s}(F, G).$$

This is quite a useful property which may lead to some non-trivial inequalities. For example, if $F_1, ..., F_r$ are r independent identically distributed random functions from $\mathcal{M}$ to itself, and if F^* denotes a uniformly distributed random function, we have

$$\operatorname*{BestAdv}_{\text{Cl}^d_a}(F_1 \circ ... \circ F_r, F^*) \leq 2^{r-1}.\left(\operatorname*{BestAdv}_{\text{Cl}^d_a}(F_1, F^*)\right)^r.$$

There is a simple link between $\|.\|_s$ and $\|.\|_a$. Actually, if for any random permutation C on $\mathcal{M}$ we let $\bar{C}$ be defined from $\{0, 1\} \times \mathcal{M}$ to $\mathcal{M}$ by

$$\bar{C}(0, x) = C(x) \quad \text{and} \quad \bar{C}(1, x) = C^{-1}(x).$$

Obviously we have

$$\|[C_1]^d - [C_2]^d\|_s = \|[\bar{C}_1]^d - [\bar{C}_2]^d\|_a. \tag{1}$$

Finally we recall the following lemma.

Lemma 2 ([25]). *Let d be an integer, $F_1, \ldots, F_r$ be r independent random function oracles, and $C_1, \ldots, C_s, D_1, \ldots, D_t$ be $s+t$ independent random permutation oracles. We let $\Omega^{F_1, \ldots, F_r, C_1, \ldots, C_s, D_1, \ldots, D_t}$ be an oracle which can access to the previous oracles and from each query x defines an output $G(x)$. We assume that Ω is such that the number of queries to F_i and C_j is limited to some integer a_i and b_j respectively, and the number of queries to D_k or D_k^{-1} is limited to c_k in total for any $i = 1, \ldots, r$, $j = 1, \ldots, s$ and $k = 1, \ldots, t$. We let the F_i^* (resp. C_j^*, D_k^*) be independent uniformly distributed random functions (resp. permutations) on the same range than F_i (resp. C_j, D_k) and we let G^* the function defined by $\Omega^{F_1^*, \ldots, F_r^*, C_1^*, \ldots, C_s^*, D_1^*, \ldots, D_t^*}$. We have*

$$\mathrm{DecF}^d_{||\cdot||_a}(G) \leq \sum_{i=1}^{r} \mathrm{DecF}^{a_i d}_{||\cdot||_a}(F_i) + \sum_{j=1}^{s} \mathrm{DecP}^{b_j d}_{||\cdot||_a}(C_j) + \sum_{k=1}^{t} \mathrm{DecP}^{c_k d}_{||\cdot||_a}(D_k) +$$
$$\mathrm{DecF}^d_{||\cdot||_a}(G^*).$$

In addition, if the Ω construction defines a permutation G, assuming that computing G^{-1} leads to the same a_i, b_j and c_k limits, we have

$$\mathrm{DecF}^d_{||\cdot||_s}(G) \leq \sum_{i=1}^{r} \mathrm{DecF}^{a_i d}_{||\cdot||_s}(F_i) + \sum_{j=1}^{s} \mathrm{DecP}^{b_j d}_{||\cdot||_s}(C_j) + \sum_{k=1}^{t} \mathrm{DecP}^{c_k d}_{||\cdot||_s}(D_k) +$$
$$\mathrm{DecF}^d_{||\cdot||_s}(G^*).$$

This lemma actually separates the problem of studying the decorrelation bias of a construction scheme into the problem of studying the decorrelation biases of its internal primitives F_i, C_j and D_k and studying the decorrelation bias of an ideal version G^* with truly random functions inside.

2   Randomness of Cryptographic Primitives Designs

2.1   The Luby–Rackoff Result

Many previous papers on cryptography investigated how much randomness provides such or such design scheme. The Luby–Rackoff result [8] addressed the Feistel construction [5] with truly random rounds. By translating it into our formalism, it turns into the following lemma.

Lemma 3 (Luby–Rackoff 1986 [8]). *Let $F_1^*, F_2^*, F_3^*, F_4^*$ be four independent random function on $\{0,1\}^{\frac{m}{2}}$ with uniform distribution. We have*

$$\mathrm{DecF}^d_{||\cdot||_a}(\Psi(F_1^*, F_2^*, F_3^*)) \leq 2d^2 . 2^{-\frac{m}{2}}$$
$$\mathrm{DecP}^d_{||\cdot||_a}(\Psi(F_1^*, F_2^*, F_3^*)) \leq 2d^2 . 2^{-\frac{m}{2}}$$
$$\mathrm{DecP}^d_{||\cdot||_s}(\Psi(F_1^*, F_2^*, F_3^*, F_4^*)) \leq 2d^2 . 2^{-\frac{m}{2}}.$$

The results hold for Feistel schemes defined from any (quasi)group operation.

(Here $\Psi(F_1, \ldots, F_r)$ is the standard notation introduced by Luby and Rackoff in order to denote a Feistel scheme in which the ith round function is F_i.) This result can be used together with Lemma 2 in order to study the decorrelation bias of a Feistel scheme with independent rounds. This is the basis of the Peanut construction over which the AES candidate DFC [2,6,7] is based.

2.2 General Approach

Many other similar results (and extensions) followed the Luby–Rackoff paper. For instance Zheng-Matsumoto-Imai [27] and Patarin's Thesis [13]. We show here how we can make this kind of result easier by using a technique inspired by both Patarin's "H coefficient method" [13,14] and Maurer [10].

We aim to upper bound the decorrelation bias of a given random function. Our paradigm consists in first proving a combinatorial lemma by using the structure of the random function then using a standard proof given by the following lemma.

Lemma 4. *Let d be an integer. Let F be a random function from a set $\mathcal{M}_1$ to a set $\mathcal{M}_2$. We let $\mathcal{X}$ be the subset of $\mathcal{M}_1^d$ of all $(x_1, \ldots, x_d)$ with pairwise different entries. We let F^* be a uniformly distributed random function from $\mathcal{M}_1$ to $\mathcal{M}_2$. We know that for all $x \in \mathcal{X}$ and $y \in \mathcal{M}_2^d$ the value $[F^*]_{x,y}^d$ is a constant $p_0 = (\#\mathcal{M}_2)^{-d}$. We assume there exists a subset $\mathcal{Y} \subseteq \mathcal{M}_2^d$ and two positive real values ϵ_1 and ϵ_2 such that*

- $|\mathcal{Y}| p_0 \geq 1 - \epsilon_1$
- $\forall x \in \mathcal{X} \quad \forall y \in \mathcal{Y} \quad [F]_{x,y}^d \geq p_0(1 - \epsilon_2)$.

Then we have $\mathrm{DecF}_{||.||_a}^d(F) \leq 2\epsilon_1 + 2\epsilon_2$.

This lemma intuitively means that if $[F]_{x,y}^d$ is close to $[F^*]_{x,y}^d$ for all x and almost all y, then the decorrelation bias of F is small.

We have a twin lemma for the $||.||_s$ norm. Here, since we can query y as well, the approximation must hold for all x and y.

Lemma 5. *Let d be an integer. Let C be a random permutation on a set $\mathcal{M}$. We let $\mathcal{X}$ be the subset of $\mathcal{M}^d$ of all $(x_1, \ldots, x_d)$ with pairwise different entries. We let F^* be a uniformly distributed random function on $\mathcal{M}$. We let C^* be a uniformly distributed random permutation on $\mathcal{M}$. We have*

- *if $[C]_{x,y}^d \geq [C^*]_{x,y}^d(1 - \epsilon)$ for all x and y in $\mathcal{X}$ then $\mathrm{DecP}_{||.||_s}^d(F) \leq 2\epsilon$*
- *if $[C]_{x,y}^d \geq [F^*]_{x,y}^d(1 - \epsilon)$ for all x and y in $\mathcal{X}$ then $\mathrm{DecP}_{||.||_s}^d(F) \leq 2\epsilon + 2d^2(\#\mathcal{M})^{-1}$.*

Proof. We use the characterization of $\mathrm{DecF}_{||.||_a}^d$ in term of best adaptive distinguisher. We let $\mathcal{A}$ be one d-limited distinguisher between F and F^* with maximum advantage. We can assume w.l.o.g. that all queries to the oracle are pairwise different (we can simulate the distinguisher by replacing repeated queries by dummy queries). The behavior of $\mathcal{A}$ is deterministically defined by the initial random tape ω and the oracle responses $y = (y_1, \ldots, y_d)$. We let x_i denotes the ith query defined by (ω, y). It actually depends on ω and $y_1, \ldots, y_{i-1}$ only. We let $x = (x_1, \ldots, x_d)$ which is assumed to be in $\mathcal{X}$. We let A be the set of all (ω, y) for which $\mathcal{A}$ outputs 0. It is straightforward that

$$\text{Adv}_{\mathcal{A}}(F, F^*) = - \sum_{(\omega, y) \in A} \Pr[\omega] \left([F]_{x,y}^d - [F^*]_{x,y}^d\right).$$

Next we have

$$\text{Adv}_{\mathcal{A}}(F, F^*) \leq \sum_{\substack{(\omega, y) \in A \\ y \in \mathcal{Y}}} \Pr[\omega] \epsilon_2 [F^*]_{x,y}^d + \sum_{\substack{(\omega, y) \in A \\ y \notin \mathcal{Y}}} \Pr[\omega][F^*]_{x,y}^d.$$

By relaxing the (ω, y) in the first sum, we observe that it is upper bounded by ϵ_2. (We just have to sum the y_js backward, starting by summing all y_ds, then y_{d-1}, ...) For the second sum, we recall that all x_is are pairwise different, so $[F^*]_{x,y}^d$ is always equal to p_0. This sum is thus less than ϵ_1.

For the $||.||_s$ norm, we simply use the $\bar{F}$ function as in Equation (1). In the case where we approximate $[C]_{x,y}^d$ by $[F^*]_{x,y}^d$, we just notice that

$$[C^*]_{x,y}^d - [F^*]_{x,y}^d \leq [F^*]_{x,y}^d d^2 (\#\mathcal{M})^{-1}.$$

$\square$

As an example of application (which will be used later on) we prove the following lemma.

Lemma 6. *For a random uniformly distributed function F^* and a random uniformly distributed permutation C^* defined over $\{0,1\}^m$, we have*

$$\text{DecP}^d(F^*) = \text{DecF}^d(C^*) \leq d(d-1)2^{-m}.$$

Proof. We use Lemma 4 for $\text{DecF}^d(C^*)$. We let $\mathcal{Y}$ be equal to the full set of pairwise different outputs (we have $\epsilon_1 = 0$). We have

$$1 - \frac{[C^*]_{x,y}^d}{p_0} = 1 - \frac{1}{(1 - 2^{-m}) \ldots (1 - (d-1).2^{-m})} \leq \frac{d(d-1)}{2} 2^{-m}$$

which gives ϵ_2.

$\square$

We can now prove Lemma 3 by using Lemma 4, 5 and 6 but with a quite compact proof.

Proof (Lemma 3). Following the Feistel scheme $F = \Psi(F_1^*, F_2^*, F_3^*)$, we let

$$x_i = (z_i^0, z_i^1)$$
$$z_i^2 = z_i^0 + F_1^*(z_i^1)$$
$$y_i = (z_i^4, z_i^3)$$

We let E be the event $z_i^3 = z_i^1 + F_2^*(z_i^2)$ and $z_i^4 = z_i^2 + F_3^*(z_i^3)$ for $i = 1, \ldots, d$. We thus have $[F]_{x,y}^d = \Pr[E]$. We now define

$$\mathcal{Y} = \left\{ (y_1, \ldots, y_d); \forall i < j \quad z_i^3 \neq z_j^3 \right\}.$$

(This is a set of non-pathologic outputs when computing $[F]_{x,y}^d$.) We can easily check that $\mathcal{Y}$ fulfill the requirements of Lemma 4. Firstly we have

$$|\mathcal{Y}| \geq \left(1 - \frac{d(d-1)}{2} 2^{-\frac{m}{2}} \right) 2^{md}$$

thus we let $\epsilon_1 = \frac{d(d-1)}{2} 2^{-\frac{m}{2}}$. Second, for $y \in \mathcal{Y}$ and any x (with pairwise different entries), we need to consider $[F]_{x,y}^d$. Let E^2 be the event that all z_i^2s are pairwise different over the distribution of F_1^*. We have

$$[F]_{x,y}^d \geq \Pr[E/E^2] \Pr[E^2].$$

For computing $\Pr[E/E^2]$ we know that z_i^3s are pairwise different, as for the z_i^2s. Hence $\Pr[E/E^2] = 2^{-md}$. It is then straightforward that $\Pr[E^2] \geq 1 - \frac{d(d-1)}{2} 2^{-\frac{m}{2}}$ which is $1 - \epsilon_2$. We thus obtain from Lemma 4 that $\mathrm{DecF}_{||.||_a}^d(F) \leq 2d(d-1) 2^{-\frac{m}{2}}$. From this and Lemma 6 we thus obtain $\mathrm{DecP}_{||.||_a}^d(F) \leq 2d^2 2^{-\frac{m}{2}}$ for $d \leq 2^{1+\frac{m}{2}}$. Since DecF is always less than 2, it also holds for larger d.

Thanks to Lemma 5, the $||.||_s$-norm case with $C = \Psi(F_1^*, F_2^*, F_3^*, F_4^*)$ is fairly similar. We let

$$x_i = (z_i^0, z_i^1)$$
$$z_i^2 = z_i^0 + F_1^*(z_i^1)$$
$$y_i = (z_i^5, z_i^4)$$
$$z_i^3 = z_i^5 - F_4^*(z_i^4)$$

for $i = 1, \ldots, d$. We have

$$[F]_{x,y}^d = \Pr \left[\begin{matrix} z_i^3 = z_i^1 + F_2^*(z_i^2) \\ z_i^4 = z_i^2 + F_3^*(z_i^3) \end{matrix} ; i = 1, \ldots, d \right].$$

Let E be this event. If we let E^2 (resp. E^3) be the event that all z_i^2 (resp. z_i^3) are pairwise different, then

$$[F]_{x,y}^d \geq \Pr[E/E^2, E^3] \Pr[E^2, E^3].$$

The E^2 and E^3 events are independent, with probability greater than $1 - \frac{d(d-1)}{2} 2^{-\frac{m}{2}}$ which is $1 - \epsilon$. The probability of E when E^2 and E^3 hold is obviously $2^{-md} = [F^*]_{x,y}^d$. So, $\mathrm{DecP}_{||.||_s}^d(F) \leq 2d^2 . 2^{-\frac{m}{2}}$. $\square$

2.3 Random Permutations Collection

We collect here a few results taken from the literature, which can be proven by our techniques.

First of all, Zheng-Matsumoto-Imai investigate the pseudorandomness of generalized Feistel transformations. Originally with two branches, they consider having k branches. The first generalization ("type-1 transformation") $\psi_1(f_1, \ldots, f_r)$ is defined by

$$\psi_1(f_1, \ldots, f_r)(x_1, \ldots, x_k) =$$
$$\psi_1(f_2, \ldots, f_r)(f_1(x_1) + x_2, x_3, x_4, \ldots, x_k, x_1).$$

The second generalization ("type-2 transformation") $\psi_2(f_1, \ldots, f_r)$ is similarly defined for k even and r multiple of $\frac{k}{2}$ by

$$\psi_2(f_1, \ldots, f_r)(x_1, \ldots, x_k) =$$
$$\psi_2(f_{\frac{k}{2}+1}, \ldots, f_r)(f_1(x_1) + x_2, x_3, f_2(x_3) + x_4, x_5, \ldots, f_{\frac{k}{2}}(x_{k-1}) + x_k, x_1).$$

And finally the "type-3 transformation" $\psi_3(f_1, \ldots, f_r)$ for r multiple of k:

$$\psi_3(f_1, \ldots, f_r)(x_1, \ldots, x_k) =$$
$$\psi_3(f_k, \ldots, f_r)(f_1(x_1) + x_2, f_2(x_2) + x_3, \ldots, f_{k-1}(x_{k-1}) + x_k, x_1).$$

(We also define $\Psi_\ell()(x) = y$ for $\ell = 1, 2, 3$.)

Lemma 7 (Zheng-Matsumoto-Imai 1989 [27]). *We consider the previous generalizations of Feistel schemes with k branches on $\{0,1\}^m$. For independent uniformly distributed random functions $F_1^*, \ldots, F_{2k-1}^*$ and an integer d, we have*

$$\mathrm{DecP}^d_{\|.\|_a}(\Psi_1(F_1^*, \ldots, F_{2k-1}^*)) \leq 2(k-1)d^2.2^{-\frac{m}{k}}$$
$$\mathrm{DecP}^d_{\|.\|_a}(\Psi_2(F_1^*, \ldots, F_{k^2-1}^*)) \leq \frac{k^2}{2}d^2.2^{-\frac{m}{k}}$$
$$\mathrm{DecP}^d_{\|.\|_a}(\Psi_3(F_1^*, \ldots, F_{k^2-1}^*)) \leq (k^2 - 2k + 2)d^2.2^{-\frac{m}{k}}.$$

The results hold for generalized Feistel schemes defined from any (quasi)group operation.[1]

(Note that they all generalize Lemma 3 for which $k = 2$.)

Proof (Sketch). We use Lemma 4 for evaluating DecF.

For Ψ_1 we let $\mathcal{Y}$ be the set of all $y = (y_1, \ldots, y_d)$ where $y_i = (y_i^1, \ldots, y_i^k)$ such that we have $y_i^j = y_{i'}^j$ for no $j > 1$ and $i < i'$. We get $\epsilon_1 = (k-1)\frac{d(d-1)}{2}2^{-\frac{m}{k}}$.

---

[1] Here we slightly improved the original result for Ψ_3 from [27] which is

$$\mathrm{DecP}^d_{\|.\|_a}(\Psi_3(F_1^*, \ldots, F_{k^2-1}^*)) \leq k(k-1)d^2.2^{-\frac{m}{k}}.$$

We then consider the event in which the first entry after the $k-1$th round takes pairwise different values for $x_1, \ldots, x_d$. Upper bounding the probability when this event occurs we get $\epsilon_2 = (k-1)\frac{d(d-1)}{2}2^{-\frac{m}{k}}$. Thus $\mathrm{DecF}^d(F) \leq 2(k-1)d(d-1)2^{-\frac{m}{k}}$.

Similarly, for Ψ_2 we let $\mathcal{Y}$ be the set of all y such that we have $y_i^j = y_{i'}^j$ for no even j and $i < i'$. We get $\epsilon_1 = \frac{k}{2} \times \frac{d(d-1)}{2}2^{-\frac{m}{k}}$. We consider the event in which all odd entries after the $k-1$th round takes pairwise different values for $x_1, \ldots, x_d$. We get $\epsilon_2 = \frac{k}{2}(k-1) \times \frac{d(d-1)}{2}2^{-\frac{m}{k}}$. Thus $\mathrm{DecF}^d(F) \leq \frac{k^2}{2}d(d-1)2^{-\frac{m}{k}}$.

For Ψ_3 we let $\mathcal{Y}$ be the set of all y such that we have $y_i^k = y_{i'}^k$ for no $i < i'$. We get $\epsilon_1 = \frac{d(d-1)}{2}2^{-\frac{m}{k}}$. We consider the event in which all first $k-1$ entries after the $k-1$th round take pairwise different values for $x_1, \ldots, x_d$. We get $\epsilon_2 = (k-1)^2 \frac{d(d-1)}{2}2^{-\frac{m}{k}}$. Thus $\mathrm{DecF}^d(F) \leq (k^2 - 2k + 2)d(d-1)2^{-\frac{m}{k}}$.

In the three cases, ϵ_2 is evaluated as the number of unexpected equalities between two outputs from a single circuit of depth $k-1$ with k inputs and internal F_j^* and additions times the probability it occurs, which is at most the depth $k-1$ times $2^{-\frac{m}{k}}$.

Now to get DecP from DecF, from Lemma 6 and the triangular inequality we have

$$\mathrm{DecP}^d(F) \leq \mathrm{DecF}^d(F) + \mathrm{DecP}^d(F^*) \leq \mathrm{DecF}^d(F) + d^2 2^{-m}.$$

We then notice that the obtained upper bounds for DecF are always $\mathrm{DecF}^d(F) \leq Ad(d-1)2^{-\frac{m}{k}}$ for some $A \geq 2$. For $d \leq A2^{m-\frac{m}{k}}$ we thus obtain $\mathrm{DecP}^d(F) \leq Ad^2 2^{-\frac{m}{k}}$. For larger d, this bound is greater than $A^3 2^{m(2-\frac{3}{k})}$ which is greater than 8 since $m \geq k \geq 2$. Since $\mathrm{DecP}^d(F)$ is always less than 2, the bound is thus still valid. $\qquad\square$

One problem proposed by Schnorr in the late 80's was to extend the Luby–Rackoff result with a single random function. This has been first solved by Pieprzyk when he show that $\Psi(F^*, F^*, F^*, F^* \circ F^*)$ is pseudorandom [16]. In [14], Patarin improved this result by using less rounds by showing that $\Psi(F^*, F^*, F^* \circ \zeta \circ F^*)$ is pseudorandom when ζ is a special fixed function (like for instance the bitwise circular rotation).

Lucks [9] adopted a different approach for reducing the number of random function in Lemma 3: instead of making all functions depend on each other, he tried to "derandomize" the first and last function and to use imbalance Feistel schemes (with branches with different size). This work was followed by Naor and Reingold's [12] who proved the following result.

Lemma 8 (Naor–Reingold 1999 [12]). *Let C_1 and C_2 be two random permutations such that $\mathrm{DecP}^1(C_1) = 0$ and $\mathrm{DecP}^2_{|||.|||_\infty}(C_i) \leq \delta_i$ for $i = 1, 2$. Let F^* be a uniformly distributed random function. We assume that C_1, C_2 and F^* are independent. For any integer d, we have*

$$\mathrm{DecP}^d_{||.||_s}(C_2^{-1} \circ \Psi(F^*, F^*) \circ C_1) \leq d(d-1)\frac{\delta_1 + \delta_2}{2} + 4d^2 \cdot 2^{-\frac{m}{2}}.$$

The result holds for Feistel schemes defined from any (quasi)group operation.[2]

The $\mathrm{DecP}^1(C_1) = 0$ requirement is quite easy to achieve since for any random C_0 we can construct $C_1(x) = C_0(x) + K$ with an independent uniformly distributed K and we get the requirement. (For any quasigroup addition.)

Proof (sketch). For $C_2^{-1} \circ \Psi(F^*, F^*) \circ C_1(x_i) = y_i$ we let u_i and v_i denote the inputs of the two F^* functions, which are the right half of $C_1(x_i)$ and the left half of $C_2(y_i)$ respectively. Applying Lemma 5 we consider the event that all u_i and v_j are pairwise different. We get

$$\epsilon = \sum_{i<j} \Pr[u_i = u_j] + \sum_{i<j} \Pr[v_i = v_j] + \sum_{i,j} \Pr[u_i = v_j] + d^2 . 2^{-m}.$$

Let us first consider $\Pr[u_i = u_j]$. We have

$$\Pr[u_i = u_j] = \sum_{\xi_i \neq \xi_j} \Pr[x_i = \xi_i, x_j = \xi_j] \Pr[u(\xi_i) = u(\xi_j)]$$

where $u(\xi)$ denotes the right half of $C_1(\xi)$ and the probabilities $\Pr[x_i = \xi_i, x_j = \xi_j]$ are taken over the distribution of the distinguisher and the oracle. Hence we have

$$\Pr[u_i = u_j] \leq \max_{\xi_i \neq \xi_j} \Pr[u(\xi_i) = u(\xi_j)].$$

Now we can see that $u(\xi_i) = u(\xi_j)$ defines a distinguisher for C_1 with two non adaptive deterministic queries ξ_i and ξ_j. Hence From $\mathrm{DecP}^2(C_1)_{|||.|||_\infty} \leq \delta_1$ we can get $\Pr[u_i = u_j] \leq \frac{1}{2}\delta_1 + 2^{-\frac{m}{2}}$. The same holds for $\Pr[v_i = v_j]$. Similarly, since $\mathrm{DecP}^1(C_1) = 0$ and u_i and v_j are independent we have $\Pr[u_i = v_j] = 2^{-\frac{m}{2}}$. Hence

$$\epsilon \leq \frac{d(d-1)}{2} \times \frac{\delta_1 + \delta_2}{2} + d(2d-1).2^{-\frac{m}{2}} + d^2.2^{-m}.$$

Thus the result holds when $2d \leq 2^{\frac{m}{2}}$. Since the bound is greater than $2^{\frac{m}{2}}$ in the other case, it also holds for $m > 1$. The trivial $m = 1$ case is later solved by inspection. $\square$

---

[2] The original result from [12] was not stated with the same hypothesis. The first result corresponds with $\delta_1 = \delta_2 = 0$ and states

$$\mathrm{DecP}^d_{||.||_\bullet}(C_2^{-1} \circ \Psi(F^*, F^*) \circ C_1) \leq 4d^2.2^{-\frac{m}{2}} + 2d^2.2^{-m}.$$

The second result needs two independent random functions F_1^* and F_2^* and the hypothesis that for any fixed $x \neq y$ the right (resp. left) halves of $C_1(x)$ and $C_1(y)$ (resp. $C_2(x)$ and $C_2(y)$) are equal with probability less than δ. It states

$$\mathrm{DecP}^d_{||.||_\bullet}(C_2^{-1} \circ \Psi(F_1^*, F_2^*) \circ C_1) \leq 2d^2\delta + 2d^2.2^{-m}.$$

In addition Naor and Reingold presented an alternate scheme with parallel branches which achieves better decorrelation bias.

Other alternate permutation constructions were investigated, for instance by Sugita [17] based on MISTY-like permutations and in [24] based on IDEA-like permutations.

3 Mode of Operation and Similar Constructions

3.1 CBC-MAC

Another family of results was originated by Bellare-Kilian-Rogaway [4]. This paper consider the regular CBC-MAC construction which transforms a block cipher function C into a message authentication code MAC by

$$\text{MAC}(m_1, \ldots, m_\ell) = C(C(\ldots C(m_1) + m_2 \ldots) + m_\ell).$$

We do not really need invertibility for C, so we can use a random function F instead of C. This construction is provably secure when ℓ is fixed, provided that F is already secure. This informal result can be quantified as follows.

Lemma 9 (Bellare-Kilian-Rogaway 1994 [4]). *For any fixed integer ℓ, we consider the function MAC defined on ℓ m-bit blocks from a uniformly distributed random function F^* as above. For any d we have $\text{DecF}_{||\cdot||_a}^d(\text{MAC}) \leq d\ell(d\ell - 1)2^{-m}$.[3]*

When F is pseudorandom it becomes trivial from this and Lemma 2 that

$$\text{DecF}_{||\cdot||_a}^d(\text{MAC}) \leq \text{DecF}_{||\cdot||_a}^{d\ell}(F) + d\ell(d\ell - 1)2^{-m}.$$

Proof (sketch). By using Lemma 4 again, we take $\mathcal{Y}$ equal to full set of $y = (y_1, \ldots, y_d)$. Thus we have $\epsilon_1 = 0$. Let us define the random variable $U_{i,j}$ as the input of the jth F^* computation on the message $x_i = (m_{i,1}, \ldots, m_{i,\ell})$ respectively. We consider the event E that all $U_{i,j}$ for any i and $j < \ell$ are different from the $U_{k,\ell}$ for any k, and all $U_{k,\ell}$ are pairwise different. As usual we obtain

$$\Pr[\text{MAC}(x_i) = y_i; i = 1, \ldots, d] \geq 2^{-md} \Pr[E]$$

since $\Pr[\text{MAC}(x_i) = y_i; i = 1, \ldots, d/E] = 2^{-md}$. Thus we can take $\epsilon_2 = \Pr[\bar{E}]$.

Let Coll be the event that we have an F^*-collision $F^*(U_{i,j}) = F^*(U_{r,s})$ with $U_{i,j} \neq U_{r,s}$ for some values of i, j, r, s with $j < \ell$ and $r < \ell$. Since the number of queries to F^* for this is $d(\ell - 1)$, we have

$$\Pr[\text{Coll}] \leq \frac{d(\ell - 1)(d(\ell - 1) - 1)}{2} 2^{-m}.$$

Now we have

$$\Pr[\bar{E}] \leq \Pr[\text{Coll}] + \Pr[\bar{E} \text{ and } \overline{\text{Coll}}].$$

---

[3] The original result from [4] is $\text{DecF}_{||\cdot||_a}^d(\text{MAC}) \leq 6d^2\ell^2 2^{-m}$.

Let us study the $U_{i,j} = U_{k,\ell}$ event with $j \leq \ell$ and $i \neq k$ if $j = \ell$ when $\overline{\text{Coll}}$ holds. We let r be the smallest positive integer such that $m_{j-r} \neq m_{\ell-r}$ (with the convention that $m_0 \neq m_{\ell-j}$; we notice that if $j = \ell$, since the ith and the kth messages are different we must have $r < \ell$). Since we have no collision on F^* we must have

- either $F^*(U_{i,j-r-1}) + m_{i,j-r} = F^*(U_{k,\ell-r-1}) + m_{k,\ell-r}$ if $r < j - 1$
- or $m_{i,1} = F^*(U_{k,\ell-r-1}) + m_{k,\ell-r}$ if $r = j - 1 < \ell - 1$
- or $F^*(U_{k,\ell-r}) = 0$ if $r = j$.

Since we have no collision all these events hold with probability at most 2^{-m}. Therefore

$$\Pr[\bar{E}] \leq \frac{d(\ell-1)(d(\ell-1)-1)}{2} 2^{-m} + \left((\ell-1)d^2 + \frac{d(d-1)}{2} \right) 2^{-m}.$$

Thus we can take $\epsilon_2 = \frac{d\ell(d\ell-1)}{2} 2^{-m}$. $\square$

3.2   Similar Results

Lemma 9 means that we can make one Fixed-Input-Length (FIL) MAC from one Single-Block-Input (SBI) MAC. There are other related results. Namely we can consider making Variable-Input-Length (VIL) MAC from FIL-MAC (see [1]), or directly from SBI-MAC. We can consider making VIL-encryption from SBI-encryption[4] (see [3]), SBI-MAC from SBI-encryption, and so on. The technique is basically the same and straightforward from our presentation of Lemma 9.

In order to define decorrelation biases for VIL-MAC, we need to face to the problem of having infinite sets. Let for instance F be a random function defined from $\mathcal{M}_1^*$ to $\mathcal{M}_2$ ($\mathcal{M}_1^*$ is the set of all finite sequences with entries in $\mathcal{M}_1$). We define the $[F]^{q_1,\ldots,q_d}$ matrix with rows defined on $\mathcal{M}_1^{q_1} \times \ldots \times \mathcal{M}_1^{q_d}$ and columns defined on $\mathcal{M}_2^d$. Next we define $\text{DecF}_D^{q_1,\ldots,q_d}(F)$ as the D-distance between $[F]^{q_1,\ldots,q_d}$ and $[F^*]^{q_1,\ldots,q_d}$. We can easily check that all previous theorems remain valid for these definitions. Additionally, we can define

$$\text{DecF}_D^{d,q}(F) = \max_{q_1+\ldots+q_d=q} \text{DecF}_D^{q_1,\ldots,q_d}(F)$$

and still check that $\text{BestAdv}_{\text{Cl}_a^{d,q}}(F, F^*) = \frac{1}{2}\text{DecF}_{||\cdot||_a}^{d,q}(F)$ where $\text{Cl}_a^{d,q}$ is the class of adaptive attacks limited to d queries with a total length of q blocks.

Here is for instance a VIL-MAC from FIL-MAC construction which improves the An-Bellare result [1].

Lemma 10. *Let F_1 and F_2 be two independent random functions defined from $\{0,1\}^{m+b}$ to $\{0,1\}^b$. For any ℓ and any $(m_1,\ldots,m_\ell) \in (\{0,1\}^m)^\ell$ we define*

$$\text{MAC}(m_1,\ldots,m_\ell) = F_2(F_1(\ldots F_1(F_1(0,m_1),m_2)\ldots,m_\ell),\ell)$$

---

[4] This is usually referred to as "mode of operation" in the literature.

where 0 means a b-bit zero string, and ℓ means an m-bit string which represents the ℓ value. Considering distinguishers limited to d queries and a total length of qm bits we have

$$\mathrm{DecF}^{d,q}_{\|\cdot\|_a} \leq \mathrm{DecF}^q_{\|\cdot\|_a}(F_1) + \mathrm{DecF}^d_{\|\cdot\|_a}(F_2) + q(q-1)2^{-m}.$$

Proof (sketch). As in the proof of Lemma 9 we consider the event E where all F_2 inputs are pairwise different. In Lemma 4 we have $\epsilon_1 = 0$ and $\epsilon_2 = \Pr[\bar{E}]$. If we have two equal F_2 inputs, we must have a collision on F_1, thus $\Pr[\bar{E}] \leq \frac{q(q-1)}{2}2^{-m}$. □

Note that we can still use

$$\mathrm{MAC}(m_1, \ldots, m_\ell) = F_2(F_1(\ldots F_1(F_1(0, m_1), m_2) \ldots, m_\ell))$$

with F_2 define on $\{0,1\}^m$ (the same construction, but without the message length) and obtain

$$\mathrm{DecF}^{d,q}_{\|\cdot\|_a} \leq \mathrm{DecF}^q_{\|\cdot\|_a}(F_1) + \mathrm{DecF}^d_{\|\cdot\|_a}(F_2) + q(q+1)2^{-m}.$$

(In the proof, any equal F_2 inputs lead to a collision on F_1 or a preimage of 0.) We state an ultimate similar result with the CBC-MAC construction.

Lemma 11 ([26]). *Let C_1 and C_2 be two independent uniformly distributed random permutations on $\{0,1\}^m$. For any ℓ and any $(m_1, \ldots, m_\ell) \in (\{0,1\}^m)^\ell$ we define*

$$\mathrm{MAC}(m_1, \ldots, m_\ell) = C_2(C_1(\ldots C_1(C_1(m_1) + m_2) \ldots + m_{\ell-1}) + m_\ell).$$

Considering distinguishers limited to d queries and a total length of qm bits we have

$$\mathrm{DecF}^{d,q}(\mathrm{MAC}) \leq d(d-1)2^{-m} + q(q+1)(1 + q2^{-m})2^{-m}.$$

The result holds for any (quasi)group addition.

Proof (sketch). Using Lemma 4, let $\mathcal{Y}$ be the set of all $y = (y_1, \ldots, y_d)$ with different y_is. We thus have $\epsilon_1 = \frac{d(d-1)}{2}2^{-m}$. Now for any collection of $x_i = (m_{i,1}, \ldots, m_{i,\ell_i})$ we let

$$U_{i,j} = C_1(\ldots C_1(C_1(m_{i,1}) + m_{i,2}) \ldots + m_{i,j-1}) + m_j.$$

We consider the event E that all U_{i,ℓ_i} are pairwise different. We have

$$[\mathrm{MAC}]^{\ell_1, \ldots, \ell_d}_{x,y} \geq 2^{-md}(1 - \Pr[\bar{E}])$$

therefore we can take $\epsilon_2 = \Pr[\bar{E}] = \Pr[\exists i < r; U_{i,\ell_i} = U_{r,\ell_r}]$.

The $U_{i,\ell_i} = U_{r,\ell_r}$ event reduces to a collision $U_{i,\ell_i - t_{i,r}} = U_{r,\ell_r - t_{i,r}}$ or to a preimage of 0 for C_1. Let Inv be the event that $C_1(U_{i,j}) = 0$ for some i, j, and let Coll be the event that we have $U_{i,j} = U_{r,s}$ for some i, j, r, s such that $(m_{i,1}, \ldots, m_{i,j}) \neq (m_{r,1}, \ldots, m_{r,s})$. We have $\epsilon_2 \leq \Pr[\text{Inv}] + \Pr[\text{Coll}]$.

The probability that any adaptive attack against C_1 finds a preimage of 0 after q queries is obviously less than $\frac{q}{2^m - q}$. Thus $\Pr[\text{Inv}] \leq \frac{q}{2^m - q}$.

We let $\mathcal{I}$ be the set of all $((i, j), (r, s))$ pairs such that $1 \leq j \leq \ell_i$ and $1 \leq s \leq \ell_r$ and $(m_{i,1}, \ldots, m_{i,j}) \neq (m_{r,1}, \ldots, m_{r,s})$. This is the set of all potential $U_{i,j} = U_{r,s}$ collision indices. We define $c(i, j, r, s)$ equal to the set of all (i, j') and (r, s') such that $j' \leq j$ and $s' \leq s$. We define an ordering on $\mathcal{I}$ by

$$((i, j), (r, s)) \leq ((i', j'), (r', s')) \iff c(i, j, r, s) \subseteq c(i', j', r', s').$$

For $((i, j), (r, s)) \in \mathcal{I}$ we let $\text{Coll}_{i,j,r,s}$ be the event that $((i, j), (r, s))$ is the minimal pair in $\mathcal{I}$ such that $U_{i,j} = U_{r,s}$. We have

$$\Pr[\text{Coll}] \leq \frac{q(q-1)}{2} \max_{((i,j),(r,s)) \in \mathcal{I}} \Pr[\text{Coll}_{i,j,r,s}].$$

For $((i, j), (r, s)) \in \mathcal{I}$, let us consider the $\text{Coll}_{i,j,r,s}$ event. We assume without loss of generality that $s \leq j$. Since we have no previous collision we must have $m_{i,j} \neq m_{r,s}$. Furthermore we must have $U_{i,j-1} \neq U_{r,s-1}$ and $j > 1$, and we need to consider the event

$$C_1(U_{i,j-1}) + m_{i,j} = U_{r,s}.$$

If $U_{i,j-1}$ is equal to some $U_{i',j'}$ with $(i, j-1) \neq (i', j')$ and $(i', j') \in c(i, j, r, s)$, we must have $((i, j-1), (i', j')) \notin \mathcal{I}$ (otherwise we have a previous collision) which means $j' = j - 1$ and $i' = r \neq i$ and $(m_{i,1}, \ldots, m_{i,j-1}) = (m_{r,1}, \ldots, m_{r,j-1})$. If $s < j$ we have $U_{i,j} = U_{r,s} = U_{i,s}$ with $((i, j), (i, s)) \in \mathcal{I}$ which contradicts the minimality of the collision. Thus $s = j$, but this contradicts $U_{i,j-1} \neq U_{r,s-1}$. Hence $U_{i,j-1}$ is not equal to some $U_{i',j'}$ with any previously used (i', j'). The distribution of $C_1(U_{i,j-1})$ is thus uniform among a set of at least $2^m - q$ elements and independent of $U_{r,s}$. Hence $\Pr[\text{Coll}_{i,j,r,s}] \leq \frac{1}{2^m - q}$.

Finally we obtain

$$\epsilon_2 \leq \frac{q}{2^m - q} + \frac{q(q-1)}{2} \times \frac{1}{2^m - q} \leq \frac{q(q+1)}{2}(1 + q2^{-m})2^{-m}.$$

$\square$

4   Conclusion

We have shown how to make several results on conventional cryptography more systematic by using decorrelation theory. The new presentation of the results helps to understand and improve them. Namely, we improved a few of the pseudorandomness results. We also improved results related to MAC constructions, and we proved that the encrypted CBC-MAC construction is secure.

References

1. J. H. An, M. Bellare. Constructing VIL-MACs from FIL-MACs: Message Authentication under Weakened Assumptions. In *Advances in Cryptology CRYPTO'99*, Santa Barbara, California, U.S.A., Lectures Notes in Computer Science 1666, pp. 252–269, Springer-Verlag, 1999.
2. O. Baudron, H. Gilbert, L. Granboulan, H. Handschuh, R. Harley, A. Joux, P. Nguyen, F. Noilhan, D. Pointcheval, T. Pornin, G. Poupard, J. Stern, S. Vaudenay. DFC Update. In Proceedings from the Second Advanced Encryption Standard Candidate Conference, National Institute of Standards and Technology (NIST), March 1999.
3. M. Bellare, P. Rogaway. On the Construction of Variable-Input-Length Ciphers. In *Fast Software Encryption*, Roma, Italy, Lectures Notes in Computer Science 1636, pp. 231–244, Springer-Verlag, 1999.
4. M. Bellare, J. Kilian, P. Rogaway. The Security of Cipher Block Chaining. In *Advances in Cryptology CRYPTO'94*, Santa Barbara, California, U.S.A., Lectures Notes in Computer Science 839, pp. 341–358, Springer-Verlag, 1994.
5. H. Feistel. Cryptography and Computer Privacy. *Scientific American*, vol. 228, pp. 15–23, 1973.
6. H. Gilbert, M. Girault, P. Hoogvorst, F. Noilhan, T. Pornin, G. Poupard, J. Stern, S. Vaudenay. Decorrelated Fast Cipher: an AES Candidate. (Extended Abstract.) In *Proceedings from the First Advanced Encryption Standard Candidate Conference*, National Institute of Standards and Technology (NIST), August 1998.
7. H. Gilbert, M. Girault, P. Hoogvorst, F. Noilhan, T. Pornin, G. Poupard, J. Stern, S. Vaudenay. Decorrelated Fast Cipher: an AES Candidate. Submitted to the Advanced Encryption Standard process. In *CD-ROM "AES CD-1: Documentation"*, National Institute of Standards and Technology (NIST), August 1998.
8. M. Luby, C. Rackoff. How to Construct Pseudorandom Permutations from Pseudorandom Functions. *SIAM Journal on Computing*, vol. 17, pp. 373–386, 1988.
9. S. Lucks. Faster Luby–Rackoff Ciphers. In *Fast Software Encryption*, Cambridge, United Kingdom, Lectures Notes in Computer Science 1039, pp. 189–203, Springer-Verlag, 1996.
10. U. M. Maurer. A Simplified and Generalized Treatment of Luby–Rackoff Pseudorandom permutation generators. In *Advances in Cryptology EUROCRYPT'92*, Balatonfüred, Hungary, Lectures Notes in Computer Science 658, pp. 239–255, Springer-Verlag, 1993.
11. U. M. Maurer. Information-Theoretic Cryptography. Invited lecture. In *Advances in Cryptology CRYPTO'99*, Santa Barbara, California, U.S.A., Lectures Notes in Computer Science 1666, pp. 47–64, Springer-Verlag, 1999.
12. M. Naor, O. Reingold. On the Construction of Pseudorandom Permutations: Luby–Rackoff Revisited. *Journal of Cryptology*, vol. 12, pp. 29–66, 1999.
13. J. Patarin. *Etude des Générateurs de Permutations Basés sur le Schéma du D.E.S.*, Thèse de Doctorat de l'Université de Paris 6, 1991.
14. J. Patarin. How to Construct Pseudorandom and Super Pseudorandom Permutations from One Single Pseudorandom Function. In *Advances in Cryptology EUROCRYPT'92*, Balatonfüred, Hungary, Lectures Notes in Computer Science 658, pp. 256–266, Springer-Verlag, 1993.
15. J. Patarin. About Feistel Schemes with Six (or More) Rounds. In *Fast Software Encryption*, Paris, France, Lectures Notes in Computer Science 1372, pp. 103–121, Springer-Verlag, 1998.

16. J. Pieprzyk. How to Construct Pseudorandom Permutations from a Single Pseudorandom Functions. In *Advances in Cryptology EUROCRYPT'90*, Aarhus, Denemark, Lectures Notes in Computer Science 473, pp. 140–150, Springer-Verlag, 1991.

17. M. Sugita. Pseudorandomness of Block Ciphers with Recursive Structures. Technical Report of IEICE. ISEC97-9.

18. S. Vaudenay. Provable Security for Block Ciphers by Decorrelation. In *STACS 98*, Paris, France, Lectures Notes in Computer Science 1373, pp. 249–275, Springer-Verlag, 1998.

19. S. Vaudenay. Provable Security for Block Ciphers by Decorrelation. (Full Paper.) Technical report LIENS-98-8, Ecole Normale Supérieure, 1998.
   URL: ftp://ftp.ens.fr/pub/reports/liens/liens-98-8.A4.ps.Z

20. S. Vaudenay. Feistel Ciphers with L_2-Decorrelation. In *Selected Areas in Cryptography*, Kingston, Ontario, Canada, Lectures Notes in Computer Science 1556, pp. 1–14, Springer-Verlag, 1999.

21. S. Vaudenay. The Decorrelation Technique Home-Page.
   URL: http://www.dmi.ens.fr/~vaudenay/decorrelation.html

22. S. Vaudenay. *Vers une Théorie du Chiffrement Symétrique*, Dissertation for the diploma of "habilitation to supervise research" from the University of Paris 7, Technical Report LIENS-98-15 of the Laboratoire d'Informatique de l'Ecole Normale Supérieure, 1998.

23. S. Vaudenay. Resistance Against General Iterated Attacks. In *Advances in Cryptology EUROCRYPT'99*, Prague, Czech Republic, Lectures Notes in Computer Science 1592, pp. 255–271, Springer-Verlag, 1999.

24. S. Vaudenay. On the Lai-Massey Scheme. Technical report LIENS-99-3, Ecole Normale Supérieure, 1999. To appear in Asiacrypt'99, LNCS, Springer-Verlag.

25. S. Vaudenay. Adaptive-Attack Norm for Decorrelation and Super-Pseudorandomness. Technical report LIENS-99-2, Ecole Normale Supérieure, 1999. (To appear in SAC'99, LNCS, Springer-Verlag.)
   URL: ftp://ftp.ens.fr/pub/reports/liens/liens-99-2.A4.ps.Z

26. S. Vaudenay. Security of CBC-MAC. Submitted.

27. Y. Zheng, T. Matsumoto, and H. Imai. "On the Construction of Block Ciphers Provably Secure and Not Relying on Any Unproved Hypotheses. In *Advances in Cryptology CRYPTO'89*, Santa Barbara, California, U.S.A., Lectures Notes in Computer Science 435, pp. 461–480, Springer-Verlag, 1990.

Correlation Properties of the Bluetooth Combiner

Miia Hermelin and Kaisa Nyberg

Nokia Research Center, Helsinki, Finland
miia.hermelin@nokia.com, kaisa.nyberg@nokia.com

Abstract. In its intended usage the lengths of the key stream sequences produced by the Bluetooth stream cipher E_0 are strictly limited. In this paper the importance of this limitation is proved by showing that the Bluetooth stream cipher with 128 bit key can be broken in $\mathcal{O}(2^{64})$ steps given an output key stream segment of length $\mathcal{O}(2^{64})$. We also show how the correlation properties of the E_0 combiner can be improved by making a small modification in the memory update function.

1 Introduction

Bluetooth^{TM} is a standard for wireless connectivity specified by the Bluetooth^{TM} Special Interest Group in [1]. The specification defines a stream cipher algorithm E_0 to be used for point-to-point encryption between the elements of a Bluetooth network. The structure of E_0 is a modification of a summation bit generator with memory. In this paper we call it the Bluetooth combiner and analyze its correlation properties. A few correlation theorems originating from [4] are stated and exploited in the analysis. Also a new kind of divide-and-conquer attack is introduced, which shows the importance of limiting the lengths of produced key stream sequences.

As a consequence of these results, we propose a modification to the Bluetooth combiner. This modification could be done at no extra cost, that is, it does not increase the complexity of the algorithm. But, on the other hand, it would improve the correlation properties of the Bluetooth combiner to some extent. However, as long as no practical attack is known against the current version of the Bluetooth combiner, the results given in this paper remain theoretical.

2 Correlation Theorems

2.1 Definitions and Notation

Let us introduce the notation to be used throughout this paper. We shall consider the field $GF(2^n)$ as a linear space with a given fixed basis, and denote by x_t an n-dimensional vector in $GF(2^n)$ as $x_t = (x_t^1, x_t^2, \ldots, x_t^n)$. The inner product "." between two vectors $w = (w_1, w_2, \ldots, w_n)$ and $x = (x_1, x_2, \ldots, x_n)$ of the space $GF(2^n)$ is defined as

$$w \cdot x = w_1 x_1 \oplus w_2 x_2 \oplus \ldots \oplus w_n x_n.$$

JooSeok Song (Ed.): ICISC'99, LNCS 1787, pp. 17–29, 2000.
© Springer-Verlag Berlin Heidelberg 2000

The linear function $L_u(x)$ is then

$$L_u(x) = u \cdot x, \ u, x \in GF(2^n).$$

We use the same definition of correlation between two Boolean functions as in [3], where it is also referred to as "normalized correlation".

Definition 1. *Let* $f, g : GF(2^n) \to GF(2)$ *be Boolean functions. The correlation between* f *and* g *is*

$$c(f, g) = 2^{-n}(\#\{x \in GF(2^n) \,|\, f(x) = g(x)\} - \#\{x \in GF(2^n) \,|\, f(x) \neq g(x)\}).$$

Sometimes the notation $c_x(f(x), g(x))$ is used to emphasize the variable with respect to which the correlation is to be calculated.

Finally, we recall Parseval's theorem, which implies, in particular, that any Boolean function is correlated to some linear functions.

Theorem 2. *(Parseval's Theorem)*

$$\sum_{w \in GF(2^n)} c(f, L_w)^2 = 1.$$

2.2   Correlation Theorems

Iterated structures and combinations of transformations with common input are frequently seen building blocks of cryptographic algorithms. The following correlation theorems are useful in the analysis of propagation of correlations over such structures. The proofs of the theorems can be found in [4].

Theorem 3. *Given functions* $f : GF(2^n) \times GF(2^k) \to GF(2)$ *and* $g : GF(2^m) \to GF(2^k)$ *we set*

$$h(x, y) = f(x, g(y)), \ x \in GF(2^n), \ y \in GF(2^m).$$

Then, for all $u \in GF(2^n)$, $v \in GF(2^m)$,

$$c_{x,y}(h(x, y), u \cdot x \oplus v \cdot y) = \sum_{w \in GF(2^k)} c_{x,z}(f(x, z), u \cdot x \oplus w \cdot z)c_y(w \cdot g(y), v \cdot y).$$

We note that Theorem 3 can be considered as a generalization of Lemma 2 of [3]. In the second correlation theorem a Boolean function, which is a sum of two functions with partially common input, is considered.

Theorem 4. *Let* $f : GF(2^n) \times GF(2^k) \to GF(2)$ *and* $g : GF(2^k) \times GF(2^m) \to GF(2)$ *be Boolean functions. Then, for all* $u \in GF(2^n)$, $w \in GF(2^m)$,

$$c_{x,y,z}(f(x, y) \oplus g(y, z), u \cdot x \oplus w \cdot z)$$
$$= \sum_{v \in GF(2^k)} c_{x,y}(f(x, y), u \cdot x \oplus v \cdot y)c_{y,z}(g(y, z), v \cdot y \oplus w \cdot z).$$

If here the two functions f and g, and the two linear combinations u and w are the same, we have the following corollary.

Corollary 5. *Let* $f : GF(2^n) \times GF(2^k) \to GF(2)$ *be a Boolean function. Then, for all* $u \in GF(2^n)$,

$$c_{x,y,\xi}(f(x, y) \oplus f(\xi, y), u \cdot (x \oplus \xi)) = \sum_{v \in GF(2^k)} c_{x,y}(f(x, y), u \cdot x \oplus v \cdot y)^2.$$

3   Combination Generators

In [3] an example of a summation bit generator with one bit of memory is introduced and analyzed. The combiner of the Bluetooth key stream generator can be considered as a variation of the thoroughly analyzed basic summation bit generator, see [3] and [2]. A general class of combination generators with memory giving the generators of [3] and [1] as special cases is defined as follows:

$$z_t = \bigoplus_{i=1}^{n} x_t^i \oplus c_t^0 \tag{1}$$

$$c_t = f(x_{t-1}, c_{t-1}, \ldots, c_{t-d}). \tag{2}$$

Here $x_t = (x_t^1, \ldots, x_t^n) \in GF(2^n)$ is the fresh input to the combiner at time t and $c_t^0 \in GF(2)$ is the one-bit input from the memory, $t = 0, 1, 2, \ldots$. The fresh input is formed by n independent sequences $x^i = (x_0^i, x_1^i, \ldots)$, $i = 1, 2, \ldots, n$, which are typically generated by n linear feedback shift registers.

The memory constitutes of md bits arranged as a register of d consecutive cells of m bits each. The memory is updated by computing a new m-bit $c_t = (c_t^0, \ldots, c_t^m)$ using a function f from the fresh input and from the contents of the memory saving the new c_t in the memory and discarding c_{t-d}. The output bit z_t is computed as an xor-sum of the fresh input x_t and the previously computed update c_t of the memory.

Correlation attacks aimed at recovering the keys, which determine the generation of the fresh input, are based on correlations between a number of fresh input bits and the output bits. For the type of generators defined by (1) such correlations relations relations can be derived from correlations between consecutive "carry" bits c_t^0.

Such correlations are usually found by exhaustive search. This is the case also with the Bluetooth combiner which is such a relatively small system that this kind of "trial and error"-search is possible. In larger systems, however, some more sophisticated means for finding these correlation equations should be used. One such method is presented in [2].

4   Bluetooth Combiner

Bluetooth chips are small components capable of short range communication with each other. The Bluetooth specification is given in [1]. In the security part of [1] an encryption algorithm E_0 is specified to be used for protection of the confidentiality of the Bluetooth communication.

The algorithm E_0 is of the form specified by (1) and (2). It consists of four LFSRs of length 128 in total, a non-linear memory update function f, which is a composition of a nonlinear f_1 and a linear mapping T.

The functions define the following recursive equations. The output key sequence z_t, used to encipher the plaintext, is

$$z_t = x_t^1 \oplus x_t^2 \oplus x_t^3 \oplus x_t^4 \oplus c_t^0,$$

where $(x_t^1, x_t^2, x_t^3, x_t^4)$ is the fresh input at time t produced by the four LFSRs. Non-linearity is represented in the sequence s_t, defined by the following formula, where "+" means the ordinary integer sum:

$$s_{t+1} = (s_{t+1}^1, s_{t+1}^0) = f_1(x_t, c_t) = \left\lfloor \frac{x_t^1 + x_t^2 + x_t^3 + x_t^4 + 2c_t^1 + c_t^0}{2} \right\rfloor.$$

The function f_1 introduces the necessary non-linearity in the system, as integer summation is non-linear in $GF(2)$. The memory bits c_t are then defined with the aid of s_t as

$$c_{t+1} = (c_{t+1}^1, c_{t+1}^0) = T(s_{t+1}, c_t, c_{t-1}) = T_0(s_{t+1}) \oplus T_1(c_t) \oplus T_2(c_{t-1}).$$

Here T_0, T_1 and T_2 are linear transformations. Although non-linearity is crucial for security, the choice of the linear mapping T has also certain influence to the security of the E_0 algorithm, as we will see later.

4.1   The Mapping T

The linear mapping T of E_0 mix the old bits from the memory to the new updated memory bits. The main focus of this work is to investigating its role in the correlation properties. For the given f_1 we see (c.f. Table 1) that $c(s_t^0, c_{t-1}^0 \oplus u \cdot x_t) = 0$, for all $u \in GF(2)$. Hence these are not useful in correlation attacks. On the other hand, $c(s_t^i, c_{t-1}^1 \oplus v^0 c_{t-1}^0 \oplus u \cdot x_t') \neq 0$, $i = 0, 1$.

The mapping T consists of three mappings, T_0, T_1 and T_2, as

$$c_{t+1} = T(s_{t+1}, c_t, c_{t-1})$$
$$= T_0(s_{t+1}) \oplus T_1(c_t) \oplus T_2(c_{t-1}).$$

In matrix form, $T_0 = T_1 = I$, where I is a 2×2 identity matrix. Further,

$$T_2 = \begin{pmatrix} 0 & 1 \\ 1 & 1 \end{pmatrix}$$

This means, the bits of $c_t = (c_t^1, c_t^0)$ are

$$c_t^1 = s_t^1 \oplus c_{t-1}^1 \oplus c_{t-2}^0 \tag{3}$$
$$c_t^0 = s_t^0 \oplus c_{t-1}^0 \oplus c_{t-2}^1 \oplus c_{t-2}^0. \tag{4}$$

With different choices of T_0, T_1 and T_2 the correlation properties of the system become different. This shall be analyzed in section 5.

4.2   Correlation Analysis of the Bluetooth Combiner

The memory in Bluetooth has four bits, two bits for each two consecutive time steps t and $t - 1$. The function, which is used to form a new term z_t of the keys stream, is linear. The non-linearity is gained from the function f_1, which is used to calculate s_t. As argued in [3], and in more general terms in [2], there remain always some correlations in such a system. They shall be analyzed next.

The analysis exploits the correlations of the form

$$c(w \cdot s_{t+1}, u \cdot x_t \oplus v \cdot c_t)$$

where $u \in GF(2^4)$ and $v = (v^1, v^0) \in GF(2^2)$. Different choices of u and v correspond to different linear combinations of $x_t^1, x_t^2, x_t^3, x_t^4, c_t^1$ and c_t^0. In the following Table 1 all the correlations are presented.

v^1	v^0	weight of u	s_{t+1}^1	s_{t+1}^0	$s_{t+1}^1 \oplus s_{t+1}^0$
0	0	0	0	0	$-\frac{5}{8}$
		1	$\frac{1}{4}$	0	0
		2	0	0	$\frac{1}{8}$
		3	0	0	0
		4	0	0	$-\frac{1}{8}$
0	1	0	$\frac{1}{4}$	0	0
		1	0	0	$\frac{1}{8}$
		2	0	0	0
		3	0	0	$-\frac{1}{8}$
		4	$-\frac{1}{4}$	0	0
1	0	0	$\frac{5}{8}$	$-\frac{1}{4}$	0
		1	0	0	$\frac{1}{4}$
		2	$-\frac{1}{8}$	$\frac{1}{4}$	0
		3	0	0	0
		4	$\frac{1}{8}$	$-\frac{1}{4}$	0
1	1	0	0	0	$\frac{1}{4}$
		1	$-\frac{1}{8}$	$\frac{1}{4}$	0
		2	0	0	0
		3	$\frac{1}{8}$	$-\frac{1}{4}$	0
		4	0	0	$-\frac{1}{4}$

Table 1. The correlations for s_{t+1}^1, s_{t+1}^0 and $s_{t+1}^1 \oplus s_{t+1}^0$.

We note that, since the system is symmetric with respect to each x_t^j, only the Hamming weight of u is of importance. We also see that for s_{t+1}^0, the correlation is zero, if $v_1 = 0$. Next we present derivation of the strongest correlation relation we found within the Bluetooth combiner.

Add c_{t-3}^0 to the both sides of (4) and rearrange the terms to get

$$c_t^0 \oplus c_{t-1}^0 \oplus c_{t-3}^0 = s_t^0 \oplus c_{t-2}^1 \oplus c_{t-2}^0 \oplus c_{t-3}^0. \tag{5}$$

Next we use Theorem 3 to get

$$c(c_t^0 \oplus c_{t-1}^0 \oplus c_{t-3}^0, 0) = c(s_t^0, c_{t-2}^1 \oplus c_{t-2}^0 \oplus c_{t-3}^0)$$
$$= \sum_{w \in GF(2^2)} c(s_t^0, w \cdot c_{t-1}) c(w \cdot c_{t-1}, c_{t-2}^1 \oplus c_{t-2}^0 \oplus c_{t-3}^0),$$

with $u = 0$, $v = (0, 0, 1, 1, 1, 0)$ and $y = (s_{t-1}^0, s_{t-1}^1, c_{t-2}^0, c_{t-2}^1, c_{t-3}^0, c_{t-3}^1)$. Now from Table 1 we know, that the terms of the sum are zero for $w = (0, 1)$, $w =$

$(1, 1)$ and $w = (0, 0)$. Only the term with $w = (1, 0)$ remains. So, the correlation equation is simplified to

$$c(c_t^0 \oplus c_{t-1}^0 \oplus c_{t-3}^0, 0)$$
$$= c(s_t^0, c_{t-1}^1)c(c_{t-1}^1, c_{t-2}^0 \oplus c_{t-2}^1 \oplus c_{t-3}^0)$$
$$= c(s_t^0, c_{t-1}^1)c(s_{t-1}^1, c_{t-2}^0).$$

Here the last equation is obtained by moving back in time for one step in equation (3), so that

$$c_{t-1}^1 = c_{t-1}^1 \oplus s_{t-1}^1 \oplus c_{t-3}^0.$$

Using the values of Table 1, we finally get

$$c(c_t^0 \oplus c_{t-1}^0 \oplus c_{t-3}^0, 0) = -\frac{1}{4} \cdot \frac{1}{4} = -\frac{1}{16}. \tag{6}$$

After this we notice that

$$z_t \oplus z_{t-1} \oplus z_{t-3} = \bigoplus_1^4 x_t^i \oplus \bigoplus_1^4 x_{t-1}^i \oplus \bigoplus_1^4 x_{t-3}^i \oplus c_t^0 \oplus c_{t-1}^0 \oplus c_{t-3}^0.$$

We conclude by equation (6) that

$$c\left(z_t \oplus z_{t-1} \oplus z_{t-3}, \bigoplus_1^4 x_t^i \oplus \bigoplus_1^4 x_{t-1}^i \oplus \bigoplus_1^4 x_{t-3}^i\right) = -\frac{1}{16}. \tag{7}$$

Since the output function of the Bluetooth combiner is XOR, it is maximum order correlation immune. Hence divide and conquer attacks in their standard form are not useful for determining the initial states of the LFSR's. In section 6 it is shown how the achieved correlation relation can be utilized to determine a theoretical upper-bound of the level of the security of the Bluetooth combiner.

5   Alternative Mappings for Mixing the Carry Bits

The goal of this section is to investigate, how the choice of the mapping T affects the correlation properties of the Bluetooth combiner. In particular, we show that the mapping T can be selected in such a way that more than two linear approximations are needed when establishing a correlation relation between consecutive carry bits.

Our method exploits a matrix which makes it possible to consider all possible linear approximations of the function f_1 simultaneously.

Let T_0, T_1 and T_2 be arbitrary 2×2 matrices:

$$T_0 = \begin{pmatrix} t_0^1 & t_0^2 \\ t_0^3 & t_0^4 \end{pmatrix}, \quad T_1 = \begin{pmatrix} t_1^1 & t_1^2 \\ t_1^3 & t_1^4 \end{pmatrix}, \quad \text{and} \quad T_2 = \begin{pmatrix} t_2^1 & t_2^2 \\ t_2^3 & t_2^4 \end{pmatrix}.$$

Then

$$c_t = T_0 s_t \oplus T_1 c_{t-1} \oplus T_2 c_{t-2}. \tag{8}$$

We write $T_1 = A \oplus B$. Here the analyst can choose A and B in which way ever is convenient, as long as their sum is T_1. The equation (8) can be written as

$$c_t = T_0 s_t \oplus A c_{t-1} \oplus B c_{t-1} \oplus T_2 c_{t-2}. \tag{9}$$

We perform one iteration by inserting equation (8), applied for $t - 1$ instead of t, to the equation (9) and get

$$c_t = T_0 s_t \oplus B c_{t-1} \oplus A T_0 s_{t-1} \oplus A T_1 c_{t-2} \oplus A T_2 c_{t-3} \oplus T_2 c_{t-2}. \tag{10}$$

Now, A may be chosen. Let D be a matrix of the form

$$D = \begin{pmatrix} d_1 & d_2 \\ 0 & d_4 \end{pmatrix}.$$

As the analyst wishes to minimize the number of correlation approximations, she wants c_t^0 not to depend on c_{t-3}^1. Therefore, she chooses $A T_2 = D$. If we assume that T_2 is invertible then such a choice is always possible. Inserting $B = T_1 \oplus A$ into equation (10), as well as $A = D T_2^{-1}$, we have

$$c_t = T_0 s_t \oplus (T_1 \oplus D T_2^{-1}) c_{t-1} \oplus (D T_2^{-1} T_0) s_{t-1} \oplus (D T_2^{-1} T_1 \oplus T_2) c_{t-2} \oplus D c_{t-3}. \tag{11}$$

In order to take the correlation approximations into consideration, we write them in matrix form as

$$s_t = X_t c_{t-1},$$

where

$$X_t = \begin{pmatrix} e_t^1 & e_t^2 \\ e_t^3 & e_t^4 \end{pmatrix},$$

and s_t and c_{t-1} are taken as vertical vectors.

We see from Table 1, that if $e_t^3 = 0$ the correlations for s_t^0

$$c(s_t^0, e_t^3 \cdot c_{t-1}^1 \oplus e_t^4 \cdot c_{t-1}^0 \oplus u \cdot x_t),$$

are always zero. Therefore we can presume $e_t^3 = 1$. The choice of u does not affect the best non-zero values of the correlations. Therefore, we shall drop $u \cdot x_t$ and merely study the combinations of s_{t+1}^i and c_t^j.

We approximate twice by inserting $s_t = X_t c_{t-1}$ into equation (11), which yields

$$c_t = (T_0 X_t \oplus T_1 \oplus D T_2^{-1}) c_{t-1} \oplus (D T_2^{-1} T_0 X_{t-1} \oplus D T_2^{-1} T_1 \oplus T_2) c_{t-2} \oplus D c_{t-3}. \tag{12}$$

In Bluetooth the generated key-sequence z_t does not depend on c_t^1 but merely on c_t^0. Hence, similarly as above in section 4.2, we aim at establishing a correlation relation between zero components of c_t.

Theorem 6. *Let in the Bluetooth combiner generator $T_0 = T_1 = I$ and T_2 an arbitrary invertible 2×2 matrix. If $t_2^3 = 1$, then two correlation approximations suffices to establish a correlation between the input and output.*

Proof. Substitute $T_0 = T_1 = I$ and the general form of T_2 into (12) and obtain the following correlation relation for c_t^0:

$$c_t^0 = (1 \oplus d_4, e_t^4 \oplus 1 \oplus d_4 t_2^1) \cdot c_{t-1}$$
$$\oplus (d_4 e_{t-1}^1 \oplus d_4 t_2^1 \oplus d_4 \oplus 1, d_4 e_{t-1}^2 \oplus d_4 t_2^1 e_{t-1}^4 \oplus d_4 t_2^1 \oplus t_2^4) \cdot c_{t-2}$$
$$\oplus d_4 c_{t-3}^0$$

To have only two approximations means that there must be neither c_{t-1}^1 nor c_{t-2}^1 in the equation above, i.e.

$$1 \oplus d_4 = 0 \tag{13}$$

and

$$d_4 e_{t-1}^1 \oplus d_4 t_2^1 \oplus d_4 \oplus 1 = 0. \tag{14}$$

If $d_4 = 0$, then the other approximation will cancel out, so that equation (11) transforms into the initial equation

$$c_t^0 = s_t^0 \oplus c_{t-1}^0 \oplus c_{t-2}^1 \oplus c_{t-2}^0.$$

As this is not what the analyst wants, she chooses $d_4 = 1$ and equation (13) is true. From (14) we then get that

$$e_{t-1}^1 \oplus t_2^1 = 0.$$

Now we check from Table 1 that $c(s_{t-1}^1, c_{t-2}^0 \oplus u \cdot x_{t-2}) \neq 0$ for some u. Hence it is possible to use this correlation if $t_2^1 = 0$. Similarly, if $t_2^1 = 1$, we see that $e_{t-1}^1 = 1$ is possible, as $c(s_{t-1}^1, c_{t-2}^1 \oplus v^0 \cdot c_{t-2}^0 \oplus u \cdot x_{t-2}) \neq 0$ for some choice of u and v^0. $\square$

In the case of the initial choice of Bluetooth T_2, we have $t_2^3 = 1$. So, as we saw earlier in section 4.2, only two iteration approximations are needed. The approximation matrices X_t in the case of (6) were

$$X_{t-1} = \begin{pmatrix} 0 & 0 \\ 1 & 0 \end{pmatrix} \quad X_t = \begin{pmatrix} 0 & 1 \\ 1 & 0 \end{pmatrix}.$$

From Table 1 we see that $c(s_{t-1}^1, c_{t-2}^1) = \frac{5}{8}$. Hence

$$T_2 = \begin{pmatrix} 1 & 1 \\ 1 & 0 \end{pmatrix}$$

with $t_2^1 = 1$ would have been still a weaker choice than the current T_2 in Bluetooth. Next we show that a stronger choice would have been possible.

Theorem 7. *Let $t_2^3 = 0$. Then at least three approximation rounds are needed.*

Proof. If $t_2^3 = 0$, and T_2 is invertible as assumed, then $t_2^1 = t_2^4 = 1$. Also $T_2^{-1} = T_2$. Let

$$D = \begin{pmatrix} d_1 & d_2 \\ d_3 & d_4 \end{pmatrix}.$$

Then, as in (12) we have

$$c_t = (X_t \oplus I \oplus DT_2)c_{t-1} \oplus (DT_2 X_{t-1} \oplus DT_2 X_{t-1} \oplus DT_2 \oplus T_2)c_{t-2},$$

and further,

$$
\begin{aligned}
c_t^0 = {} & (1 \oplus d_3, e_t^4 \oplus 1 \oplus t_2^2 d_3 \oplus d_4) \cdot c_{t-1} \\
& \oplus [d_3 e_{t-1}^1 \oplus e_{t-1}^3 (t_2^2 d_3 \oplus d_4) \oplus d_3, \\
& d_3 e_{t-1}^2 \oplus e_{t-1}^4 (t_2^2 d_3 \oplus d_4) \oplus t_2^2 d_3 \oplus d_4 \oplus 1] \cdot c_{t-2} \\
& \oplus (d_3, d_4) \cdot c_{t-3}
\end{aligned}
$$

Now, if $d_3 = 1$, then we have c_{t-3}^1 in the equation, so we need to do at least one more approximation, hence two approximations is not enough. If $d_3 = 0$, then c_{t-1}^1 is within the equation of c_t^0 and again more than two approximations are needed.   □

An example of a matrix T_2, which requires at least three approximations to get correlation relations between the carry bits c_t^0 from different time instances, is $T_2 = I$. We consider, for example, the correlation between c_t^0 and c_{t-4}^0. The correlation relation could involve some x_t variables at appropriate moments t, but in what follows we restrict to the case where the Hamming weight of u is always zero.

Corresponding to the equations (3) and (4) we now have

$$
\begin{aligned}
c_t^1 &= s_t^1 \oplus c_{t-1}^1 \oplus c_{t-2}^1 \\
c_t^0 &= s_t^0 \oplus c_{t-1}^0 \oplus c_{t-2}^0.
\end{aligned}
$$

Since no x_{t-1} is involved, we have by Theorem 3 and with the aid of Table 1

$$
\begin{aligned}
c(c_t^0, c_{t-4}^0) &= c(s_t^0, c_{t-1}^0 \oplus c_{t-2}^0 \oplus c_{t-4}^0) \\
&= \sum_w c(s_t^0, w \cdot c_{t-1}) c(w \cdot c_{t-1}, c_{t-1}^0 \oplus c_{t-2}^0 \oplus c_{t-4}^0) \\
&= c(s_t^0, c_{t-1}^1) c(c_{t-1}^1, c_{t-1}^0 \oplus c_{t-2}^0 \oplus c_{t-4}^0).
\end{aligned}
$$

The second equality follows from Theorem 3, and in the third we noted that the only non-zero correlation for $c(s_t^0, w \cdot c_{t-1})$ is due to $w = (1, 0)$. We continue in the same manner to obtain:

$$
\begin{aligned}
& c(c_t^0, c_{t-4}^0) \\
={} & -\frac{1}{4} c(s_{t-1}^1 \oplus s_{t-1}^0, c_{t-2}^1 \oplus c_{t-3}^1 \oplus c_{t-3}^0 \oplus c_{t-4}^0) \\
={} & -\frac{1}{4} \sum_w c(s_{t-1}^1 \oplus s_{t-1}^0, w \cdot c_{t-2}) c(w \cdot c_{t-2}, c_{t-2}^1 \oplus c_{t-3}^1 \oplus c_{t-3}^0 \oplus c_{t-4}^0) \\
={} & -\frac{1}{4} (c(s_{t-1}^1 \oplus s_{t-1}^0, 0) c(c_{t-2}^1 \oplus c_{t-3}^1 \oplus c_{t-3}^0 \oplus c_{t-4}^0, 0) \\
& + c(s_{t-1}^1 \oplus s_{t-1}^0, c_{t-2}^1 \oplus c_{t-2}^0) c(c_{t-2}^0, c_{t-3}^1 \oplus c_{t-3}^0 \oplus c_{t-4}^0)).
\end{aligned}
\tag{15}
$$

From the two terms inside the brackets only the last one is non-zero. To see this we calculate the first term

$$c(c_{t-2}^1 \oplus c_{t-3}^1 \oplus c_{t-3}^0 \oplus c_{t-4}^0, 0)$$
$$= c(s_{t-2}^1, c_{t-3}^0 \oplus c_{t-4}^1 \oplus c_{t-4}^0)$$
$$= \sum_w c(s_{t-2}^1, w \cdot c_{t-3}) c(w \cdot c_{t-3}, c_{t-3}^0 \oplus c_{t-4}^1 \oplus c_{t-4}^0)$$
$$= c(s_{t-2}^1, c_{t-3}^0) c(c_{t-4}^1 \oplus c_{t-4}^0, 0) + c(s_{t-2}^1, c_{t-3}^1) c(c_{t-3}^1, c_{t-3}^0 \oplus c_{t-4}^1 \oplus c_{t-4}^0).$$

In the last equality the first term in the sum is zero, since the memory bits on the same moment are assumed to be statistically independent. We continue with the second part of the sum:

$$c(c_{t-3}^0 \oplus c_{t-3}^1, c_{t-4}^1 \oplus c_{t-4}^0)$$
$$= c(s_{t-3}^1 \oplus s_{t-3}^0, c_{t-5}^1 \oplus c_{t-5}^0)$$
$$= \sum_w c(s_{t-3}^1 \oplus s_{t-3}^0, w \cdot c_{t-4}) c(w \cdot c_{t-4}, c_{t-5}^1 \oplus c_{t-5}^0)$$
$$= -\frac{5}{8} (c(s_{t-3}^1 \oplus s_{t-3}^0, 0) c(c_{t-5}^1 \oplus c_{t-5}^0, 0)$$
$$+ c(s_{t-3}^1 \oplus s_{t-3}^0, c_{t-4}^1 \oplus c_{t-4}^0) c(c_{t-4}^1 \oplus c_{t-4}^0, c_{t-5}^1 \oplus c_{t-5}^0))$$
$$= -\frac{5}{8} \cdot \frac{1}{4} c(c_{t-4}^1 \oplus c_{t-4}^0, c_{t-5}^1 \oplus c_{t-5}^0).$$

These calculations are easy to generalize to any moment $t - j$, $j = 0, 1, 2, \ldots$, so actually we have

$$c(c_{t-3}^0 \oplus c_{t-3}^1, c_{t-4}^1 \oplus c_{t-4}^0)$$
$$= \left(-\frac{5}{32}\right)^k c(c_{t-3-k}^1 \oplus c_{t-3-k}^0, c_{t-4-k}^1 \oplus c_{t-4-k}^0).$$

Hence, infinitely many approximations should be done, and so the correlation can be regarded as zero:

$$c(c_{t-3}^0 \oplus c_{t-3}^1, c_{t-4}^1 \oplus c_{t-4}^0) = \lim_{k \to \infty} \left(-\frac{5}{32}\right)^k = 0.$$

Let us now return to the equation 15. We have, that

$$c(c_t^0, c_{t-4}^0)$$
$$= -\frac{1}{4} \cdot \frac{1}{4} c(c_{t-2}^0, c_{t-3}^1 \oplus c_{t-3}^0 \oplus c_{t-4}^0)$$
$$= -\frac{1}{16} c(s_{t-2}^0, c_{t-3}^1) = -\frac{1}{64},$$

which is significantly smaller than $\frac{1}{16}$ in equation 6. In this manner, the correlation can be calculated for other weights of u, too. The resulting values degenerate to a single product, as above, so that the product of the correlations is always significantly smaller than $\frac{1}{16}$.

The Bluetooth combination generator is a strengthened version of the basic summation bit generator. By increasing the size of the memory the correlations have been reduced. We have shown that with the same memory size, by making a small modification in the memory update function, it would be possible to further reduce the correlations.

6   Ultimate Divide and Conquer

In this section it is shown that divide and conquer attack becomes possible if the length of the given keystream is longer than the period p of the shortest (say, the first) LFSR used in the key stream generation. Assume that there is a relation with a non-zero correlation ρ between a linear combination of the shift register output bits

$$(u_0^1 x_t^1 \oplus \ldots \oplus u_0^n x_t^n) \oplus \ldots \oplus (u_d^1 x_{t-d}^1 \oplus \ldots \oplus u_d^n x_{t-d}^n)$$

and the key stream bits

$$w_0 z_t \oplus w_1 z_{t-1} \oplus \ldots \oplus w_d z_{t-d},$$

over a number of $d+1$ time steps. Then it follows by Corollary 5 that we have a correlation relation between a linear combination of the keystream bits

$$w_0(z_t \oplus z_{t+p}) \oplus w_1(z_{t-1} \oplus z_{t+p-1}) \oplus \ldots \oplus w_d(z_{t-d} \oplus z_{t+p-d})$$

and a linear combination of the LFSR output bits

$$u_0^2(x_t^2 \oplus x_{t+p}^2) \oplus \ldots \oplus u_0^n(x_t^n \oplus x_{t+p}^n) \oplus \ldots$$
$$\oplus u_d^2(x_{t-d}^2 \oplus x_{t+p-d}^2) \oplus \ldots \oplus u_d^n(x_{t-d}^n \oplus x_{t+p-d}^n),$$

where the output bits from the first (the shortest) shift register cancel, since they are equal.

By Corollary 5, the strength of the correlation over the period p is at least ρ^2. Further, Corollary 5 shows how this lower bound can be improved. We state this result in a form of a theorem as follows.

Theorem 8. *For a combination generator, assume that we have the following correlation*

$$c(w_0 z_t \oplus w_1 z_{t-1} \oplus \ldots \oplus w_d z_{t-d},$$
$$(u_0^1 x_t^1 \oplus \ldots \oplus u_0^n x_t^n) \oplus \ldots + (u_d^1 x_{t-d}^1 \oplus \ldots \oplus u_d^n x_{t-d}^n)) = \rho \neq 0.$$

Let the lengths of the registers be $L_1, \ldots, L_n$ and the periods $p_1, \ldots, p_n$. Then given a keystream of length $p_1 p_2 \cdots p_k + \frac{1}{\rho^4} + d$ one can do exhaustive search over the $L_{k+1} + \ldots + L_n$ bits which form the initial contents of $n-k$ registers.

If the LFSR registers have primitive feedback polynomials, then $p_i = 2^{L_i} - 1$. In most applications n is even and the lengths L_i are about the same. Then given a sufficiently strong correlation between the input bits and the output bits of a combination generator, the complexity to determine the complete initial state of length L is about $\mathcal{O}(2^{L/2})$. In other words, by generating key stream of length $\mathcal{O}(2^{L/2})$ one can successfully carry out exhaustive search over $L/2$ bits of the initial state.

6.1   Periodic Correlations in Bluetooth

Computation of the correlations for the Bluetooth E_0 combiner is somewhat complicated due to multiple iteration. We make use of the relation $c_t^0 + c_{t-1}^0 + c_{t-3}^0 = 0$, see (6). Applying Theorem 3 we get

$$c(c_t^0 \oplus c_{t-1}^0 \oplus c_{t-3}^0 \oplus c_{t+p}^0 \oplus c_{t+p-1}^0 \oplus c_{t+p-3}^0, 0)$$
$$= c(s_t^0 \oplus s_{t+p}^0, c_{t-2}^0 \oplus c_{t-2}^1 \oplus c_{t-3}^0 plus c_{t+p-2}^0 \oplus c_{t+p-2}^1 + c_{t+p-3}^0)$$
$$= \sum_{w,w' \in GF(2^2)} c(s_t^0 \oplus s_{t+p}^0, w \cdot c_{t-1} \oplus w' \cdot c_{t+p-1})$$
$$\cdot c(w \cdot c_{t-1} \oplus w' \cdot c_{t+p-1}, c_{t-2}^0 \oplus c_{t-2}^1 \oplus c_{t-3}^0 \oplus c_{t+p-2}^0 \oplus c_{t+p-2}^1 \oplus c_{t+p-3}^0).$$

Now we apply Theorem 4 to the first correlation in the product and get

$$c(s_t^0 \oplus s_{t+p}^0, w \cdot c_{t-1} \oplus w' \cdot c_{t+p-1})$$
$$= \sum_{u \in GF(2^2)} c(s_t^0, w \cdot c_{t-1} \oplus u \cdot x) c(s_t^0, w' \cdot c_{t-1} \oplus u \cdot x).$$

Here x has one, two, or three coordinates, depending on whether p is the least common period of of one, two, or three LFSRs, respectively.

Let us now consider the case where p is the least common period of two LFSRs. From Table 1 we see that these correlations are nonzero if and only if $u = (0, 0)$ and $w = w' = (1, 0)$, or $u = (1, 1)$ and $w = w' = (1, 0)$, or $u = (0, 1)$ and $w = w' = (1, 1)$, or finally, $u = (1, 0)$ and $w = w' = (1, 1)$.

The value $w = w' = (1, 1)$ leads to a longer correlation relation extending over at least two rounds, and hence are expected be of less in amount, but still non-negative. Therefore, we discard the corresponding terms, and get a lower bound to the correlation from the remaining terms with $w = w' = (1, 0)$ as follows

$$c(c_t^0 \oplus c_{t-1}^0 \oplus c_{t-3}^0 \oplus c_{t+p}^0 \oplus c_{t+p-1}^0 \oplus c_{t+p-3}^0, 0)$$
$$\geq (c(s_t^0, c_{t-1}^1)^2 + c(s_t^0, c_{t-1}^1 \oplus x_t^1 \oplus x_t^2)^2)$$
$$\cdot c(c_{t-1}^1 \oplus c_{t+p-1}^0, c_{t-2}^0 \oplus c_{t-2}^1 \oplus c_{t-3}^0 \oplus c_{t+p-2}^0 \oplus c_{t+p-2}^1 \oplus c_{t+p-3}^0)$$
$$= (c(s_t^0, c_{t-1}^1)^2 + c(s_t^0, c_{t-1}^1 \oplus x_t^1 \oplus x_t^2)^2) \cdot \sum_{u \in GF(2^2)} c(s_{t-1}^1, c_{t-2}^0 \oplus u \cdot x)^2$$
$$= ((-1/4)^2 + (1/4)^2)(1/4)^2 = 2^{-7},$$

using the correlation values given in Table 1.

It should be stressed, however, that the presented ultimate divide and conquer attack is of theoretical nature, and practical only if the analyzer is given access to key stream extending over periods of partial input. For example, the Bluetooth E_0 algorithm in its intended use generates only short segments of keystream to encrypt each plaintext frame starting from a new independent initial state.

7   Conclusions

We have seen how the correlations in the Bluetooth combiner could be reduced by making a small modification in its memory update function. This improvement is, however, rather theoretical in nature, but quite interesting as such. The methods used in finding this modification are specific to Bluetooth, but could be easily adapted to other similar combiner generators. The technique involves a matrix describing potential approximations based on known non-zero linear correlations over the non-linear part of the memory update function.

We also showed how any significant correlations over a combiner can be used to launch a divide and conquer attack against any combiner generator provided that sufficient amount of the output keystream is given. If the input to the combiner is produced using a certain number of LFSRs with primitive feedback polynomials, and the number of bits of the total initial state is L, then the complexity of this attack is upper bounded by $\mathcal{O}(2^{L/2})$. This will require the amount of same magnitude $\mathcal{O}(2^{L/2})$ of the output bits.

We conclude that if the effective key length of a combiner generator is required to be about the same magnitude as the size of the initial state, then the usage of the generator must be restricted in such a way that the length of any keystream block ever produced by this generator never exceeds the shortest period of the input sequences.

References

1. Bluetooth[TM] SIG. *Bluetooth Specification*, Version 1.0 A, 24 July 1999.
2. J. Dj. Golić. Correlation properties of a general binary combiner with memory, *Journal of Cryptology*, Vol 9 Number 2, 1996, pp. 111-126.
3. W. Meier and O. Staffelbach. Correlation properties of combiners with memory in stream ciphers, *Journal of Cryptology* , Vol 5 Number 1, 1992, pp. 67-86.
4. K. Nyberg. Correlation theorems in cryptanalysis (submitted)

Preventing Double-Spent Coins from Revealing User's Whole Secret

DaeHun Nyang and JooSeok Song

Department of Computer Science, Yonsei University
SeodaemunGu ShinchonDong 134, Seoul 120-749, Korea
{nyang,jssong}@emerald.yonsei.ac.kr

Abstract. Most digital cash systems have the identity revelation capability under the condition of double-spending, but the capability may be misused as a framing tool by Bank. We present a method that provides both the identity revelation capability and the framing prevention property.

Keywords : Cryptography, Digital Cash, Zero-Knowledge Interactive Proof, Blind Signature

1 Motivation

Digital cash is the most versatile tool for the electronic commerce, and it may be an alternative of the paper money. Chaum introduced quite a new concept of digital money in his paper [3]. Besides its conveniences in its usage, Chaum's digital cash provides its user anonimity. In 1993, Brands designed an efficient off-line cash system, which includes not only all merits that Chaum's system has, but also other good functionalities such as the framing prevention[1].

One of the most intriguing properties that Brands' scheme has is that User is computationally protected against Bank's framing, or Bank cannot compute a proof of double-spending if User follows the protocols and does not double-spend. Anonimoty control and concern about Bank's framing attempts has led researches such as [5], [6], [7] and [10]. But, they did not treat the following problem: what if User spends a coin more than twice? Does the fact guarantee that User is guilty for $n(> 2)$-spending? When Bank insists that $n(> 2)$-spending occurred in a digital cash system, there are actually two possibilities. One is Bank's framing of User. Since Bank knew already user's secret when double-spending occurred, it is possible to make another illegal copy of the double-spent coin, if needed, with the collusion of Shop. The other possibility is that User actually spent one coin more than twice. Besides $n(> 2)$-spending problem, u_1 revealed may be misused by Bank in various ways(e.g. Bank can withdraw digital coins using u_1).

This unavoidable problem comes from the lack of the user's secrecy that would not be revealed even when she double-spends a coin. If she has a key that would maintain its secrecy after the double-spending revelation and the key is required to perform the payment protocol, User cannot assert Bank's framing while she spent a coin more than twice. Owing to the existence of the secret that only User knows, Bank can make Judge sure that $n(> 2)$-spending has really occurred by User, not counterfeited by Bank itself.

JooSeok Song (Ed.): ICISC'99, LNCS 1787, pp. 30–37, 2000.

However, when double-spending occurs, the maintenance of secrecy or framing prevention looks incompatible with the identity revelation. In this paper, we solve the problem by introducing another secret into Brands' scheme and using the normal 3-move zero-knowledge interactive proof(ZKIP).

2 Features on Brands' Scheme

First, we briefly describe Brands' scheme. Refer to [1] for further understanding.

- The setup of the system: Bank generates at random a generator-tuple (g, g_1, g_2) and a number $x \in_R Z_q^*$. Also, she chooses two collision-intractable hash functions H, H_0. H is used for the signature generation/verification and H_0 is for the computation of challenges. The generator-tuple and two hash functions are her public-key, and x is her private-key.
- Opening an account: User's identity is $I = g_1^{u_1} g_2$(User is computationally protected from Bank's framing) or $I = g_1^{u_1} g_2^{u_2}$(the framing attempts of Bank have negligible probability of success, regardless of computing power). Because the case of $I = g_1^{u_1} g_2^{u_2}$ is easily adapted from the case of $I = g_1^{u_1} g_2$, we proceed for the latter case.
  User generates randomly $u_1 \in_R Z_q$ and computes $I = g_1^{u_1}$. If $Ig_2 \neq 1$, User transmits I to Bank, and keeps u_1 secret. Bank keeps I with User's identifying information in her database. Bank computes $z = (Ig_2)^x$, and transmits it to User.
- The withdrawal protocol: After proving ownership of her account, User perform the following withdrawal protocol.
  1. Bank generates $w \in_R Z_q$, and sends $a = g^w$ and $b = (Ig_2)^w$ to User.
  2. User Generates $s \in_R Z_q^*, x_1, x_2 \in_R Z_q$, and computes $A = (Ig_2)^s, B = g_1^{x_1} g_2^{x_2}$ and $z' = z^s$. User generates $u, v \in_R Z_q$ and computes $a' = a^u g^v, b' = b^{su} A^v$. User then computes the challenge $c' = H(A, B, z', a', b')$, and sends the blinded challenge $c = c'/u \bmod q$ to Bank.
  3. Bank sends the response $r = cx + w \bmod q$ to User, and debits the account of User.
  4. User accepts iff $g^r = h^c a$ and $(Ig_2)^r = z^c b$. If this holds, User computes $r' = ru + v \bmod q$.
  Now User has a coin that looks like $(A, B, Sign(A, B) = (z', a', b', r'))$.
- The payment protocol:
  1. User sends $(A, B, Sign(A, B))$ to Shop.
  2. If $A \neq 1$, then Shop computes and sends the challenge $d = H_0(A, B, I_S, date/time)$.
  3. User computes and sends the responses $r_1 = d(u_1 s) + x_1 \bmod q$ and $r_2 = ds + x_2 \bmod q$.
  4. Shop accepts iff $Sign(A, B)$ is valid, and $g_1^{r_1} g_2^{r_2} = A^d B$.
- The deposit protocol: Shop performs the same procedure as that in the payment protocol with Bank. Bank accepts iff $Sign(A, B)$ is valid, and $g_1^{r_1} g_2^{r_2} = A^d B$ and A has not been stored before. If A is already in the deposit database, Bank can find the double-spender's identity by computing $g_1^{(r_1 - r_1')/(r_2 - r_2')}$.

The role of u_1 in Brands' scheme is both to identify each user's coins when double-spending occurs and to prevent Bank from framing User on double-spending. Thus, the revelation of u_1 means not only double-spending detection, but also the possibility of Bank's framing. After Bank's double-spending detection, Bank comes to know u_1 and s. With u_1 and s, Bank can make another forged transcript of dealing with the double-spent coin. In this way, Bank can frame $n(> 2)$-spending on User for an arbitrary n. Besides that, u_1 revealed may be misused by Bank in various ways(e.g. Bank can withdraw digital coins using u_1). If needed, Bank may ask Shop to accept the forged transcript of dealing as true in front of Judge. Confronted with this case, User has no way to insist that she has spent the coin not $n(> 2)$ times, but only twice.

3 Protocol Design Concept

As mentioned before, u_1, the one secret that User has plays two roles in Brands' scheme. With only one secret, we cannot obtain both the identity revelation and the framing prevention when double-spending occurs. As a natural consequence, we separate u_1 into u_1 and u_2 such that each secret might play only one role. That is, u_1 reveals User's identity and u_2 prevents Bank from framing User on $n(> 2)$-spending when double-spending occurs. u_2 would not be revealed even when User double-spends a coin. Separation of u_1 into u_1 and u_2 according to their roles enables us to achieve both functionalities.

Note that Brands' payment protocol does not include the commitment move by the prover, whereas 3-move ZKIP systems such as [4, 8] include the commitment stage. In that point of view, Brands' scheme separates the commitment stage from the payment protocol and includes it in the withdrawal protocol. As known, committed values must be different whenever the prover proves its knowledge in the 3-move ZKIP. If the same committed value is used again, the verifier can easily compute the prover's secret. The usage of the same committed value during the proof of the same knowledge in ZKIP corresponds exactly to the double-spending of the same coin in Brands' setting. And the exposure of the secret in ZKIP corresponds to the exposure of User's identity in Brands' scheme.

The knowledge of the first secret(u_1) is proved in the same way as that of Brands', whereas the proof of knowledge of the second secret(u_2) is performed by the normal 3-move ZKIP. This setting allows that u_1 will be revealed but u_2 will be kept secret whenever double-spending occurs. Thus, the payment protocol now, looks the normal ZKIP for u_2.

Our approach has another merit that User does not need to perform the account opening procedure again. Unlike Brands' scheme, we can keep using the revealed u_1 together with u_2. Suppose that u_1 is known to Bank before the withdrawal protocol starts owing to the previous double-spending. After the withdrawal ends, Bank cannot compute $A = (Ig_2)^s$ without knowing s, which is randomly and secretly selected by User. That is, Bank's knowledge of u_1 does not help Bank to know what the blinded message A of $m = Ig_2$ is. Consequently, Bank cannot link User with a coin that is returned after the deposit protocol. Even if Bank tries to do exhaustive search for every user's identity with coins after the deposit protocol, she cannot link User's identity with a one-spent coin $(r_1, r_2, A, B, sign(A, B), I_S, date/time)$. It can be easily seen as following:

$$r_1 = d(u_1 s) + x_1 \bmod q$$

$$r_2 = ds + x_2 \bmod q$$

In the above simultaneous equation, there are four unknown variables s, x_1, x_2, u_1. Thus, even when Bank correctly guesses u_1, there are infinitely many solutions of the equation and she has no way of validating of her guess. However, Bank may be able to frame User by making a forged transcript by the known u_1 unless User has the unknown u_2.

4   Privacy Enhaced Digital Cash System

Our digital cash system consists of system setup, opening an account, the withdrawal protocol, the payment protocol and the deposit protocol. All the conventions are the same as that of Brands' one. As mentioned already, we augment Brands' scheme such that it has two secrets for User and the added secret is treated as the secret of the normal 3-move ZKIP technique.

- System setup: This part is the same as that of Brands' one, but domains of two hash functions H and H_0 are corrected such that it may accommodate the added functionalities.
  $H : G_q \times G_q \times G_q \times G_q \times G_q \times G_q \times G_q \times G_q \to Z_q^*$
  $H_0 : G_q \times G_q \times G_q \times G_q \times SHOP - ID \times DATE/TIME \to Z_q$
- Opening an account: Besides u_1, User has another secret u_2 and corresponding account information I_2. Also, Bank must return account information restricted by x, the Bank's secret. Only adding another secret u_2 is different from Brands'.

$$I_1 = g_1^{u_1}, I_2 = g_1^{u_2}$$

$$z_1 = (I_1 g_2)^x, z_2 = (I_2 g_2)^x$$

- The withdrawal protocol: User blinds account information by s_1 and s_2 that is related to u_1 and u_2, respectively. However, the number that will be used as a commitment for u_2 in the payment protocol is not prepared in this phase. Instead, the commitment will be generated during each payment. After the completion of the withdrawal protocol, User has a coin $(A, B, C, Sign(A, B, C))$.
  1. Bank generates $w \in_R Z_q$, and sends $a = g^w$ and $b_1 = (I_1 g_2)^w, b_2 = (I_2 g_2)^w$ to User.
  2. User Generates $s_1, s_2 \in_R Z_q^*, x_1, x_2 \in_R Z_q$, and computes

     $$A = (I_1 g_2)^{s_1}, B = g_1^{x_1} g_2^{x_2}, C = (I_2 g_2)^{s_2} \text{ and } z_1' = z_1^{s_1}, z_2' = z_2^{s_2}.$$

     User generates $u, v \in_R Z_q$ and computes

     $$a' = a^u g^v, b_1' = b_1^{s_1 u} A^v, b_2' = b_2^{s_2 u} C^v$$

User then computes the challenge

$$c' = H(A, B, C, z'_1, z'_2, a', b'_1, b'_2),$$

and sends the blinded challenge $c = c'/u \bmod q$ to Bank.

3. Bank sends the response $r = cx + w \bmod q$ to User, and debits the account of User.

4. User accepts iff

$$g^r = h^c a \text{ and } (I_1 g_2)^r = z_1^c b_1, (I_2 g_2)^r = z_2^c b_2$$

If this holds, User computes

$$r' = ru + v \bmod q$$

Proposition 1. *A and C(the blinded numbers of $I_1 g_2$ and $I_2 g_2$, respectively) and their signatures are unconditionally untraceable to any specific execution of the withdrawal protocol.*

Proof Sketch: Given any $h \neq 1$, for each pair $(A, B, C, Sign(A, B, C) = (z'_1, z'_2, a', b'_1, b'_2, r'))$ and the information that Bank gets during the execution of any protocol in which User accepts, there are exactly q possible random choices of sets (s_1, s_2, t, u, v) that could have been made by User that bring about the link. $\square$

Proposition 2. *Under the discrete logarithm assumption, there is no way that has noneglibible probability of success for User such that she ends up with a pair $(A, B, C, sign(A, B, C))$ for which she knows two different representations of (A, B, C) with respect to (g_1, g_2).*

Proof Sketch: Refer to Proposition 12 in [2]. $\square$

Proposition 2 means that User cannot double-spend a coin without changing the internal structure of the coin, that is the account information. With the propositions, we obtain the following assumption.

Assumption 1.*The withdrawal protocol is a restrictive blind signature protocol (blinded numbers are A and C) with blinding invariant functions IV_1 and IV_2 with respect to (g_1, g_2) defined by $IV_1(a_1, a_2) = IV_2(a_1, a_2)a_1/a_2 \bmod q$.*

Now, Bank's attempt to link a coin with User is always frustrated by the following proposition.

Proposition 3. *Under the discrete logarithm assumption, Bank's prior knowledge of u_1 does not help her compute the coin.*

Proof: It's trivial since s_1, s_2, x_1, x_2 are chosen secretly to make a coin $(A = (I_1 g_2)^{s_1}, B = g_1^{x_1} g_2^{x_2}, C = (I_2 g_2)^{s_2}, sign(A, B, C))$ by User. Bank cannot compute them. $\square$

- The payment protocol: User can prove the knowledge of the representation of A, B, C with respect to $g_1 g_2$ with the payment protocol. A, B is for the knowledge proof of u_1, and C is for that of u_2. In the protocol, $D = g_1^{deal_1} g_2^{deal_2}$ corresponds to the commitment numbers of the normal 3-move ZKIP.

1. User generates $deal_1, deal_2 \in_R Z_q^*$ and computes

$$D = g_1^{deal_1} g_2^{deal_2}$$

   User sends $(A, B, C, D, Sign(A, B, C))$ to Shop.
2. If $A \neq 1$, then Shop computes and sends the challenge

$$d = H_0(A, B, C, D, I_S, date/time)$$

3. User computes and sends the responses

$$r_1 = d(u_1 s_1) + x_1 \bmod q, r_2 = d s_1 + x_2 \bmod q$$

$$r_3 = d(u_2 s_2) + deal_1 \bmod q, r_4 = d s_2 + deal_2 \bmod q$$

4. Shop accepts iff $Sign(A, B, C)$ is valid, and

$$g_1^{r_1} g_2^{r_2} = A^d B, g_1^{r_3} g_2^{r_4} = C^d D$$

Proposition 4. *During the payment protocol, User does not reveal more information than Brands' does.*

Proof Sketch: The information for the proof of knowledge of representation of A with respect to $g_1 g_2$ is the same as that for Brands'. Added information is for the the proof of knowledge of representation of C, but it is Schnorr's identification scheme and does not reveal any information on u_2. $\square$

- The deposit protocol: First, Bank checks the validity of the coin by scrutinizing the signature and the transcript submitted by Shop. Double-spending is examined by comparing (A, C) with entries in Bank's database. If found match, Bank can extract account information by computing

$$I_1 = g_1^{(r_1 - r_1')/(r_2 - r_2')}.$$

Since $deal_1 \neq deal_1'$ and $deal_2 \neq deal_2'$, $g_1^{(r_3 - r_3')/(r_4 - r_4')}$ does not reveal I_2 even if double-spending occurs. Because Bank cannot compute any forged transcript of dealing with the double-spent coin without knowing u_2, User is computationally protected from Bank's framing attempt after the double-spending.

1. Shop sends the transcript

$$(A, B, C, D, date/time, Sign(A, B, C), r_1, r_2, r_3, r_4)$$

2. Bank accepts iff $Sign(A, B, C)$ is valid, and

$$g_1^{r_1} g_2^{r_2} = A^d B, g_1^{r_3} g_2^{r_4} = C^d D$$

and (A, C) has not been stored before. If (A, C) is already in the deposit database, Bank can find the double-spender's identity by computing

$$g_1^{(r_1 - r_1')/(r_2 - r_2')}$$

After the double-spending is detected, there are possibly two choices regarding the use of (u_1, u_2). One is that User throws away (u_1, u_2) and performs the whole procedure of account-opening again, and the other is that User keeps using (u_1, u_2). As mentioned before, the revelation of u_1 does not give any hint for Bank to link a coin with User's identity. Thus, (u_1, u_2) can be re-used without any modification. This feature is very useful, since it eliminates the need of re-opening an account and updating the database of Bank.

5 Conclusion

We presented Bank's framing problem that might arise after double-spending in digital cash systems, and propose a digital cash system that satisfies both the double-spending detection capability and the framing prevention property. The idea is that our digital cash system uses two secrets: one is for double-spending detection and the other would not be revealed even when double-spending occurs. Though we present a digital cash system based on Brands' scheme, the idea may be applied to existing and forth-coming digital cash systems.

In summary, our digital cash system has the following properties:

1. If $n(\geq 2)$-spending occurs, User cannot deny it.
2. Bank cannot frame User on $n(> 2)$-spending, while User only double-spends($n = 2$).
3. One-spent coin does not reveal User's identity.
4. After the double-spending detection, account-reopening procedure is not needed.

Though we did not present the observer-based cash protocol, it is easy to augment the proposed protocol such that it might have that pre-restriction capability on double-spending. Other than that, our technique can be easily embeded to recently published digital cash system in [5].

References

1. S. Brands, Untraceable off-Line cash in wallets with observers, Advances in Cryptology : Proceedings of Crypto '93, Lecture Notes in Computer Science, Springer-Verlag, (1994) pp. 302-318.
2. S. Brands, An Efficient Off-line Electronic Cash System Based on the representation problem, CWI Technical Report CS-R9323, (1993).

3. D. Chaum, Security Without Identification Transaction Systems to Make Big Brother Obsolete, *Communications of the ACM*, Vol. 28, No. 10, 1985, pp. 1030–1044.
4. Amos Fiat and Adi Shamir, How To Prove Yourself: Practical Solutions to Identification and Signature Problems, Advances in Cryptology : Proceedings of Crypto 86, Lecture Notes in Computer Science, Springer-Verlag, (1987) pp. 186–194.
5. Y. Frankel, Y. Tsiounis and M. Yung, Indirect discourse proof: achieving efficient fair off-line e-cash, Advances in Cryptology : Proceedings of AsiaCrypt 96, Lecture Notes in Computer Science, Springer-Verlag, (1996) pp. 286–300.
6. M. Jakobsson and M. Yung, Revokable and versatile electronic money, The 3rd ACM Conference on Computer and Communications Security, (1996) pp. 76–87.
7. S. Miyazaki and K. Sakurai, A more efficient untraceable e-cash system with partially blind signatures based on the discrete logarithm problem, Financial Cryptography, (1998).
8. J.J. Quisquater and L.S. Guillou, A Paradoxical Identity Based Signature Scheme Resulting from Zero Knowledge, Advances in Cryptology : Proceedings of EuroCrypt 88, Lecture Notes in Computer Science, Springer-Verlag, (1988) pp. 77–86.
9. C.P. Schnorr, Efficient Identification and Signatures for Smart cards, Advances in Cryptology : Proceedings of Crypto 89, Lecture Notes in Computer Science, Springer-Verlag, New York, (1989) pp. 239–251.
10. B. Schoenmakers, An efficient electronic payment system withstanding parallel attacks, Technical Report, CWI, (1995), Available in "http://www.cwi.nl/cwi/publications/".

On the Optimal Diffusion Layers with Practical Security against Differential and Linear Cryptanalysis

Ju-Sung Kang[1], Choonsik Park[1], Sangjin Lee[2], and Jong-In Lim[2]

[1] Section 0710, Electronics and Telecommunications Research Institute
Yusong P.O. Box 106, Taejon, 305-350, Korea
{jskang,csp}@etri.re.kr
[2] Department of Mathematics, Korea University
Chochiwon, 347-700, Korea
{sangjin,jilim}@tiger.korea.ac.kr

Abstract. In this works we examine the diffusion layers of some block ciphers referred to as substitution-permutation networks. We investigate the practical security of these diffusion layers against differential and linear cryptanalysis by using the notion of active S-boxes. We show that the minimum number of differentially active S-boxes and that of linearly active S-boxes are generally not identical and propose some special conditions in which those are identical. Moreover, we apply our results to analyze three diffusion layers used in the block ciphers E2, CRYPTON and Rijndael, respectively. It is also shown that these all diffusion layers have achieved optimal security according to their each constraints of using operations.

1 Introduction

Shannon suggested that practical secure product ciphers may be constructed using a mixing transformation consisting of a number of layers or rounds of "confusion" and "diffusion"[19]. The confusion component is a nonlinear substitution on a small subblock and the diffusion component is a linear mixing of the subblock connections.

The Substitution-Permutation Networks(SPN) structure is directly based on the concepts of confusion and diffusion. One round of an SPN structure generally consists of three layers of substitution, permutation, and key addition. Substitution layer is made up of small nonlinear substitutions referred to as S-boxes easily implemented by table lookup for confusion effect. Permutation layer is a linear transformation in order to diffuse the cryptographic characteristics of substitution layer. Key addition layer is to implant round subkeys of the cipher and the position of this layer is variable according to ciphers. A typical example of one round of an SPN structure is given in Figure 1.

Due to memory requirements, most block cipher designers use small S-boxes, e.g. with 4 or 8 input bits. Thus the diffusion of S-box outputs by permutation

JooSeok Song (Ed.): ICISC'99, LNCS 1787, pp. 38–52, 2000.
© Springer-Verlag Berlin Heidelberg 2000

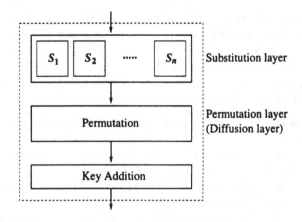

Fig. 1. One round of an SPN structure

layer plays a great role in providing immunity against various attacks including differential and linear cryptanalysis.

On the other hand, permutation layers of most modern block ciphers are not simple bitwise position permutations or transpositions but linear transformations on some vector spaces over various finite fields. Hence in this paper, we call permutation layer as "diffusion layer" for the distinctness.

Diffusion layers of modern block ciphers of SPN structure are linear transformations on Z_2^n over some finite fields such as $GF(2)$ or $GF(2^n)$ and have one-to-one correspondence to appropriate matrix. That is, most diffusion layers have appropriate matrix representations. In this work, with these matrix representations we study the practical security against differential and linear cryptanalysis for the diffusion layers of three AES 1 round candidate algorithms E2, CRYPTON and Rijndael. The diffusion effects of the diffusion layers constructed with simple transpositions and an appropriate linear transformation were well studied in [7]. However, the diffusion layers of E2, CRYPTON and Rijndael are different from that of [7].

2   Practical Security against DC and LC

2.1   Background

The most well-known method of analyzing block ciphers today is differential cryptanalysis(DC), proposed by Biham and Shamir[2,3] in 1990. DC is a chosen plaintext attack in which the attacker chooses plaintexts of certain well-considered differences. Biham and Shamir used the notion of "characteristic", while Lai, Massey and Murphy[11] showed that the notion "differential" strictly reflects the strength of a cipher against DC. Roughly speaking, a differential is a collection of characteristics.

Another method of analyzing block ciphers is linear cryptanalysis(LC), published by Matsui[13] in 1993. The attacks based on LC are known plaintext attacks and the attack on the DES is faster than the attack by DC. The first version of LC applied "linear approximation" to an attack of block ciphers, but Nyberg[15] has considered a collection of linear approximation, which she called a "linear hull", for strict evaluation of the strength against LC.

Kanda et al.[8] classified four measures to evaluate the security of a cipher against DC and LC as follows:

- Precise measure: The maximum average of differential and linear hull probabilities[11,15].
- Theoretical measure: The upper bounds of the maximum average of differential and linear hull probabilities[14,16,1,9].
- Heuristic measure: The maximum average of differential characteristic and linear approximation probabilities[2,3,13].
- Practical measure: The upper bounds of the maximum average of differential characteristic and linear approximation probabilities[10,18,5].

DC and LC are the most powerful attacks to most symmetric block ciphers. Accordingly, it is a basic requisite for the designer to evaluate the security of any new proposed cipher against DC and LC, and to prove that it's sufficiently resistant against them. In this paper, we consider practical measure out of the above four measures since practical measure is feasible to evaluate while others are not practical.

2.2   Differentially and Linearly Active S-Boxes

Let S be an S-box with m input and output bits, i.e., $S : Z_2^m \to Z_2^m$. Differential and linear probabilities of S are defined as the following definition.

Definition 1 *For any given Δx, Δy, a, $b \in Z_2^m$, define differential and linear probabilities of S by*

$$DP^S(\Delta x \to \Delta y) = \frac{\#\{x \in Z_2^m \; : \; S(x) \oplus S(x \oplus \Delta x) = \Delta y\}}{2^m}$$

and

$$LP^S(a \to b) = \left(\frac{\#\{x \in Z_2^m \; : \; <a, x> = <b, S(x)>\}}{2^{m-1}} - 1 \right)^2 ,$$

respectively, where $< \alpha, \beta >$ denotes the parity(0 or 1) of bitwise product of α and β.

DP^S and LP^S for a strong S-box S should be small enough for any input difference $\Delta x \neq 0$ and output mask value $b \neq 0$. So we define parameters represent immunity of an S-box and each substitution layer of SPN structure against DC and LC as follows:

Definition 2 *The maximum differential and linear probabilities of S are defined by*

$$DP_{\max}^S = \max_{\Delta x \neq 0, \; \Delta y} DP^S(\Delta x \to \Delta y)$$

and

$$LP^S_{\max} = \max_{a,\, b \neq 0} LP^S(a \to b) \,,$$

respectively.

Definition 3 *Assume that each substitution layer of a SPN structure consist of* n *S-boxes* $S_1, S_2, \cdots, S_n$. *The maximum differential and linear probability of the substitution layer are defined by*

$$p = \max_{1 \leq i \leq n} DP^{S_i}_{\max} \,, \quad q = \max_{1 \leq i \leq n} LP^{S_i}_{\max} \,,$$

respectively.

Evaluation of security for a block cipher of SPN structure by practical measure begins with the concept of active S-box.

Definition 4 *Differentially active S-box is defined as an S-box given a nonzero input difference and linearly active S-box as an S-box given a nonzero output mask value.*

By computing the minimum number of differentially and linearly active S-boxes, we can evaluate security of a block cipher on the viewpoint of practical security against DC and LC[10, 18, 5]. We can obtain upper bounds of the maximum differential characteristic and linear approximation probabilities from the minimum number of active S-boxes. Thus in the case of SPN structure, it is important to analyze the increasing amounts of minimum number of active S-boxes by considering diffusion layer in consecutive two rounds.

Note that we can omit the key addition layer to compute the number of active S-boxes since this layer has no influence under the assumption that the key addition layer is performed by bitwise EXORs. Define the SDS function with three layers of substitution-diffusion-substitution for analyzing the role of diffusion layer to rise the number of active S-boxes in consecutive two rounds of a SPN structure(Figure 2).

Throughout this paper we assume that all S-boxes in the substitution layer are bijective. If an S-box is bijective and differentially/linearly active, then it has a non-zero output difference/input mask value[14]. So when all S-boxes in substitution layer are bijective, we can define the minimum number of active S-boxes of the SDS function. Set diffusion layer of the SDS function as D, input difference of D as $\Delta x = x \oplus x^*$, output difference as $\Delta y = y \oplus y^* = D(x) \oplus D(x^*)$, and input and output mask value as a and b, respectively.

Definition 5 *The minimum number of differentially and linearly active S-boxes of the SDS function are defined by*

$$\beta_d(D) = \min_{\Delta x \neq 0} \{H_c(\Delta x) + H_c(\Delta y)\}$$

and

$$\beta_l(D) = \min_{b \neq 0} \{H_c(a) + H_c(b)\} \,,$$

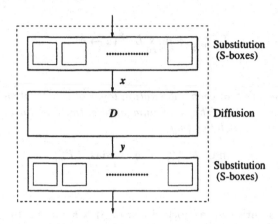

Fig. 2. SDS function

respectively, where for each $x = (x_1, x_2, \cdots, x_n) \in Z_2^{mn}$, $\forall x_i \in Z_2^m$, component Hamming weight of x is defined by

$$H_c(x) = \#\{1 \leq i \leq n \ : \ x_i \neq 0\} \ .$$

Theorem 1 *Let $\beta_d(D)$ and $\beta_l(D)$ be the minimum differentially and linearly active S-boxes in the SDS function, respectively. Then the maximum differential and linear probabilities p_{sds} and q_{sds} of the SDS function hold for*

$$p_{sds} \leq p^{\beta_d(D)} \quad and \quad q_{sds} \leq q^{\beta_l(D)} \ .$$

The above theorem is obtained easily by the maximality of p(or q) and the minimality of $\beta_d(D)$(or $\beta_l(D)$). Evaluation of practical security against DC and LC is based on this theorem.

2.3   Matrix Representation of Diffusion Layer

Most diffusion layers of modern block ciphers of SPN structure are linear transformations on Z_2^n over some finite fields such as $GF(2)$ or $GF(2^n)$ and have one-to-one correspondence to appropriate matrix. That is, most diffusion layers have appropriate matrix representations. If we use this matrix representation for a diffusion layer, then we obtain the relationship between input and output differences(or mask values) as the following theorem.

Theorem 2 *Assume that the diffusion layer D of the SDS function is represented as a matrix M. Then the matrix for relationship between input and output differences is represented as the same matrix M, while the matrix for relationship between output and input mask values is represented as the transposed matrix M^t. That is,*

$$\Delta y = M \Delta x \ , \quad a = M^t b \ .$$

Proof: This theorem is proven in [4] by using the notion of correlation matrix.

□

It is possible that we compute the minimum number of differentially and linearly active S-boxes($\beta_d(D)$ and $\beta_l(D)$) of the SDS function by using the matrix representation of Theorem 2. However the minimum number of differentially and linearly active S-boxes are not identical in general. In the next section, we will show that $\beta_d(D) \neq \beta_l(D)$ by proposing a counterexample. On the other hand, the minimum number of differentially and linearly active S-boxes are identical for the special types of representation matrix M_D as the following two theorems.

Theorem 3 *Let the diffusion layer D of the SDS function be represented as $n \times n$ matrix M. If M is a symmetric or orthogonal matrix, then $\beta_d(D) = \beta_l(D)$.*

Proof: By Theorem 2 and Definition 5, $\beta_d(D)$ and $\beta_l(D)$ are expressed as follows:

$$\beta_d(D) = \min_{\Delta x \neq 0} \{H_c(\Delta x) + H_c(M\Delta x)\},$$

$$\beta_l(D) = \min_{b \neq 0} \{H_c(M^t b) + H_c(b)\}.$$

From this, we can easily see that $\beta_d(D) = \beta_l(D)$ if M is a symmetric matrix which $M^t = M$.

Meanwhile, if M is an orthogonal matrix that $M^{-1} = M^t$, then $a = M^t b$ implies that $b = Ma$, and the condition $b = Ma \neq 0$ is identical to $a \neq 0$ since M is an invertible matrix. Thus

$$\beta_l(D) = \min_{a \neq 0} \{H_c(a) + H_c(Ma)\},$$

and $\beta_d(D) = \beta_l(D)$.

□

Theorem 4 *If M^t is obtained from M by applying operations of exchanging row or column vectors, then $\beta_d(D) = \beta_l(D)$, where M is the representation matrix of the diffusion layer D of the SDS function.*

Proof: The operation of exchanging row vectors of M results in changing the order of components of output difference Δy, and this operation does not affect to the component Hamming weight $H_c(\Delta y)$. On the other hand, it is clear that $H_c(\Delta y)$ is determined by column vectors of M but unconcerned to their location. Thus the operation of exchanging column vectors of M also does not affect to the component Hamming weight $H_c(\Delta y)$. Since a row(column) vector of M_D is a column(row) vector of M^t, operations of exchanging row or column vectors of M doesn't affect to the component Hamming weight $H_c(a)$ also. Therefore, if M^t is obtained from M by those operations, $\beta_d(D) = \beta_l(D)$.

□

3   Diffusion Layer of E2

E2 is a 128-bit block cipher designed by NTT of Japan, which is one of the fifteen candidates in the first round of the AES(Advanced Encryption Standard) project[17]. Overall structure of E2 is Feistel network as DES and adopt round function to what is called "2-round SPN structure"[8]. However 2-round SPN structure without key addition layer is exactly equal to the SDS function.

In the round function of E2, diffusion layer is constructed with just bitwise EXORs and expressed as an 8×8 matrix with only 0 and 1 entries. Kanda et al.[8] described the relationship between the matrix representation and the actual construction of the diffusion layer and proposed a search algorithm for constructing the optimal diffusion layer. Furthermore, they have shown that the round function of Feistel structure with the 2-round SPN structure requires one-fourth as many rounds as the "1-round SPN structure", which is composed of one substitution and one permutation layer, to achieve the same differential and linear probabilities. That is, the round function using the 2-round SPN structure is twice as efficient as that using the 1-round SPN structure.

Let each S-box of substitution layers have m input and output bits, and the number of S-boxes in each substitution layer be n. Assume that inputs of the SDS function are linearly transformed to outputs per m-bit and the diffusion layer is constructed with just bitwise EXORs. Then the diffusion layer is represented as an $n \times n$ matrix M which all entries are zero or one as follows:

$$y_i = \bigoplus_{j=1}^{n} \mu_{ij} x_j = \bigoplus_{\mu_{ij}=1} x_j ,$$

where $x = (x_1, x_2, \cdots, x_n) \in (Z_2^m)^n$ is an input, $y = (y_1, y_2, \cdots, y_n)$ is the output, and $M = (\mu_{ij})$.

Kanda et al.[8] studied diffusion property of the diffusion layer with this matrix representation. Their study was based on the relationship between the matrix for differential characteristic and linear approximation. However they made two conjectures to unfold their theory. The Conjecture 1 of [8] is correct since this is a special case of Theorem 2, but the Conjecture 2 of [8] is a wrong opinion. We disprove this conjecture by proposing a counterexample.

Conjecture 2 of [8]. *In the SDS function, the minimum number of differentially active S-boxes is equal to the minimum number of lineally active S-boxes. That is, $\beta_d(D) = \beta_l(D)$, where M is the representation matrix of the diffusion layer D.*

Counterexample for the Conjecture 2 of [8]: Suppose that the diffusion layer of SDS function with $n = 4$ be represented by the following invertible matrix:

$$M = \begin{pmatrix} 1 & 1 & 1 & 1 \\ 1 & 0 & 0 & 1 \\ 0 & 1 & 0 & 1 \\ 0 & 0 & 1 & 0 \end{pmatrix} , \quad M^t = \begin{pmatrix} 1 & 1 & 0 & 0 \\ 1 & 0 & 1 & 0 \\ 1 & 0 & 0 & 1 \\ 1 & 1 & 1 & 0 \end{pmatrix} .$$

If $H_c(\Delta x) = 1$, then $H_c(\Delta y) \geq 2$ since $H_c(\Delta y)$ is determined by a column vector of M and Hamming weight of each column vector is at least 2. $H_c(\Delta y)$ is determined by the EXORs between any different two column vectors if $H_c(\Delta x) = 2$. Any the EXOR between two column vectors has Hamming weight at least 1. Thus the minimum number of differentially active S-boxes is $\beta_d(D) = 3$.

On the other hand, by Theorem 2, relationship between output and input mask values is represented as the transpose matrix M^t of M. Note that Hamming weight of the fourth column vector of M^t is 1. Consider the output mask value of the form $b = (0, 0, 0, b_4)$, $b_4 \neq 0$,

$$\begin{pmatrix} 1\,1\,0\,0 \\ 1\,0\,1\,0 \\ 1\,0\,0\,1 \\ 1\,1\,1\,0 \end{pmatrix} \begin{pmatrix} 0 \\ 0 \\ 0 \\ b_4 \end{pmatrix} = \begin{pmatrix} 0 \\ 0 \\ b_4 \\ 0 \end{pmatrix}$$

then corresponding input mask value $a = (0, 0, b_4, 0)$. From this we can obtain that $\beta_l(D) = 2$. Consequently we know that $\beta_d(D) \neq \beta_l(D)$ for the above 4×4 matrix M. $\qquad \square$

As a matter of convenience, we abuse our notation and use $\beta_d(M)$(or $\beta_l(M)$) for $\beta_d(D)$(or $\beta_l(D)$) henceforth, where M is the representation matrix of the diffusion layer D. In the block cipher E2, designers considered the SDS function with $n = 8$. Kanda et al.[8] suggested a method of determining an 8×8 matrix $M = P$ yielding the maximum value of $\beta_d(P)$ using the search algorithm. Using this search algorithm, they found that there is no matrix with $\beta_d(P) \geq 6$, and that there are some candidate matrices with $\beta_d(P) = 5$. Here, we give theoretical proof for the fact that $\beta_d(P) = 5$ is optimal and also that $\beta_l(P) \leq 5$, where P is a 8×8 invertible matrix.

Theorem 5 *Assume that the number of S-boxes in the substitution layer of the SDS function is 8($n = 8$). If the representation matrix P of the diffusion layer is an 8×8 invertible matrix, then $\beta_d(P)$, $\beta_l(P) \leq 5$.*

Proof: Since P is an 8×8 invertible matrix, eight column vectors $P_1, P_2, \cdots, P_8$ are linearly independent. Thus the number of columns with the Hamming weight 8 is at most one. Note that $\beta_d(P)$ is closely related to the Hamming weights of column vectors of P. We separate the proof into four cases. Here, the Hamming weight $H_c(P_j)$ of a column vector P_j is the number of entries with 1 in P_j.
Case 1 If $\min_{1 \leq j \leq 8} H_c(P_j) = 7$, for any two column vectors P_j and P_k, $H_c(P_j \oplus P_k) \leq 2$. By considering Δx such that $H_c(\Delta x) = 2$, we obtain that $\beta_d(P) \leq 2 + 2 = 4$.
Case 2 Suppose that $\min_{1 \leq j \leq 8} H_c(P_j) = 6$. If there exists a column vector with Hamming weight 8, then $H_c(\Delta y) \leq 2$ for some Δx such that $H_c(\Delta x) = 2$. If there exists a column vector with Hamming weight 7, the minimum value of $H_c(\Delta y)$ at most 3, since we can consider the EXORs between column with Hamming weight 6 and 7, where $H_c(\Delta x) = 2$. At last, if the Hamming weight of all column vectors is 6, then although some different four column vectors include

0 entries in distinct rows, the Hamming weight of EXORs between one of this four columns and another fifth column vector is 2. Consequently, we obtain that $\beta_d(P) \leq 5$.

Case 3 Assume that $\min_{1 \leq j \leq 8} H_c(P_j) = 5$. If there exists a column vector with the Hamming weight 8, then $H_c(\Delta y) \leq 3$ for some Δx so that $H_c(\Delta x) = 2$. If there exists a column vector with the Hamming weight 7 or 6, by similar analysis to *Case 2*, we can obtain what we want. In the case that the Hamming weight of each column vectors is 5, although 0 entries of some different five column vectors are arranged optimally, another sixth column vector and one of this five columns have in common at least two 0 entries at the same rows. Thus $H_c(\Delta y)$ is at most 2 where $H_c(\Delta x) = 2$. Therefore $\beta_d(P) \leq 5$ also holds in this case.

Case 4 Assume that $\min_{1 \leq j \leq 8} H_c(P_j) \leq 4$. Consider the only Δx such that $H_c(\Delta x) = 1$. Then we obtain easily $\beta_d(P) \leq 5$, since there exists a column with the Hamming weight 4.

By *Case 1 - 4*, we obtain that $\beta_d(P) \leq 5$ always holds whenever P is 8×8 invertible matrix. On the other hand, by Theorem 2, $\beta_l(P)$ is related to P^t. Thus we can also obtain the same result for $\beta_l(P)$ by considering the Hamming weight of row vectors instead of column vectors of P.   □

We can see that each case in the proof of the above theorem depends only on the Hamming weight property of column vectors. Kanda et al.[8] found 10080 candidate matrices with $\beta_d(P) = 5$ by searching algorithm and for all candidate matrices, the total Hamming weight is 44 with 4 column(row) vectors of six Hamming weight and 4 column(row) vectors of five Hamming weight. One of these candidate matrices which is easy to determining construction is used in the block cipher E2 as follows:

$$P = \begin{pmatrix} 0&1&1&1&1&1&1&0 \\ 1&0&1&1&0&1&1&1 \\ 1&1&0&1&1&0&1&1 \\ 1&1&1&0&1&1&0&1 \\ 1&1&0&1&1&1&0&0 \\ 1&1&1&0&0&1&1&0 \\ 0&1&1&1&0&0&1&1 \\ 1&0&1&1&1&0&0&1 \end{pmatrix} , \quad P^t = \begin{pmatrix} 0&1&1&1&1&1&0&1 \\ 1&0&1&1&1&1&1&0 \\ 1&1&0&1&0&1&1&1 \\ 1&1&1&0&1&0&1&1 \\ 1&0&1&1&1&0&0&1 \\ 1&1&0&1&1&1&0&0 \\ 1&1&1&0&0&1&1&0 \\ 0&1&1&1&0&0&1&1 \end{pmatrix} .$$

It is easy to see that the above matrix P^t is obtained from P by applying the operations of exchanging row or column vectors. Therefore, by Theorem 4, we obtain that

$$\beta_d(P) = \beta_l(P) .$$

That is, the Conjecture 2 of [8] is correct for the above matrix P.

4   Diffusion Layer of CRYPTON

The block cipher CRYPTON[12] is one of the fifteen candidates in the first round of the AES. Overall structure of CRYPTON is the SPN structure influenced by SQUARE[5] which using round function allowing parallel processing on the whole data block. The round function of CRYPTON consists of four parallelizable steps: byte-wise substitution, column-wise bit transformation, column-to-row transposition, and key addition. Out of these four steps the column-wise bit transformation and the column-to-row transposition are included in the diffusion layer. In this section, we concentrate our discussion on the column-wise bit transformation.

The bit transformation π mixes four bytes in each byte column of a 4×4 byte array and π is expressed as $\pi = (\pi_0, \pi_1, \pi_2, \pi_3)$, where π_i is the bit transformation of the i-th column. In order to analyze the diffusion effect of π, it suffices to consider π_0 since another π_i's$(i = 1, 2, 3)$ are obtained from π_0 by the simple byte transposition.

For any $x = (x_1, x_2, x_3, x_4) \in (Z_2^8)^4$, $y = \pi_0(x) = (y_1, y_2, y_3, y_4) \in (Z_2^8)^4$ is defined by

$$y_1 = (x_1 \wedge m_1) \oplus (x_2 \wedge m_2) \oplus (x_3 \wedge m_3) \oplus (x_4 \wedge m_4)$$
$$y_2 = (x_1 \wedge m_2) \oplus (x_2 \wedge m_3) \oplus (x_3 \wedge m_4) \oplus (x_4 \wedge m_1)$$
$$y_3 = (x_1 \wedge m_3) \oplus (x_2 \wedge m_4) \oplus (x_3 \wedge m_1) \oplus (x_4 \wedge m_2)$$
$$y_4 = (x_1 \wedge m_4) \oplus (x_2 \wedge m_1) \oplus (x_3 \wedge m_2) \oplus (x_4 \wedge m_3) ,$$

where for $a, b \in Z_2^8$, $a \wedge b$ is the bitwise AND, and

$$m_1 = 0xfc = 11111100_{(2)}$$
$$m_2 = 0xf3 = 11110011_{(2)}$$
$$m_3 = 0xcf = 11001111_{(2)}$$
$$m_4 = 0x3f = 00111111_{(2)} .$$

Now we consider the matrix representation of the bit transformation π_0. The bit transformation π_0 can be implemented by using bitwise EXOR and AND logic, and is a linear transformation on the vector space Z_2^{32} over the finite field $GF(2)$. Hence π_0 has the unique matrix representation under the appropriate basis. Let the standard basis be given for the vector space Z_2^{32}. Then the 32×32 matrix Q over $GF(2)$ corresponding to π_0 has expression as follows: For $1 \leq i \leq 4$, set

$$MI_i = \begin{pmatrix} m_{i1} & 0 & 0\,0\,0\,0\,0 & 0 \\ 0 & m_{i2} & 0\,0\,0\,0\,0 & 0 \\ \multicolumn{4}{c}{\cdots\cdots\cdots\cdots\cdots} \\ \multicolumn{4}{c}{\cdots\cdots\cdots\cdots\cdots} \\ \multicolumn{4}{c}{\cdots\cdots\cdots\cdots\cdots} \\ 0 & 0 & 0\,0\,0\,0\,0 & m_{i8} \end{pmatrix} ,$$

where $m_i = (m_{i1}, m_{i2}, \cdots, m_{i8})$, then

$$Q = \begin{pmatrix} MI_1 & MI_2 & MI_3 & MI_4 \\ MI_2 & MI_3 & MI_4 & MI_1 \\ MI_3 & MI_4 & MI_1 & MI_2 \\ MI_4 & MI_1 & MI_2 & MI_3 \end{pmatrix} .$$

That is,

$$(\pi_0(x))^t = Qx^t .$$

In [12], the author examined the diffusion property of π_0 closely by using the notion of "diffusion order". However the definition of diffusion order is exactly the same as that of "branch number" used in the former document of block cipher SHARK[18].

Definition 6 *For any transformation L, the branch number of L is defined by*

$$\mathcal{B}(L) = min_{a \neq 0}\{H_c(a) + H_c(L(a))\} .$$

It is shown that $\mathcal{B}(\pi_0) = 4$ and there are only 204 values among 2^{32} possible values that achieve the branch number 4[12]. On the other hand, we can theoretically show that $\mathcal{B}(\pi_0) \leq 4$ by the similar process of the proof of Theorem 5 since Q is an invertible matrix. Thus we obtain the fact that the diffusion effect of π_0 is optimal under the condition that the transformation is consisted of only bitwise EXOR and AND logic.

The branch number of diffusion layer is closely related to the minimum number of differentially and linearly active S-boxes of this layer. In the case of π_0, it is easily seen that $\beta_d(\pi_0) = \mathcal{B}(\pi_0)$ since

$$\pi_0(x \oplus x^*) = \pi_0(x) \oplus \pi_0(x^*) .$$

Moreover Q is a symmetric matrix($Q = Q^t$), thus $\beta_d(\pi_0) = \beta_l(\pi_0)$ by Theorem 3. Consequently, for the bit transformation π_0 in the diffusion layer of CRYPTON, we obtain that

$$\beta_d(\pi_0) = \beta_l(\pi_0) = \mathcal{B}(\pi_0) = 4$$

and this is an optimal number of the linear transformation constructed with only bitwise EXORs and ANDs.

5   Diffusion Layer of Rijndael

Rijndael[6] is a block cipher, designed as a candidate algorithm for the AES. Recently, NIST announced the five AES finalist candidates for round 2[20]. The block cipher Rijndael was included in the five AES finalists. The design of Rijndael was strongly influenced by the design of the former block cipher SQUARE[5]. In this section, we study the diffusion layer which used commonly in SQUARE and Rijndael.

The diffusion layers of SQUARE and Rijndael are consisted of row-wise(or column-wise) bit transformations and bytewise transpositions that operate on a 4×4 array of bytes. The bit transformations that used in the two block ciphers are mathematically identical with the exception of the fact that it's a row-wise operation in the case of SQUARE and column-wise in the case of Rijndael. For the convenience, we use the notation θ of SQUARE for the bit transformation.

Let $\xi = (\xi_0, \xi_1, \xi_2, \xi_3) \in GF(2^8)^4$ be 4-byte input of θ and $\theta(\xi) = \zeta = (\zeta_0, \zeta_1, \zeta_2, \zeta_3) \in GF(2^8)^4$ be the corresponding output. Then ξ and θ correspond to polynomials in $GF(2^8)[x]$. That is, they can be denoted by

$$\xi(x) = \xi_0 + \xi_1 x + \xi_2 x^2 + \xi_3 x^3$$

and

$$\zeta(x) = \zeta_0 + \zeta_1 x + \zeta_2 x^2 + \zeta_3 x^3 ,$$

respectively. Defining $c(x) = c_0 + c_1 x + c_2 x^2 + c_3 x^3$ we can describe θ as a modular polynomial multiplication:

$$\zeta = \theta(\xi) \Leftrightarrow \zeta(x) = c(x)\xi(x) \pmod{1 + x^4} . \tag{1}$$

The matrix representation of (1) is as follows:

$$\begin{pmatrix} \zeta_0 \\ \zeta_1 \\ \zeta_2 \\ \zeta_3 \end{pmatrix} = \begin{pmatrix} c_0 & c_3 & c_2 & c_1 \\ c_1 & c_0 & c_3 & c_2 \\ c_2 & c_1 & c_0 & c_3 \\ c_3 & c_2 & c_1 & c_0 \end{pmatrix} \begin{pmatrix} \xi_0 \\ \xi_1 \\ \xi_2 \\ \xi_3 \end{pmatrix} . \tag{2}$$

Let C be the 4×4 matrix of (2), then (2) can be written simply as

$$\zeta^t = C\xi^t .$$

Note that the bit transformation θ is a linear transformation on the vector space $(GF(2^8))^4$ over the finite field $GF(2^8)$. It is shown in [5] that if $\Delta\xi(x) = \xi(x) \oplus \xi^*(x)$ is an input difference of θ, the output difference is

$$\Delta\zeta(x) = c(x)\xi(x) \oplus c(x)\xi^*(x) \pmod{1 + x^4}$$
$$= c(x)\Delta\xi(x) \pmod{1 + x^4} .$$

This can be written equivalently as

$$(\Delta\zeta)^t = C(\Delta\xi)^t \tag{3}$$

which is the formula appeared in Theorem 2. From this we obtain that

$$\beta_d(\theta) = \mathcal{B}(\theta) . \tag{4}$$

Let $a(x)$ be an input mask value and $b(x)$ be the corresponding output mask value on the view point of LC. Then $a(x)$ and $b(x)$ are satisfied with

$$a(x) = c(x^{-1})b(x) \pmod{1 + x^4} \tag{5}$$

and (5) can be written as

$$a_0 = c_0 \bullet b_0 \oplus c_1 \bullet b_1 \oplus c_2 \bullet b_2 \oplus c_3 \bullet b_3$$
$$a_1 = c_3 \bullet b_0 \oplus c_0 \bullet b_1 \oplus c_1 \bullet b_2 \oplus c_2 \bullet b_3$$
$$a_2 = c_2 \bullet b_0 \oplus c_3 \bullet b_1 \oplus c_0 \bullet b_2 \oplus c_1 \bullet b_3$$
$$a_3 = c_1 \bullet b_0 \oplus c_2 \bullet b_1 \oplus c_3 \bullet b_2 \oplus c_0 \bullet b_3$$

where "•" means the multiplication on the finite field $GF(2^8)$, $a(x) = a_0 + a_1 x + a_2 x^2 + a_3 x^3$ and $b(x) = b_0 + b_1 x + b_2 x^2 + b_3 x^3$. The matrix representation of the above formula is

$$
\begin{pmatrix} a_0 \\ a_1 \\ a_2 \\ a_3 \end{pmatrix} = \begin{pmatrix} c_0 & c_1 & c_2 & c_3 \\ c_3 & c_0 & c_1 & c_2 \\ c_2 & c_3 & c_0 & c_1 \\ c_1 & c_2 & c_3 & c_0 \end{pmatrix} \begin{pmatrix} b_0 \\ b_1 \\ b_2 \\ b_3 \end{pmatrix}
$$

and equivalently

$$
a^t = C^t b^t \tag{6}
$$

that is also appeared in Theorem 2.

In fact, two matrices C and C^t of (3) and (6), respectively, used commonly in the block ciphers SQUARE and Rijndael are given as follows:

$$
C = \begin{pmatrix} 2 & 3 & 1 & 1 \\ 1 & 2 & 3 & 1 \\ 1 & 1 & 2 & 3 \\ 3 & 1 & 1 & 2 \end{pmatrix}, \quad C^t = \begin{pmatrix} 2 & 1 & 1 & 3 \\ 3 & 2 & 1 & 1 \\ 1 & 3 & 2 & 1 \\ 1 & 1 & 3 & 2 \end{pmatrix}.
$$

At this point, it is easy to see that the matrix C can be obtained from C^t by appropriate transpositions of row and column vectors. Therefore, by Theorem 4 and (4), we obtain the fact that

$$
\beta_d(\theta) = \beta_l(\theta) = \mathcal{B}(\theta) .
$$

On the other hand, the branch number $\mathcal{B}(\theta) = 5$ and this is the maximal branch number. In [18] it was shown how a linear transformation on $(GF(2^m))^n$ optimal branch number $\mathcal{B}(\mathcal{B} = n+1)$ can be constructed from a maximal distance separable code. The polynomial multiplication with $c(x)$ corresponds to a special subset of the maximal distance separable codes.

However the fact that $\mathcal{B}(\theta) = 5$ also can be shown by the similar methods used in the proof of Theorem 5. Since the additive operation of $GF(2^8)$ is the bitwise EXOR, the Hamming weights of EXORs among column vectors of the matrix C are reflected to compute the branch number $\mathcal{B}(\theta)$.

The bit transformation θ has the best diffusion effect but its computational efficiency is relatively of low grade since the multiplication in $GF(2^8)$ is complicated. The authors of SQUARE and Rijndael insisted that computational efficiency of θ improved by using

$$
c(x) = 02 + 01 \cdot x + 01 \cdot x^2 + 03 \cdot x^3
$$

since the multiplication in $GF(2^8)$ can be implemented by one bit shift operations and bitwise EXORs. There is some truth in this assertion if we consider only encryption process since the coefficients $01, 02, 03$ of $c(x)$ are relatively small. If we consider decryption process, the inverse of θ is needed and θ^{-1} corresponds to the polynomial

$$
d(x) = 0E + 09 \cdot x + 0D \cdot x^2 + 0B \cdot x^3 .
$$

However the coefficients $09, 0B, 0D, 0E$ of $d(x)$ are large and computational efficiency of θ^{-1} is relatively low. Therefore we obtain that the bit transformation of SQUARE and Rijndael has the best diffusion effect but its computational efficiency is inferior to another diffusion layers.

6   Comparison of Diffusion Layers of E2, CRYPTON, and Rijndael

While the diffusion layers of E2 and CRYPTON are linear transformations over $GF(2)$, the diffusion layer of Rijndael is a linear transformations over the finite field $GF(2^8)$. The diffusion layer of Rijndael is the best out of these three algorithms on the view point of diffusion effect. But on the view point of computational efficiency, the order of excellency is E2, CRYPTON, and Rijndael. Therefore it is instructive that security hard to be compatible with computational efficiency even when we consider only diffusion layer. The result of comparing briefly for the diffusion layers used in the three block ciphers is given in the table 1.

Table 1. Comparison of diffusion layers

Cipher	E2	CRYPTON	Rijndael
Diffusion Layer	$P : 8 \times 8$ matrix (LT over $GF(2)$)	$\pi_0 : GF(2)^{32} \to GF(2)^{32}$ (LT over $GF(2)$)	$\theta : GF(2^8)^4 \to GF(2^8)^4$ (LT over $GF(2^8)$)
Operations	EXORs	EXORs ANDs	EXORs Mul. in $GF(2^8)$ (Shifts, EXORs)
$\beta_d = \beta_l$ (Maximum)	5 (9)	4 (5)	5 (5)
Diffusion effect	56 %	80 %	100 %

7   Conclusion

We examined the diffusion layers of some block ciphers referred to as substitution-permutation networks. We investigated the practical security of these diffusion layers against differential and linear cryptanalysis by using the notion of active S-boxes. It was shown that the minimum number of differentially active S-boxes and that of linearly active S-boxes are generally not identical in Section 3 and we proposed some special conditions in which those are identical(See Theorem 3 and 4). Moreover, we applied our results to analyze three diffusion layers used in the block ciphers E2, CRYPTON and Rijndael, respectively. It was also shown that these all diffusion layers have achieved optimal security according to each their constraints of using operations.

References

1. K. Aoki and K. Ohta, *Strict evaluation of the maximum average of differential probability and the maximum average of linear probability*, IEICE TRANS. FUNDAMENTALS, No. 1, 1997, pp. 2-8.
2. E. Biham and A. Shamir, *Differential cryptanalysis of DES-like Cryptosystems*, Advances in Cryptology - CRYPTO'90, LNCS 537, Springer-Verlag, 1990, pp. 2-21.
3. E. Biham and A. Shamir, *Differential cryptanalysis of DES-like Cryptosystems*, Journal of Cryptology, 1991, 4, pp. 3-72.
4. J. Daemen, R. Govaerts and J. Vandewalle, *Correlation Matrices*, Fast Software Encryption, LNCS 1008, Springer-Verlag, 1994, pp. 275-285.
5. J. Daemen, L. R. Knudsen and V. Rijmen, *The block cipher SQUARE*, Fast Software Encryption, LNCS 1267, Springer-Verlag, 1997, pp. 149-165.
6. J. Daemen and V. Rijmen, *The Rijndael block cipher*, AES Proposal, 1998.
7. H. M. Heys and S. E. Tavares, *Substitution-permutation networks resistant to differential and linear cryptanalysis*, Journal of Cryptology, 1996, 9, pp.1-19.
8. M. Kanda, Y. Takashima, T. Matsumoto, K. Aoki, and K. Ohta, *A strategy for constructing fast round functions with practical security against differential and linear cryptanalysis*, Selected Areas in Cryptography, LNCS 1556, 1999, pp. 264-279.
9. Y. Kaneko, F. Sano and K. Sakurai, *On provable security against differential and linear cryptanalysis in generalized Feistel ciphers with multiple random functions*, Proceedings of SAC'97, 1997, pp. 185-199.
10. L. R. Knudsen, *Practically secure Feistel Ciphers*, Fast Software Encryption, LNCS 809, 1994, pp. 211-221.
11. X. Lai, J. L. Massey, and S. Murphy, *Markov Ciphers and Differential Cryptanalysis*, Advances in Cryptology - Eurocrypt'91, LNCS 547, Springer-Verlag, 1991, pp. 17-38.
12. C. H. Lim, *CRYPTON : A new 128-bit block cipher*, AES Proposal, 1998.
13. M. Matsui, *Linear cryptanalysis method for DES cipher*, Advances in Cryptology - Eurocrypt'93, LNCS 765, Springer-Verlag, 1993, pp. 386-397.
14. M. Matsui, *New Structure of Block Ciphers with Provable Security against Differential and Linear Cryptalaysis*, Fast Software Encryption, LNCS 1039, Springer-Verlag, 1996, pp. 205-218.
15. K. Nyberg, *Linear Approximation of Block Ciphers*, Advances in Cryptology - Eurocrypt'94, LNCS 950, Springer-Verlag, 1994, pp. 439-444.
16. K. Nyberg and L. R. Knudsen, *Provable Security against Differential Cryptanalysis*, Journal of Cryptology, 1995, No. 8, (1), pp. 27-37.
17. NTT-Nippon Telegraph and Telephone Corporation, *E2 : Efficient Encryption algorithm*, AES Proposal, 1998.
18. V. Rijmen, J. Daemen, B. Preneel, A. Bosselaers and E. D. Win, *The cipher SHARK*, Fast Software Encryption, LNCS 1039, Springer-Verlag, 1996, pp. 99-112.
19. C. E. Shannon, *Communication theory of secrecy systems*, Bell System Technical Journal, 28, 1949, pp. 656-715.
20. http://csrc.nist.gov/encryption/aes/aeshome.htm

Non-linear Complexity of the Naor–Reingold Pseudo-random Function

William D. Banks[1], Frances Griffin[2],
Daniel Lieman[3], and Igor E. Shparlinski[4]

[1] Department of Mathematics, University of Missouri
Columbia, MO 65211, USA
Phone: [1 - 573] 882 2393   Fax: [1 - 573] 882 1869
bbanks@math.missouri.edu
[2] Department of Mathematics, Macquarie University
Sydney, NSW 2109, Australia
Phone: [61 - 1] 9850 8923   Fax: [61 - 1] 9850 8114
fgriffin@ics.mq.edu.au
[3] Department of Mathematics, University of Missouri
Columbia, MO 65211, USA
Phone: [1 - 573] 882 4305   Fax: [1 - 573] 882 1869
lieman@math.missouri.edu
[4] Department of Computing, Macquarie University
Sydney, NSW 2109, Australia
Phone: [61 - 1] 9850 9585   Fax: [61 - 1] 9850 9551
igor@ics.mq.edu.au

Abstract. We obtain an exponential lower bound on the non-linear complexity of the new pseudo-random function, introduced recently by M. Naor and O. Reingold. This bound is an extension of the lower bound on the linear complexity of this function that has been obtained by F. Griffin and I. E. Shparlinski.

1 Introduction

Let p and l be primes with $l|p-1$ and let $n \geq 1$ be an integer.

Denote by $\mathbb{F}_p$ the finite field of p elements which we identify with the set $\{0, \ldots, p-1\}$. Select an element $g \in \mathbb{F}_p^*$ of multiplicative order l, that is,

$$g^i \neq 1, \ 1 \leq i \leq l-1, \qquad g^l = 1.$$

Then for each n-dimensional vector $\mathbf{a} = (a_1, \ldots, a_n) \in (\mathbb{F}_l^*)^n$ one can define the function

$$f_{\mathbf{a}}(X) = g^{a_1^{x_1} \cdots a_n^{x_n}} \in \mathbb{F}_p,$$

where $X = x_1 \ldots x_n$ is the bit representation of an n-bit integer X, $0 \leq X \leq 2^n - 1$, with some extra leading zeros if necessary. Thus, given $\mathbf{a} = (a_1, \ldots, a_n) \in (\mathbb{F}_l^*)^n$, for each $X = 0, \ldots, 2^n - 1$ this function produces a certain element of $\mathbb{F}_p$. After that it can be continued periodically.

JooSeok Song (Ed.): ICISC'99, LNCS 1787, pp. 53–59, 2000.
© Springer-Verlag Berlin Heidelberg 2000

For a randomly chosen vector $\mathbf{a} \in (\mathbb{F}_l^*)^n$, M. Naor and O. Reingold [4] have proposed the function $f_{\mathbf{a}}(X)$ as an efficient pseudo-random function (it is assumed in [4] that n is the bit length of p but similar results hold in much more general settings).

It is shown in [4] that the function $f_{\mathbf{a}}(X)$ has some very desirable security properties, provided that certain standard cryptographic assumptions about the hardness of breaking the Diffie-Hellman cryptosystem hold. It is also shown in [4] that this function can be computed in parallel by threshold circuits of bounded depth and polynomial size.

The distribution properties of this function have been studied in [8] and it has been proved that the statistical distribution of $f_{\mathbf{a}}(X)$ is exponentially close to uniform for almost all $\mathbf{a} \in (\mathbb{F}_l^*)^n$.

For the elliptic curve version of this generator similar results have been obtained in [9].

The linear complexity, which is an important cryptographic characteristic of this sequence, has been estimated in [2].

Here we study the more general question of non-linear complexity.

Given an integer $d \geq 1$ and an N-element sequence $W_1, \ldots, W_N$ over a ring $\mathcal{R}$, we define the *degree d complexity*, $L(d)$, as the smallest number L such that there exists a polynomial $F(Z_1, \ldots, Z_L)$ over $\mathcal{R}$ of degree at most d in L variables such that

$$W_{X+L} = F(W_X, \ldots, W_{X+L-1}), \qquad X = 1, \ldots, N - L.$$

The case $d = 1$ is closely related to the notion of the *linear complexity*, $\widetilde{L}$, the only distinction being that in the traditional definition of linear complexity only homogeneous linear polynomials are considered. However, this distinction is not very important since one can easily verify that $L(1) \leq \widetilde{L} \leq L(1) + 1$.

Linear complexity is an essential cryptographic characteristic that has been studied in many works, see [1, 3, 5–7]. Since non-linear complexity is harder to study, hence much less is known about this characteristic, even though it is of ultimate interest as well, see [1, 5].

In this paper we extend the method of [2] and obtain an exponential lower bound on the degree d complexity, $L_{\mathbf{a}}(d)$, of the sequence $f_{\mathbf{a}}(X)$, $X = 0, \ldots, 2^n - 1$, which holds for almost all $\mathbf{a} \in (\mathbb{F}_l^*)^n$.

Throughout the paper, $\log z$ denotes the binary logarithm of z.

2   Preparations

We need some statements about the distribution in $\mathbb{F}_l^*$ of products of the form

$$\mathbf{b}^{\mathbf{z}} = b_1^{z_1} \ldots b_m^{z_m}, \qquad \mathbf{z} = (z_1, \ldots, z_m) \in \{0, 1\}^m,$$

which are of independent interest.

Denote

$$\mathbf{i} = (1, \ldots, 1) \in \{0, 1\}^m,$$

so that $\mathbf{b}^{\mathbf{i}} = b_1 \ldots b_m$.

Lemma 1. *For all but at most*

$$N_{m,d} \leq \sum_{r=1}^{d} r \binom{2^m + r - 2}{r} (l-1)^{m-1}$$

vectors $\mathbf{b} = (b_1, \ldots, b_m) \in (\mathbb{F}_l^*)^m$

$$\mathbf{b}^{\mathbf{z}_1} + \ldots + \mathbf{b}^{\mathbf{z}_r} \neq \mathbf{b}^{\mathbf{i}},$$

for any choice of $r \leq d$ *vectors* $\mathbf{z}_\nu \in \{0, 1\}^m$ *with* $\mathbf{z}_\nu \neq \mathbf{i}$, $\nu = 1, \ldots, r$.

Proof. For all $r = 1, \ldots, d$, let $\mathcal{Z}_r$ denote the set of all non-equivalent r-tuples $(\mathbf{z}_1, \ldots, \mathbf{z}_r)$ with $\mathbf{z}_\nu \in \{0, 1\}^m$ and $\mathbf{z}_\nu \neq \mathbf{i}$, $\nu = 1, \ldots, r$, where two r-tuples are considered to be equivalent if one is a permutation of the other.

The cardinality $\#\mathcal{Z}_r$ of this set is equal to the number of solutions of the equation

$$\sum_{k=1}^{2^m - 1} n_k = r$$

in nonnegative integers $n_1, \ldots, n_{2^m - 1}$. Indeed, if we list the vectors

$$\mathbf{v}_k \in \{0, 1\}^m \backslash \{\mathbf{i}\}, \qquad k = 1, \ldots, 2^m - 1,$$

then every r-tuple in $\mathcal{Z}_r$ is uniquely defined by the number of times n_k that the vector $\mathbf{v}_k$ occurs in the r-tuple.

Therefore

$$\#\mathcal{Z}_r = \binom{2^m + r - 2}{r}.$$

For each r-tuple $(\mathbf{z}_1, \ldots, \mathbf{z}_r) \in \mathcal{Z}_r$ the number of solutions of the equation

$$\mathbf{b}^{\mathbf{z}_1} + \ldots + \mathbf{b}^{\mathbf{z}_r} = \mathbf{b}^{\mathbf{i}},$$

in $\mathbf{b} \in (\mathbb{F}_l^*)^m$ does not exceed $r(l-1)^{m-1}$. This can easily be proved for all $m \geq 1$ by induction in r.

It is convenient to start the induction with $r = 0$ where the statement is clearly true for all $m \geq 1$ (the equation $\mathbf{b}^{\mathbf{i}} = 0$ has no solutions).

Otherwise we select j such that the vector $\mathbf{z}_r$ has a zero jth component. This is always possible because $\mathbf{z}_r \neq \mathbf{i}$. Then the above equation can be written in the form $A = Bb_j$ where A and B do not depend on b_j. Because of our choice of j, we see that by induction, B vanishes for at most $(r-1)(l-1)^{m-2}$ vectors $(b_1, \ldots, b_{j-1}, b_{j+1}, \ldots, b_m) \in (\mathbb{F}_l^*)^{m-1}$ and in this case we have at most $l-1$ values for b_j. If $B \neq 0$ then for any vector $(b_1, \ldots, b_{j-1}, b_{j+1}, \ldots, b_m) \in (\mathbb{F}_l^*)^{m-1}$ the value of b_j is defined uniquely. Therefore the number of solutions does not exceed $(r-1)(l-1)^{m-1} + (l-1)^{m-1} = r(l-1)^{m-1}$. This completes the induction. Accordingly,

$$N_{m,d} \leq \sum_{r=1}^{d} r(l-1)^{m-1} \#\mathcal{Z}_r$$

and the bound follows. $\square$

We also need the following Lemma 2 of [2] which shows that for large m, the products $\mathbf{b}^{\mathbf{z}}$ with $\mathbf{z} \in \{0, 1\}^m$ are quite dense in $\mathbb{F}_l^*$.

Lemma 2. *Fix an arbitrary $\Delta > 0$. Then for all but at most*

$$M_m \le 2^{-m} \Delta^{-1} (l - 1)^{m+2}$$

vectors $\mathbf{b} = (b_1, \ldots b_m) \in (\mathbb{F}_l^*)^m$, the 2^m products $\mathbf{b}^{\mathbf{z}}$, $\mathbf{z} \in \{0, 1\}^m$ take at least $l - 1 - \Delta$ values from $\mathbb{F}_l^*$.

3   Lower Bound of the Degree d Complexity

Now we are prepared to prove our main result.

Theorem 1. *Assume that for some $\gamma > 0$*

$$n \ge (1 + \gamma) \log l.$$

Then for any integer $d \ge 1$ and any $\delta > 0$ the degree d complexity, $L_{\mathbf{a}}(d)$, of the sequence $f_{\mathbf{a}}(X)$, $X = 0, \ldots, 2^n - 1$, satisfies

$$L_{\mathbf{a}}(d) \ge \begin{cases} 0.5(l - 1)^{1/d - \delta/d}, & \text{if } \gamma \ge 1 + 1/d; \\ 0.5(l - 1)^{\gamma/(d+1) - \delta/d}, & \text{if } \gamma < 1 + 1/d; \end{cases}$$

for all but at most

$$N \le \left(\frac{d+1}{d!} + o(1) \right) (l - 1)^{n-\delta}, \qquad l \to \infty,$$

vectors $\mathbf{a} \in (\mathbb{F}_l^*)^n$

Proof. If $\delta \ge \max\{1, \gamma\}$ then the bound is trivial. Otherwise we put

$$t = \left\lfloor \min \left\{ \frac{1 - \delta}{d}, \frac{\gamma - \delta}{d+1} \right\} \log(l - 1) \right\rfloor, \qquad s = n - t$$

and

$$\Delta = \left\lceil (l - 1) \left(\binom{2^t + d}{d} + 1 \right)^{-1} \right\rceil - 1.$$

Therefore

$$2^{-s} = 2^{t-n} \le 2^t l^{-1-\gamma}. \tag{1}$$

From the inequality $2^{td} \le (l - 1)^{1-\delta}$ we see that

$$\sum_{r=1}^{d} r \binom{2^t + r - 2}{r} \le \left(\frac{1}{(d-1)!} + o(1) \right) (l - 1)^{1-\delta} \tag{2}$$

We also have

$$2^{t(d+1)} \leq (l-1)^{\gamma-\delta} \qquad \text{and} \qquad \Delta \geq (d! + o(1))\,(l-1)2^{-td}. \tag{3}$$

From Lemmas 1 and 2 and the bounds (1), (2) and (3) we derive

$$N_{t,d} \leq \sum_{r=1}^{d} r \binom{2^t + r - 2}{r}(l-1)^{t-1}$$

$$\leq \left(\frac{1}{(d-1)!} + o(1)\right)(l-1)^{t-\delta}$$

and

$$M_s \leq 2^{-s}\Delta^{-1}(l-1)^{s+2} \leq \left(\frac{1}{d!} + o(1)\right) 2^{t(d+1)}(l-1)^{s-\gamma}$$

$$\leq \left(\frac{1}{d!} + o(1)\right)(l-1)^{s-\delta}.$$

Let $\mathcal{A}$ be the set of vectors $\mathbf{a} \in (\mathbb{F}_l^*)^n$ such that simultaneously

$$\#\{a_1^{y_1} \ldots a_s^{y_s} \mid (y_1, \ldots, y_s) \in \{0, 1\}^s\} \geq l - 1 - \Delta$$

and

$$\sum_{\nu=1}^{d} a_{s+1}^{k_{1,\nu}} \ldots a_n^{k_{t,\nu}} \neq a_{s+1} \ldots a_n$$

for any $(k_{1,\nu}, \ldots, k_{t,\nu}) \in \{0, 1\}^t$ with $(k_{1,\nu}, \ldots, k_{t,\nu}) \neq (1, \ldots, 1)$.
Then, from the above inequalities, we derive

$$\#\mathcal{A} \geq (l-1)^n - N_{t,d}(l-1)^{n-t} - M_s(l-1)^{n-s}$$

$$\geq (l-1)^n - \left(\frac{1}{(d-1)!} + o(1)\right)(l-1)^{n-\delta}$$

$$- \left(\frac{1}{d!} + o(1)\right)(l-1)^{n-\delta}$$

$$= (l-1)^n - \left(\frac{d+1}{d!} + o(1)\right)(l-1)^{n-\delta}.$$

We show that the lower bound of the theorem holds for any $\mathbf{a} \in \mathcal{A}$, thus from $N \leq (l-1)^n - \#\mathcal{A}$ and the above inequality we obtain the desired upper bound on N.

Let us fix $\mathbf{a} \in \mathcal{A}$. Assume that $L_{\mathbf{a}}(d) \leq 2^t - 1$. Then there exists a polynomial

$$F(Z_1, \ldots, Z_{2^t-1}) \in \mathbb{F}_p[Z_1, \ldots, Z_{2^t-1}],$$

such that

$$F(f_{\mathbf{a}}(X), \ldots, f_{\mathbf{a}}(X + 2^t - 2)) = f_{\mathbf{a}}(X + 2^t - 1)$$

for all $X = 0, \ldots, 2^n - 2^t$.

Now suppose $X = 2^t Y$, where $Y = y_1 \ldots y_s$ is an s-bit integer, and let $K = k_1 \ldots k_t$ be a t-bit integer. We remark that the bits of K form the rightmost bits of the sum $X + K$. Then we have

$$f_{\mathbf{a}}(2^t Y + K) = g^{a_1^{y_1} \ldots a_s^{y_s} e_K}, \qquad Y = 0, \ldots, 2^s - 1,$$

where

$$e_K = a_{s+1}^{k_1} \ldots a_n^{k_t}, \qquad K = 0, \ldots, 2^t - 1,$$

and $K = k_1 \ldots k_t$ is the bit expansion of K.

Denote by $\Phi_{\mathbf{a}}(u)$ the following exponential polynomial

$$\Phi_{\mathbf{a}}(u) = F\left(g_0^u, \ldots, g_{2^t-2}^u\right) - g_{2^t-1}^u, \qquad u \in \mathbb{F}_l,$$

where

$$g_K = g^{e_K}, \qquad K = 0, \ldots, 2^t - 1.$$

Collecting together terms with equal values of exponents and taking into account that, because of the choice of the set $\mathcal{A}$, the value of

$$g_{2^t-1} = g^{a_{s+1} \ldots a_n}$$

is unique, we obtain that $\Phi_{\mathbf{a}}(u)$ can be expressed in the form

$$\Phi_{\mathbf{a}}(u) = \sum_{\nu=1}^{R} C_\nu h_\nu^u,$$

where

$$1 \le R \le \binom{2^t + d - 1}{d} + 1,$$

with some coefficients $C_\nu \in \mathbb{F}_p^*$ and pairwise distinct $h_\nu \in \mathbb{F}_p^*$, $\nu = 1, \ldots, R$.

Recalling that $\mathbf{a} \in \mathcal{A}$, we conclude that $\Phi_{\mathbf{a}}(u) \ne 0$ for at most Δ values of $u = 1, \ldots, l - 1$. On the other hand, from the properties of Vandermonde determinants, it is easy to see that for any $u = 1, \ldots, l - 1$, $\Phi_{\mathbf{a}}(u + v) \ne 0$ for at least one $v = 0, \ldots, R - 1$. Therefore, $\Phi_{\mathbf{a}}(u) \ne 0$ for at least

$$(l-1)/R \ge (l-1)\left(\binom{2^t + d - 1}{d} + 1\right)^{-1} > \Delta$$

values of $u = 1, \ldots, l - 1$, which is not possible because of the choice of $\mathcal{A}$. The obtained contradiction implies that $L_{\mathbf{a}}(d) \ge 2^t$. $\qquad \square$

4   Remarks

It is useful to recall that typically the bit length of p and l are of the same order as n. Thus

$$\log p \asymp \log l \asymp n.$$

In the most interesting case n is the bit length of p, that is, $n \sim \log p$. In this case Theorem 1 implies a lower bound on $L_{\mathbf{a}}(d)$ which is exponential in n, if $l \le p^{1-\varepsilon}$ for some $\varepsilon > 0$. On the other hand, it would be interesting to estimate the linear and higher degree complexity for all values of $l \le p$.

It is also an interesting open question to study the linear complexity or higher degree complexity of single bits of $f_{\mathbf{a}}(X)$. For example, one can form the sequence $\beta_{\mathbf{a}}(X)$ of the rightmost bits of $f_{\mathbf{a}}(X)$, $X = 0, \ldots, 2^n - 1$, and study its linear and higher degree complexity (as elements of $\mathbb{F}_2$). Unfortunately we do not see any approaches to this question.

Acknowledgment. This work has been motivated by a question asked during a seminar talk about the results of [2, 8, 9] given by I.S. at the Centre for Applied Cryptographic Research at the University of Waterloo, whose hospitality is gratefully acknowledged.

References

1. T. W. Cusick, C. Ding and A. Renvall, *Stream Ciphers and Number Theory*, Elsevier, Amsterdam, 1998.
2. F. Griffin and I. E. Shparlinski, 'On the linear complexity of the Naor-Reingold pseudo-random function', *Proc. 2nd Intern. Conf. on Information and Communication Security (ICICS'99), Sydney*, Lect. Notes in Comp. Sci., v.1726, Springer-Verlag, Berlin, 1999, 301–308.
3. A. J. Menezes, P. C. van Oorschot and S. A. Vanstone, *Handbook of Cryptography*, CRC Press, Boca Raton, FL, 1996.
4. M. Naor and O. Reingold, 'Number-theoretic constructions of efficient pseudo-random functions', *Proc. 38th IEEE Symp. on Foundations of Comp. Sci. (FOCS'97), Miami Beach*, IEEE, 1997, 458–467.
5. H. Niederreiter, 'Some computable complexity measures for binary sequences', *Proc. Intern. Conf. on Sequences and their Applications (SETA'98), Singapore*, C. Ding, T. Helleseth and H. Niederreiter (Eds.), Springer-Verlag, London, 1999, 67–78.
6. H. Niederreiter and M. Vielhaber, 'Linear complexity profiles: Hausdorff dimension for almost perfect profiles and measures for general profiles', *J. Compl.*, **13** (1996), 353–383.
7. R. A. Rueppel, 'Stream ciphers', *Contemporary Cryptology: The Science of Information Integrity*, IEEE Press, NY, 1992, 65–134.
8. I. E. Shparlinski, 'On the uniformity of distribution of the Naor–Reingold pseudo-random function', *Finite Fields and Their Appl.* (to appear).
9. I. E. Shparlinski, 'On the Naor–Reingold pseudo-random function from elliptic curves', *Preprint*, 1999, 1–9.

Relationships between Bent Functions and Complementary Plateaued Functions

Yuliang Zheng[1] and Xian-Mo Zhang[2]

[1] School of Comp & Info Tech, Monash University
McMahons Road, Frankston, Melbourne, VIC 3199, Australia
yuliang@pscit.monash.edu.au
http://www.pscit.monash.edu.au/links/
[2] School of Info Tech & Comp Sci
The University of Wollongong, Wollongong
NSW 2522, Australia
xianmo@cs.uow.edu.au

Abstract. We introduce the concept of complementary plateaued functions and examine relationships between these newly defined functions and bent functions. Results obtained in this paper contribute to the further understanding of profound secrets of bent functions. Cryptographic applications of these results are demonstrated by constructing highly nonlinear correlation immune functions that possess no non-zero linear structures.

Keywords: Plateaued Functions, Complementary Plateaued Functions, Bent Functions, Cryptography

1 Introduction

Bent functions achieve the maximum nonlinearity and satisfy the propagation criterion with respect to every non-zero vector. These functions, however, are neither balanced nor correlation immune. Furthermore they exist only when the number of variables is even. All these properties impede the direct applications of bent functions in cryptography. They also indicate the importance of further understanding the characteristics of bent functions in the construction of Boolean functions with cryptographically desirable properties. This extends significantly a recent paper by Zheng and Zhang [12] where a new class of functions called plateaued functions were introduced. In particular, (i) we introduce the concept of complementary plateaued functions; (ii) we establish relationships between bent and complementary plateaued functions; (iii) we show that complementary plateaued functions provide a new avenue to construct bent functions; (iv) we prove a new characteristic property of non-quadratic bent functions by the use of complementary plateaued functions; (v) As an application, we construct balanced, highly nonlinear correlation immune functions that have no non-zero linear structures.

2 Boolean Functions

Definition 1. *We consider functions from V_n to $GF(2)$ (or simply functions on V_n), V_n is the vector space of n tuples of elements from $GF(2)$. Usually we*

JooSeok Song (Ed.): ICISC'99, LNCS 1787, pp. 60–75, 2000.
© Springer-Verlag Berlin Heidelberg 2000

write a function f on V_n as $f(x)$, where $x = (x_1, \ldots, x_n)$ is the variable vector in V_n. The truth table of a function f on V_n is a $(0,1)$-sequence defined by $(f(\alpha_0), f(\alpha_1), \ldots, f(\alpha_{2^n-1}))$, and the sequence of f is a $(1,-1)$-sequence defined by $((-1)^{f(\alpha_0)}, (-1)^{f(\alpha_1)}, \ldots, (-1)^{f(\alpha_{2^n-1})})$, where $\alpha_0 = (0, \ldots, 0, 0)$, $\alpha_1 = (0, \ldots, 0, 1)$, $\ldots$, $\alpha_{2^n-1} = (1, \ldots, 1, 1)$. The matrix of f is a $(1,-1)$-matrix of order 2^n defined by $M = ((-1)^{f(\alpha_i \oplus \alpha_j)})$ where $\oplus$ denotes the addition in $GF(2)$. f is said to be balanced if its truth table contains an equal number of ones and zeros.

Given two sequences $\tilde{a} = (a_1, \cdots, a_m)$ and $\tilde{b} = (b_1, \cdots, b_m)$, their component-wise product is defined by $\tilde{a} * \tilde{b} = (a_1 b_1, \cdots, a_m b_m)$. In particular, if $m = 2^n$ and $\tilde{a}, \tilde{b}$ are the sequences of functions f and g on V_n respectively, then $\tilde{a} * \tilde{b}$ is the sequence of $f \oplus g$ where $\oplus$ denotes the addition in $GF(2)$.

Let $\tilde{a} = (a_1, \cdots, a_m)$ and $\tilde{b} = (b_1, \cdots, b_m)$ be two sequences or vectors, the scalar product of $\tilde{a}$ and $\tilde{b}$, denoted by $\langle \tilde{a}, \tilde{b} \rangle$, is defined as the sum of the component-wise multiplications. In particular, when $\tilde{a}$ and $\tilde{b}$ are from V_m, $\langle \tilde{a}, \tilde{b} \rangle = a_1 b_1 \oplus \cdots \oplus a_m b_m$, where the addition and multiplication are over $GF(2)$, and when $\tilde{a}$ and $\tilde{b}$ are $(1,-1)$-sequences, $\langle \tilde{a}, \tilde{b} \rangle = \sum_{i=1}^{m} a_i b_i$, where the addition and multiplication are over the reals.

An affine function f on V_n is a function that takes the form of $f(x_1, \ldots, x_n) = a_1 x_1 \oplus \cdots \oplus a_n x_n \oplus c$, where $a_j, c \in GF(2)$, $j = 1, 2, \ldots, n$. Furthermore f is called a linear function if $c = 0$.

A $(1,-1)$-matrix A of order m is called a Hadamard matrix if $AA^T = mI_m$, where A^T is the transpose of A and I_m is the identity matrix of order m. A Sylvester-Hadamard matrix of order 2^n, denoted by H_n, is generated by the following recursive relation

$$H_0 = 1, \ H_n = \begin{bmatrix} H_{n-1} & H_{n-1} \\ H_{n-1} & -H_{n-1} \end{bmatrix}, \ n = 1, 2, \ldots.$$

Let ℓ_i, $0 \le i \le 2^n - 1$, be the i row of H_n. It is known that ℓ_i is the sequence of a linear function $\varphi_i(x)$ defined by the scalar product $\varphi_i(x) = \langle \alpha_i, x \rangle$, where α_i is the ith vector in V_n according to the ascending alphabetical order.

The Hamming weight of a $(0,1)$-sequence ξ, denoted by $HW(\xi)$, is the number of ones in the sequence. Given two functions f and g on V_n, the Hamming distance $d(f,g)$ between them is defined as the Hamming weight of the truth table of $f(x) \oplus g(x)$.

The equality in the following lemma is called Parseval's equation (Page 416 [4]).

Lemma 1. Let f be a function on V_n and ξ denote the sequence of f. Then

$$\sum_{i=0}^{2^n-1} \langle \xi, \ell_i \rangle^2 = 2^{2n}$$

where ℓ_i is the ith row of H_n, $i = 0, 1, \ldots, 2^n - 1$.

Definition 2. *The* nonlinearity *of a function f on V_n, denoted by N_f, is the minimal Hamming distance between f and all affine functions on V_n, i.e., $N_f = \min_{i=1,2,\ldots,2^{n+1}} d(f, \varphi_i)$ where $\varphi_1, \varphi_2, \ldots, \varphi_{2^{n+1}}$ are all the affine functions on V_n.*

The following characterizations of nonlinearity will be useful (for a proof see for instance [5]).

Lemma 2. *The nonlinearity of f on V_n can be expressed by*

$$N_f = 2^{n-1} - \frac{1}{2} \max\{|\langle \xi, \ell_i \rangle|, 0 \leq i \leq 2^n - 1\}$$

where ξ is the sequence of f and $\ell_0, \ldots, \ell_{2^n-1}$ are the rows of H_n, namely, the sequences of linear functions on V_n.

The nonlinearity of functions on V_n is upper bounded by $2^{n-1} - 2^{\frac{1}{2}n-1}$.

Definition 3. *Let f be a function on V_n. For a vector $\alpha \in V_n$, denote by $\xi(\alpha)$ the sequence of $f(x \oplus \alpha)$. Thus $\xi(0)$ is the sequence of f itself and $\xi(0) * \xi(\alpha)$ is the sequence of $f(x) \oplus f(x \oplus \alpha)$. Set*

$$\Delta_f(\alpha) = \langle \xi(0), \xi(\alpha) \rangle,$$

the scalar product of $\xi(0)$ and $\xi(\alpha)$. $\Delta_f(\alpha)$ is also called the auto-correlation *of f with a shift α.*

We can simply write $\Delta_f(\alpha)$ as $\Delta(\alpha)$ if no confusion takes place.

Definition 4. *Let f be a function on V_n. We say that f satisfies the* propagation criterion *with respect to α if $f(x) \oplus f(x \oplus \alpha)$ is a balanced function, where $x = (x_1, \ldots, x_n)$ and α is a vector in V_n. Furthermore f is said to satisfy the propagation criterion of degree k if it satisfies the propagation criterion with respect to every non-zero vector α whose Hamming weight is not larger than k (see [6]).*

The *strict avalanche criterion (SAC)* [9] is the same as the propagation criterion of degree one.

Obviously, $\Delta(\alpha) = 0$ if and only if $f(x) \oplus f(x \oplus \alpha)$ is balanced, i.e., f satisfies the propagation criterion with respect to α.

Definition 5. *Let f be a function on V_n. α in V_n is called a* linear structure *of f if $|\Delta(\alpha)| = 2^n$ (i.e., $f(x) \oplus f(x \oplus \alpha)$ is a constant).*

For any function f, $\Delta(\alpha_0) = 2^n$, where α_0 is the zero vector on V_n. It is easy to verify that the set of all linear structures of a function f form a linear subspace of V_n, whose dimension is called the *linearity of f*. It is also well-known that if f has non-zero linear structure, then there exists a nonsingular $n \times n$ matrix B over $GF(2)$ such that $f(xB) = g(y) \oplus h(z)$, where $x = (y, z)$, $y \in V_p$, $z \in V_q$, g is a function on V_p and g has no non-zero linear structure, and h is a linear function on V_q. Hence q is equal to the linearity of f.

The following lemma is the re-statement of a relation proved in Section 2 of [2].

Lemma 3. *Let f be a function on V_n and ξ denote the sequence of f. Then*

$$(\Delta(\alpha_0), \Delta(\alpha_1), \ldots, \Delta(\alpha_{2^n-1}))H_n = (\langle \xi, \ell_0 \rangle^2, \langle \xi, \ell_1 \rangle^2, \ldots, \langle \xi, \ell_{2^n-1} \rangle^2)$$

where α_j is the binary representation of an integer j, $j = 0, 1, \ldots, 2^n - 1$ and ℓ_i is the ith row of H_n.

There exist a number of equivalent definitions of correlation immune functions [1,3]. It is easy to verify that the following definition is equivalent to Definition 2.1 of [1]:

Definition 6. *Let f be a function on V_n and let ξ be its sequence. Then f is called a kth-order correlation immune function if and only if $\langle \xi, \ell \rangle = 0$ for every ℓ, the sequence of a linear function $\varphi(x) = \langle \alpha, x \rangle$ on V_n constrained by $1 \le HW(\alpha) \le k$.*

For convenience sake in this paper we give the following statement.

Lemma 4. *Let f be a function on V_n and let ξ be its sequence. Then $\langle \xi, \ell_i \rangle = 0$, where ℓ_i is the ith row of H_n, if and only if $f(x) \oplus \langle \alpha_i, x \rangle$ is balanced, where α_i is the binary representation of integer i, $i = 0, 1, \ldots, 2^n - 1$.*

In fact, ℓ_i is the sequence of linear function $\varphi(x) = \langle \alpha_i, x \rangle$. This proves Lemma 4. Due to Lemma 4 and Definition 6, we conclude

Lemma 5. *Let f be a function on V_n and let ξ be its sequence. Then f is a kth-order correlation immune function if and only if $f(x) \oplus \langle \alpha, x \rangle$ where α is any vector in V_n, constrained by $1 \le HW(\alpha) \le k$.*

Definition 7. *A function f on V_n is called a bent function [7] if $\langle \xi, \ell_i \rangle^2 = 2^n$ for every $i = 0, 1, \ldots, 2^n - 1$, where ℓ_i is the ith row of H_n.*

A bent function on V_n exists only when n is even, and it achieves the maximum nonlinearity $2^{n-1} - 2^{\frac{1}{2}n-1}$. From [7] we have the following:

Theorem 1. *Let f be a function on V_n. The following statements are equivalent: (i) f is bent, (ii) the nonlinearity of f, N_f, satisfies $N_f = 2^{n-1} - 2^{\frac{1}{2}n-1}$, (iii) $\Delta(\alpha) = 0$ for any non-zero α in V_n, (iv) the matrix of f is an Hadamard matrix.*

Bent functions have following properties [7]:

Proposition 1. *Let f be a bent function on V_n and ξ denote the sequence of f. Then (i) the degree of f is at most $\frac{1}{2}n$, (ii) for any nonsingular $n \times n$ matrix B over $GF(2)$ and any vector $\beta \in V_p$, $g(x) = f(xB \oplus \beta)$ is a bent function, (iii) for any affine function ψ on V_n, $f \oplus \psi$ is a bent function, (iv) $2^{-\frac{1}{2}n}\xi H_n$ is the sequence of a bent function.*

The following is from [11] (called Theorem 18 in that paper).

Lemma 6. *Let f be a function on V_n ($n \ge 2$), ξ be the sequence of f, and p is an integer, $2 \le p \le n$. If $\langle \xi, \ell_j \rangle \equiv 0 \pmod{2^{n-p+2}}$, where ℓ_j is the jth row of H_n, $j = 0, 1, \ldots, 2^n - 1$, then the degree of f is at most $p - 1$.*

3   Plateaued Functions

3.1 rth-order Plateaued Functions

The concept of plateaued functions was first introduced in [12]. In addition to the concept, the same paper also studies the existence, properties and construction methods of plateaued functions.

Notation 1. *Let f be a function on V_n and ξ denote the sequence of f. Set $\Im_f = \{i \mid \langle \xi, \ell_i \rangle \neq 0, \ 0 \leq i \leq 2^n - 1\}$ where ℓ_i is the ith row of H_n, $i = 0, 1, \ldots, 2^n - 1$.*

We will simply write $\Im_f$ as $\Im$ when no confusion arises.

Definition 8. *Let f be a function on V_n and ξ denote the sequence of f. If there exists an even number r, $0 \leq r \leq n$, such that $\#\Im = 2^r$ and each $\langle \xi, \ell_j \rangle^2$ takes the value of 2^{2n-r} or 0 only, where $\overline{\ell_j}$ denotes the jth row of H_n, $j = 0, 1, \ldots, 2^n - 1$, then f is called a rth-order plateaued function on V_n. f is also called a plateaued function on V_n if we ignore the particular order r.*

Due to Parseval's equation, the condition $\#\Im = 2^r$ can be obtained from the condition "each $\langle \xi, \ell_j \rangle^2$ takes the value of 2^{2n-r} or 0 only, where ℓ_j denotes the jth row of H_n, $j = 0, 1, \ldots, 2^n - 1$". For convenience sake, however, both conditions are mentioned in Definition 8.

The following can be immediately obtained from Definition 8.

Proposition 2. *Let f be a function on V_n. We conclude (i) if f is a rth-order plateaued function then r must be even, (ii) f is an nth-order plateaued function if and only if f is bent, (iii) f is a 0th-order plateaued function if and only if f is affine.*

The next result is a consequence of Theorem 3 of [8].

Proposition 3. *A partially-bent function is a plateaued function.*

However, it is important to note that the converse of Proposition 3 has been shown to be false [12].

3.2 $(n-1)$th-order Plateaued Functions on V_n

Following the general results on rth-order plateaued functions on V_n [12], in this paper we examine in greater depth the properties and construction methods of $(n-1)$th-order plateaued functions on V_n. These properties will be useful in research into bent functions.

Proposition 4. *Let p be a positive odd number and g be a $(p-1)$th-order plateaued function on V_p. Then*

(i) *the nonlinearity of g, N_g, satisfies $N_g = 2^{p-1} - 2^{\frac{1}{2}(p-1)}$,*
(ii) *the degree of g is at most $\frac{1}{2}(p+1)$,*
(iii) *g has at most one non-zero linear structure,*

(iv) *for any nonsingular $p \times p$ matrix B over $GF(2)$ and any vector $\beta \in V_p$,*
$h(y) = g(yB \oplus \beta)$ is also a $(p-1)$th-order plateaued function, where $y \in V_p$,
(v) *for any affine function ψ on V_p, $g \oplus \psi$ is also a $(p-1)$th-order plateaued*
function on V_p.

Proof. Due to Lemmas 2 and 6, (1) and (ii) are obvious. We now prove (iii).
Applying Lemma 3 to function g, we have

$$(\Delta(\beta_0), \Delta(\beta_1), \ldots, \Delta(\beta_{2^p-1}))H_p = (\langle \xi, e_0 \rangle^2, \langle \xi, e_1 \rangle^2, \ldots, \langle \xi, e_{2^p-1} \rangle^2)$$

where β_j is the binary representation of an integer j, $j = 0, 1, \ldots, 2^p - 1$ and
e_i is the ith row of H_p. Multiplying the above equality by itself, we obtain
$2^p \sum_{j=0}^{2^p-1} \Delta^2(\beta_j) = \sum_{j=0}^{2^p-1} \langle \xi, e_1 \rangle^4$. Note that $\Delta(\beta_0) = 2^p$ and that g is a $(p-1)$th-order plateaued function on V_p. Hence $2^p(2^{2p} + \sum_{j=1}^{2^p-1} \Delta^2(\beta_j)) = 2^{3p+1}$. It
follows that $\sum_{j=1}^{2^p-1} \Delta^2(\beta_j) = 2^{2p}$. This proves that g has at most one non-zero
linear structure and hence (iii) is true. (iv) and (v) are easy to verify. □

Theorem 2. *Let p be a positive odd number and g be a $(p-1)$th-order plateaued*
function on V_p that has no non-zero linear structure. Then there exists a non-
singular $2^p \times 2^p$ matrix B over $GF(2)$, such that $h(y) = g(yB)$, where $y \in V_p$,
is a $(p-1)$th-order plateaued function on V_p and also a 1st-order correlation
immune function.

Proof. Set $\Omega = \{\beta | \beta \in V_p, \langle \xi, e_\beta \rangle = 0\}$, where e_β is identified with e_i and β is
the binary representation of an integer i, $0 \leq i \leq 2^p - 1$.

Since $\#\Omega = 2^{p-1}$, the rank of Ω, denoted $rank(\Omega)$, satisfies $rank(\Omega) \geq p-1$.
We now prove $rank(\Omega) = p$. Assume that $rank(\Omega) = p-1$. Since $\#\Omega = 2^{p-1}$, Ω
is identified with a $(p-1)$-dimensional linear subspace of V_p. Recall that we can
use a nonsingular affine transformation on the variables to transform a linear
subspace into any other linear subspace with the same dimension. Without loss
of the generality, we assume that Ω is composed of $\beta_0, \beta_1, \ldots, \beta_{2^{p-1}-1}$, where
each β_j is the binary representation of an integer j, $0 \leq j \leq 2^p - 1$. By using
Lemma 3, we have

$$(\langle \xi, e_0 \rangle^2, \langle \xi, e_1 \rangle^2, \ldots, \langle \xi, e_{2^p-1} \rangle^2)H_p = 2^p(\Delta_g(\beta_0), \Delta_g(\beta_1), \ldots, \Delta_g(\beta_{2^p-1}))$$

and hence

$$(0, 0, \ldots, 0, 2^{p+1}, 2^{p+1}, \ldots, 2^{p+1})H_p = 2^p(\Delta_g(\beta_0), \Delta_g(\beta_1), \ldots, \Delta_g(\beta_{2^p-1}))$$

where the number of zeros is equal to 2^{p-1}. By using the construction of H_p and
comparing the terms in the above equality, we find that $\Delta_g(\beta_{2^p-1}) = -2^p$. That
is, β_{2^p-1} is a non-zero linear structure of g. This contradicts the assumption in the
proposition, that g has no non-zero linear structure. This proves $rank(\Omega) = p$.
Hence we can choose p linearly independent vectors $\gamma_1, \ldots, \gamma_p$ from Ω.

Let μ_j denote the vector in V_p, whose jth term is one and all other terms are zeros, $j = 1, \ldots, p$. Define a $p \times p$ matrix B over $GF(2)$, such that $\gamma_j B = \mu_j$, $j = 1, \ldots, p$. Set $h(y) = g(yB^T)$, where $y \in V_p$ and B^T is the transpose of B. Due to (iv) of Proposition 4, $h(y)$ is a $(p-1)$th-order plateaued function on V_p. Next we prove that $h(y)$ is a 1st-order correlation immune function.

Note that $h(y) \oplus \langle \mu_j, y \rangle = g(yb^T) \oplus \langle \mu_j, y \rangle = g(z) \oplus \langle \mu_j, z(B^T)^{-1} \rangle$ where $z = yB^T$.

On the other hand,

$$\langle \mu_j, z(B^T)^{-1} \rangle = z(B^T)^{-1} \mu_j^T = z(B^{-1})^T \mu_j^T = z(\mu_j B^{-1})^T = z\gamma_j^T = \langle z, \gamma_j \rangle$$

It follows that $h(y) \oplus \langle \mu_j, y \rangle = g(z) \oplus \langle \gamma_j, z \rangle$ where $z = yB^T$.

Note that e_{γ_j} is the sequence of linear function $\psi_{\gamma_j} = \langle \gamma_j, y \rangle$. Since $\gamma_j \in \Omega$, $\langle \xi, e_{\gamma_j} \rangle = 0$. Due to Lemma 4, $g(z) \oplus \langle \gamma_j, z \rangle$ is balanced. Hence $h(y) \oplus \langle \mu_j, y \rangle$ is balanced. By using Lemma 5, we have proved that $h(y)$ is a 1st-order correlation immune function.   □

Theorem 3. *Let p be a positive odd integer and g be a $(p-1)$th-order plateaued function on V_p. If g has a non-zero linear structure, then there exists a non-singular $2^p \times 2^p$ matrix B over $GF(2)$, such that $g(yB) = cx_1 \oplus h(z)$ where $y = (x_1, x_2, \ldots, x_p)$, $z = (x_2, \ldots, x_n)$, each $x_j \in GF(2)$ and the function h is a bent function on V_{p-1}.*

Proof. Since g has a non-zero linear structure, there exists a nonsingular $2^p \times 2^p$ matrix B over $GF(2)$, such that $g^*(y) = g(yB) = cx_1 \oplus h(z)$ where $y = (x_1, x_2, \ldots, x_p)$, $z = (x_2, \ldots, x_n)$ and h is a function on V_{p-1}. We only need to prove that h is bent. Without loss of generality, assume that $c = 1$. Then we have $g^*(y) = x_1 \oplus h(z)$. Let η denote the sequence of h. Hence the sequence of g^*, denoted by ξ, satisfies $\xi = (\eta, -\eta)$. Let e_i denote the ith row of H_{p-1}. From the structure of Sylvester-Hadamard matrices, (e_i, e_i) is the ith row of H_p, denoted by ℓ_i, $i = 0, 1, \ldots, 2^{p-1} - 1$, and $(e_i, -e_i)$ is the $(2^{p-1} + i)$th row of H_p, denoted by $\ell_{2^{p-1}+i}$, $i = 0, 1, \ldots, 2^{p-1} - 1$. Obviously

$$\langle \xi, \ell_i \rangle = 0, \ i = 0, 1, \ldots, 2^{p-1} - 1 \tag{1}$$

Since g^* is a $(p-1)$th-order plateaued function on V_p, (1) implies

$$\langle \xi, \ell_{2^{p-1}+i} \rangle = \pm 2^{\frac{1}{2}(p+1)}, \ i = 0, 1, \ldots, 2^{p-1} - 1 \tag{2}$$

Note that $\langle \xi, \ell_{2^{p-1}+i} \rangle = 2\langle \eta, e_i \rangle$, $i = 0, 1, \ldots, 2^{p-1} - 1$. From (2), $\langle \eta, e_i \rangle = \pm 2^{\frac{1}{2}(p-1)}$, $i = 0, 1, \ldots, 2^{p-1} - 1$. This proves that h is a bent function on V_{p-1}.   □

4 Complementary $(n-1)$th-order Plateaued Functions on V_n

To explore new properties of bent functions, we propose the following new concept.

Definition 9. *Let p be a positive odd number and g_1, g_2 be two functions on V_p. Denote the sequences of g_1 and g_2 by ξ_1 and ξ_2 respectively. Then g_1 and g_2 are said to be* complementary $(p-1)$th-order plateaued functions on V_p *if they are $(p-1)$th-order plateaued functions on V_p, and satisfy the property that $\langle \xi_1, e_i \rangle = 0$ if and only if $\langle \xi_2, e_i \rangle \neq 0$, and $\langle \xi_1, e_i \rangle \neq 0$ if and only if $\langle \xi_2, e_i \rangle = 0$.*

The following Lemma can be found in [10]:

Lemma 7. *Let $k \geq 2$ be a positive integer and $2^k = a^2 + b^2$ where $a \geq b \geq 0$ and both a and b are integers. Then $a^2 = 2^k$ and $b = 0$ when k is even, and $a^2 = b^2 = 2^{k-1}$ when n is odd.*

Proposition 5. *Let p be a positive odd number and g_1, g_2 be two functions on V_p. Denote the sequences of g_1 and g_2 by ξ_1 and ξ_2 respectively. Then g_1 and g_2 are complementary $(p-1)$th-order plateaued functions on V_p if and only if $\langle \xi_1, e_i \rangle^2 + \langle \xi_2, e_i \rangle^2 = 2^{p+1}$, where e_i is the ith row of H_p, $i = 0, 1, \ldots, 2^p - 1$.*

Proof. The necessity is obvious. We now prove the sufficiency. We keep using all the notations in Definition 9. Assume that $\langle \xi_1, e_i \rangle^2 + \langle \xi_2, e_i \rangle^2 = 2^{p+1}$, where e_i is the ith row of H_p, $i = 0, 1, \ldots, 2^p - 1$. Since $p + 1$ is even, by using Lemma 7, we conclude $\langle \xi_1, e_i \rangle^2 = 2^{p+1}$ or 0, $i = 0, 1, \ldots, 2^p - 1$. Similarly $\langle \xi_2, e_i \rangle^2 = 2^{p+1}$ or 0, $i = 0, 1, \ldots, 2^p - 1$. It is easy to see that g_1 and g_2 are complementary $(p-1)$th-order plateaued functions on V_p. □

Theorem 4. *Let p be a positive odd number and g_1, g_2 be two functions on V_p. Then g_1 and g_2 are complementary $(p-1)$th-order plateaued functions on V_p if and only if for every non-zero vector β in V_p, $\Delta_{g_1}(\beta) = -\Delta_{g_2}(\beta)$.*

Proof. Applying Lemma 3 to function g_1 and g_2, we obtain

$$(\Delta_{g_1}(\beta_0) + \Delta_{g_2}(\beta_0), \Delta_{g_1}(\beta_1) + \Delta_{g_2}(\beta_1), \ldots, \Delta_{g_1}(\beta_{2^p-1}) + \Delta_{g_2}(\beta_{2^p-1}))H_p$$
$$= (\langle \xi_1, e_0 \rangle^2 + \langle \xi_2, e_0 \rangle^2, \langle \xi_1, e_1 \rangle^2 + \langle \xi_2, e_1 \rangle^2, \ldots, \langle \xi_1, e_{2^p-1} \rangle^2 + \langle \xi_2, e_{2^p-1} \rangle^2) \quad (3)$$

where β_i is the binary representation of integer i and e_i is the ith row of H_p, $i = 0, 1, \ldots, 2^p - 1$.

Assume that g_1 and g_2 are complementary $(p-1)$th-order plateaued functions on V_p. From (3), we have

$$(\Delta_{g_1}(\beta_0) + \Delta_{g_2}(\beta_0), \Delta_{g_1}(\beta_1) + \Delta_{g_2}(\beta_1), \ldots, \Delta_{g_1}(\beta_{2^p-1}) + \Delta_{g_2}(\beta_{2^p-1}))H_p$$
$$= (2^{p+1}, 2^{p+1}, \ldots, 2^{p+1}) \quad (4)$$

or

$$(\Delta_{g_1}(\beta_0) + \Delta_{g_2}(\beta_0), \Delta_{g_1}(\beta_1) + \Delta_{g_2}(\beta_1), \ldots, \Delta_{g_1}(\beta_{2^p-1}) + \Delta_{g_2}(\beta_{2^p-1}))$$
$$= 2(1, 1, \ldots, 1)H_p$$

Comparing the jth terms in the two sides of the above equality, we have $\Delta_{g_1}(\beta) + \Delta_{g_2}(\beta) = 2^{p+1}$, for $\beta = 0$, and $\Delta_{g_1}(\beta) + \Delta_{g_2}(\beta) = 0$, for $\beta \neq 0$.

Conversely, assume that $\Delta_{g_1}(\beta) + \Delta_{g_2}(\beta) = 0$, for $\beta \neq 0$. From (3), we have

$$(2^{p+1}, 0, \ldots, 0)H_p$$
$$= (\langle \xi_1, e_0 \rangle^2 + \langle \xi_2, e_0 \rangle^2, \langle \xi_1, e_1 \rangle^2 + \langle \xi_2, e_1 \rangle^2, \ldots, \langle \xi_1, e_{2^p-1} \rangle^2 + \langle \xi_2, e_{2^p-1} \rangle^2)$$

It follows that $\langle \xi_1, e_i \rangle^2 + \langle \xi_2, e_i \rangle^2 = 2^{p+1}$, $i = 0, 1, \ldots, 2^p - 1$. This proves that g_1 and g_2 are complementary $(p-1)$th-order plateaued functions on V_p. □

By using Theorem 4, we conclude

Proposition 6. *Let p be a positive odd number and g_1, g_2 be complementary $(p-1)$th-order plateaued functions on V_p. Then*

(i) β is a non-zero linear structure of g_1 if and only if β is a non-zero linear structure of g_2,

(ii) one and only one of g_1 and g_2 is balanced.

Proof. (i) can be obtained from Theorem 4.

(ii) We keep using the notations in Definition 9. From Proposition 5, $\langle \xi_1, e_0 \rangle^2 = 2^{p+1}$ if and only if $\langle \xi_2, e_0 \rangle^2 = 0$, and $\langle \xi_1, e_0 \rangle^2 = 0$ if and only if $\langle \xi_2, e_0 \rangle^2 = 2^{p+1}$. Note that e_0 is the all-one sequence hence $\langle \xi_j, e_0 \rangle = 0$ implies g_j is balanced. Hence one and only one of g_1 and g_2 is balanced. □

Proposition 7. *Let p be a positive odd number and g_1, g_2 be complementary $(p-1)$th-order plateaued functions on V_p. For any $\beta, \gamma \in V_p$, set $g_1^*(y) = g_1(y \oplus \beta)$ and $g_2^*(y) = g_2(y \oplus \gamma)$. Then $g_1^*(y)$ and $g_2^*(y)$ are complementary $(p-1)$th-order plateaued functions on V_p.*

Proof. Since g_1, g_2 are complementary $(p-1)$th-order plateaued functions on V_p, from Theorem 4, for any non-zero vector α in V_p, $\Delta_{g_1}(\alpha) = -\Delta_{g_2}(\alpha)$. On the other hand, it is easy to verify $\Delta_{g_2^*}(\alpha) = \Delta_{g_2}(\alpha)$, where α is any vector in V_p. Hence for any non-zero vector β in V_p, $\Delta_{g_1}(\alpha) = -\Delta_{g_2^*}(\alpha)$. Again, by using Theorem 4, we have proved that g_1, g_2^* are complementary $(p-1)$th-order plateaued functions on V_p. By the same reasoning, we can prove that g_1^* and g_2^* are complementary $(p-1)$th-order plateaued functions on V_p. □

Now fix β, i.e., fix g_1^* in Proposition 7, and let γ be arbitrary. We can see that there exist more than one function that can team up with g_1^* to form complementary $(p-1)$th-order plateaued functions on V_p. This shows that the relationship of complementary $(p-1)$th-order plateaued functions on V_p is not a one-to-one correspondence.

Theorem 5. *Let p be a positive odd number and ξ_1, ξ_2 be two $(1, -1)$ sequences of length 2^p. Set $\eta_1 = 2^{-\frac{1}{2}(p+1)}(\xi_1 + \xi_2)H_p$ and $\eta_2 = 2^{-\frac{1}{2}(p+1)}(\xi_1 - \xi_2)H_p$. Then ξ_1 and ξ_2 are the sequences of complementary $(p-1)$th-order plateaued functions on V_p if and only if η_1 and η_2 are the sequences of complementary $(p-1)$th-order plateaued functions on V_p.*

Proof. Assume that ξ_1 and ξ_2 are the sequences of complementary $(p-1)$th-order plateaued functions on V_p respectively. It can be verified straightforwardly that both η_1 and η_2 are $(1, -1)$ sequences. Hence both η_1 and η_2 are the sequences of functions on V_p.

Furthermore we have

$$\eta_1 H_p = 2^{\frac{1}{2}(p+1)}(\frac{1}{2}(\xi_1 + \xi_2)), \quad \eta_2 H_p = 2^{\frac{1}{2}(p+1)}(\frac{1}{2}(\xi_1 - \xi_2)) \tag{5}$$

Note that both $\frac{1}{2}(\xi_1 + \xi_2)$ and $\frac{1}{2}(\xi_1 - \xi_2)$ are $(0, 1, -1)$ sequences. From (5), $\langle \eta_1, e_i \rangle$ and $\langle \eta_2, e_i \rangle$, where e_i is the ith row of H_p, $i = 0, 1, \ldots, 2^p - 1$, take the value of $\pm 2^{\frac{1}{2}(p+1)}$ or 0 only. On the other hand, it is easy to see that the ith term of $\frac{1}{2}(\xi_1 \pm \xi_2)$ is non-zero if and only if the ith term of $\frac{1}{2}(\xi_1 \mp \xi_2)$ is zero. This proves that $\langle \eta_1, e_i \rangle \neq 0$ if and only if $\langle \eta_2, e_i \rangle = 0$, also $\langle \eta_1, e_i \rangle = 0$ if and only if $\langle \eta_2, e_i \rangle \neq 0$, $i = 0, 1, \ldots, 2^p - 1$. By using Proposition 5 η_1 and η_2 are the sequences of complementary $(p - 1)$th-order plateaued functions on V_p.

Conversely, Assume that η_1 and η_2 are the sequences of complementary $(p-1)$th-order plateaued functions on V_p. Note that $\xi_1 = 2^{-\frac{1}{2}(p+1)}(\eta_1 + \eta_2)H_p$ and $\xi_2 = 2^{-\frac{1}{2}(p+1)}(\eta_1 - \eta_2)H_p$. Inverse the above deduction, we have proved that ξ_1 and ξ_2 are the sequences of complementary $(p - 1)$th-order plateaued functions on V_p.

$\square$

In Section 5, we will prove that the existence of complementary $(n - 2)$th-order plateaued functions on V_{n-1} is equivalent to the existence of bent functions on V_n.

5   Relating Bent Functions on V_n to Complementary $(n - 2)$th-order Plateaued Functions on V_{n-1}

Lemma 8. *Let n be a positive even number and f be a function on V_n. Denote the sequence of f by $\xi = (\xi_1, \xi_2)$, where both ξ_1 and ξ_2 are of length 2^{n-1}. Let ξ_1 and ξ_2 be the sequences of functions f_1 and f_2 on V_{n-1} respectively. Then f is bent if and only if f_1 and f_2 are complementary $(n - 2)$th-order plateaued functions on V_{n-1}.*

Proof. Obviously, $\xi H_n = (\langle \xi, \ell_0 \rangle, \langle \xi, \ell_1 \rangle, \ldots, \langle \xi, \ell_{2^n-1} \rangle)$ where ℓ_j is the jth row of H_n, $j = 0, 1, \ldots, 2^n - 1$. Hence

$$(\xi_1, \xi_2) \begin{bmatrix} H_{n-1} & H_{n-1} \\ H_{n-1} & -H_{n-1} \end{bmatrix} = (\langle \xi, \ell_0 \rangle, \langle \xi, \ell_1 \rangle, \ldots, \langle \xi, \ell_{2^n-1} \rangle) \tag{6}$$

For each j, $0 \leq j \leq 2^{n-1}-1$, comparing the jth terms in the two sides of equality (6), also comparing the $2^{n-1}+j$ terms in the two sides of the equality, we find

$$\langle \xi_1, e_j \rangle + \langle \xi_2, e_j \rangle = \langle \xi, \ell_j \rangle, \quad \langle \xi_1, e_j \rangle - \langle \xi_2, e_j \rangle = \langle \xi, \ell_{2^{n-1}+j} \rangle \tag{7}$$

e_j is the jth row of H_{n-1}, $j = 0, 1, \ldots, 2^{n-1}-1$.

Assume that f is bent. From Theorem 1, $|\langle \xi, \ell_j \rangle| = 2^{\frac{1}{2}n}$ and $|\langle \xi, \ell_{2^{n-1}+j} \rangle| = 2^{\frac{1}{2}n}$, $j = 0, 1, \ldots, 2^{n-1}-1$.

Due to (7), $|\langle \xi_1, e_j \rangle + \langle \xi_2, e_j \rangle| = |\langle \xi_1, e_j \rangle - \langle \xi_2, e_j \rangle| = 2^{\frac{1}{2}n}$. This causes $\langle \xi_1, e_j \rangle = 2^{\frac{1}{2}n}$ and $\langle \xi_2, e_j \rangle = 0$ otherwise $\langle \xi_1, e_j \rangle = 0$ and $\langle \xi_2, e_j \rangle = 2^{\frac{1}{2}n}$. This proves that f_1 and f_2 are complementary $(n-2)$th-order plateaued functions on V_{n-1}.

Conversely, assume that f_1 and f_2 are complementary $(n-2)$th-order plateaued functions on V_{n-1}. ¿From Proposition 5, for each i, $0 \leq i \leq 2^{n-1}-1$, $\langle \xi_1, e_i \rangle$ and $\langle \xi_1, e_i \rangle$ take the value of $\pm 2^{\frac{1}{2}n}$ or 0 only. Furthermore $\langle \xi_1, e_i \rangle = 0$ implies $\langle \xi_2, e_i \rangle \neq 0$, and $\langle \xi_1, e_i \rangle \neq 0$ implies $\langle \xi_2, e_i \rangle = 0$. ¿From (7), $\langle \xi, \ell_j \rangle = \pm 2^{\frac{1}{2}n}$ and $\langle \xi, \ell_{2^{n-1}+j} \rangle \pm 2^{\frac{1}{2}n}$, $j = 0, 1, \ldots, 2^{n-1}-1$. Due to Theorem 1, f is bent. $\square$

Lemma 8 can be briefly restated as follows:

Theorem 6. *Let n be a positive even number and f be a function on V_n. Then f is bent if and only if the two functions on V_{n-1}, $f(0, x_2, \ldots, x_n)$ and $f(1, x_2, \ldots, x_n)$, are complementary $(n-2)$th-order plateaued functions on V_{n-1}.*

Proof. It is easy to verify that $f(x_1, \ldots, x_n) = (1 \oplus x_1)f(0, x_2, \ldots, x_n) \oplus x_1 f(1, x_2, \ldots, x_n)$. Set $f_1(x_2, \ldots, x_n) = f(0, x_2, \ldots, x_n)$ and $f_2(x_2, \ldots, x_n) = f(1, x_2, \ldots, x_n)$. Denote the sequences of f_1 and f_2 by ξ_1 and ξ_2 respectively. Obviously, the sequence of f, denoted by ξ, satisfies $\xi = (\xi_1, \xi_2)$. By using Lemma 8, we have proved the theorem. $\square$

Due to Theorem 6, the following proposition is obvious.

Proposition 8. *Let n be a positive even number and f be a function on V_n. Then f is bent if and only if the two functions on V_{n-1}, $f(x_1, \ldots, x_{j-1}, 0, x_{j+1}, \ldots, x_n)$ and $f(x_1, \ldots, x_{j-1}, 1, x_{j+1}, \ldots, x_n)$ are complementary $(n-2)$th-order plateaued functions on V_{n-1}. $j = 1, \ldots, n$.*

The following theorem follows Theorem 6 and Proposition 7.

Theorem 7. *Let n be a positive even number and f be a function on V_n. Write $x = (x_1, \ldots, x_n)$ and $y = (x_2, \ldots, x_n)$ where $x_j \in GF(2)$, $j = 1, \ldots, n$. Set $f_1(x_2, \ldots, x_n) = f(0, x_2, \ldots, x_n)$ and $f_2(x_2, \ldots, x_n) = f(1, x_2, \ldots, x_n)$. Then f is bent if and only if $g(x) = (1 \oplus x_1)f_1(y \oplus \gamma_1) \oplus x_1 f_2(y \oplus \gamma_2)$ is bent, where γ_1 and γ_2 are any two vectors in V_{n-1}.*

By using Theorem 5 and Lemma 8, we conclude

Theorem 8. *Let $\xi = (\xi_1, \xi_2)$ be a $(1, -1)$ sequence of length 2^n, where both ξ_1 and ξ_2 are of length 2^{n-1}. Then ξ is the sequence of a bent function if and only if $2^{-\frac{1}{2}n}((\xi_1 + \xi_2)H_{n-1}, (\xi_1 - \xi_2)H_{n-1})$ is the sequence of a bent function.*

Theorems 6, 7 and 8 represent new characterisations of bent functions. In addition, Theorems 7 and 8 provide methods of constructing new bent function from known bent functions.

6   Non-quadratic Bent Functions

Definition 10. *Let f be a function on V_n and W be an r-dimensional linear subspace of W. From linear algebra, V_n can be divided into 2^{n-r} disjoint cosets of W:*

$$V_n = U_0 \cup U_1 \cup \cdots \cup U_{2^{n-r}-1}$$

where $U_0 = W$, $\#U_j = 2^r$, $j = 0, 1, \ldots, 2^{n-r} - 1$, and for any two vectors γ and β in V_n, β and γ belong to the same coset U_j if and only if $\beta \oplus \gamma \in W$. The partition is unique if the order of the cosets is ignored. Each U_j can be expressed as $U_j = \gamma_j \oplus W$ where γ_j is a vector in V_n and $\gamma_j \oplus W$ denotes $\{\gamma_j \oplus \alpha | \alpha \in W\}$ however γ_j is not unique. For a coset $U = \gamma \oplus W$, define a function g on W such that $g(\alpha) = f(\gamma \oplus \alpha)$ for every $\alpha \in W$. Then g is called the restriction of f to coset $\gamma \oplus W$. g can be denoted by $f_{\gamma \oplus W}$. In particular, the restriction of f to linear subspace W, denoted by f_W, is a function h on W such that $h(\alpha) = f(\alpha)$ for every $\alpha \in W$.

Proposition 9. *Let f be a bent function on V_n and W be an arbitrary $(n-1)$-dimensional linear subspace. Let V_n divided into two disjoint cosets: $V_n = W \cup U$. Then the restriction of f to linear subspace W, f_W, and the restriction of f to coset U, f_U, are complementary $(n-2)$th-order plateaued functions on V_{n-1}.*

Proof. In fact, $W^* = \{(0, x_2, \ldots, x_n) | x_2, \ldots, x_n \in GF(2)\}$ forms an $(n-1)$-dimensional linear subspace and $U^* = \{(1, x_2, \ldots, x_n) | x_2, \ldots, x_n \in GF(2)\}$ is a coset of W. By using a nonsingular linear transformation on the variables, we can transform W into W^* and U into U^* simultaneously.. By using Theorem 6, we have proved the Proposition.   □

Proposition 9 shows that the restriction of f to any $(n-1)$-dimensional linear subspace is still cryptographically strong.

We now prove the following characteristic property of quadratic bent functions.

Lemma 9. *Let f be a bent function on V_n. Then for any $(n-1)$-dimensional linear subspace W, the restriction of f to W has a non-zero linear structure if and only if f is quadratic.*

Proof. Let f be quadratic and W be an arbitrary $(n-1)$-dimensional linear subspace. Since $n - 1$ is odd, the restriction of f to W, denoted by g, is not bent.

Hence due to (iii) of Theorem 1, there exists a non-zero vector β in W, such that $g(y) \oplus g(y \oplus \beta)$ is not balanced. On the other hand, since g is also quadratic, $g(y) \oplus g(y \oplus \beta)$ is affine. It is easy to see that any non-balanced affine function must be constant. This proves that β is a non-zero linear structure of g.

We now prove the converse: "if for any $(n-1)$-dimensional linear subspace W, the restriction of f to W has a non-zero linear structure, then f is quadratic" by induction on the dimension n.

Let $n = 2$. Bent functions on V_2 must be quadratic. For $n = 4$, from (i) of Proposition 1, bent functions on V_4 must be quadratic.

Assume that the converse is true for $4 \leq n \leq k-2$ where k is even. We now prove the converse for $n = k$.

Let f be a bent function on V_k such that for any $(k-1)$-dimensional linear subspace W the restriction of f to W has a non-zero linear structure.

It is easy to see that f can be expressed as $f(x) = x_1 g(y) \oplus h(y)$ where $y = (x_2, \ldots, x_k)$, both g and h are functions on V_{k-1}. From Theorem 6, $f(0, x_2, \ldots, x_k) = h(y)$ and $f(1, x_2, \ldots, x_k) = g(y) \oplus h(y)$ are complementary $(k-2)$th-order plateaued functions on V_{k-1}.

Since $\{(0, x_2, \ldots, x_k) | x_2, \ldots, x_k \in GF(2)\}$ forms a $(k-1)$-dimensional linear subspace, due to the assumption about f: "the restriction of f to any $(k-1)$-dimensional linear subspace has a non-zero linear structure", $f(0, x_2, \ldots, x_k) = h(y)$ has a non-zero linear structure. Without loss of generality, we can assume that the vector β in V_{k-1}, $\beta = (1, 0, \ldots, 0)$, is the non-zero linear structure of $h(y)$. It is easy to see $h(y) = cx_2 \oplus b(z)$ where c is a constant in $GF(2)$, $z = (x_3, \ldots, x_k)$ and $b(z)$ is a function on V_{k-2}. Without loss of generality, we assume that $c = 1$. From Theorem 3, $b(z)$ is a bent function on V_{k-2}.

It is easy to see $\Delta_h(\beta) = -2^{k-1}$. From Theorem 4, $\beta = (1, 0, \ldots, 0)$ is also a linear structure of $g(y) \oplus h(y)$ and $\Delta_{g \oplus h} = 2^{k-1}$. Hence $g(y) \oplus h(y)$ can be expressed as $g(y) \oplus h(y) = dx_2 \oplus p(z)$, where $z = (x_3, \ldots, x_k)$. Due to Theorem 3, $p(z)$ is a bent function on V_{k-2}. Since $\Delta_{g \oplus h}(\beta) = 2^{k-1}$, $d = 0$. Hence $g(y) = h(y) \oplus p(z) = x_2 \oplus b(z) \oplus p(z)$ and hence

$$f(x) = x_1(x_2 \oplus b(z) \oplus p(z)) \oplus x_2 \oplus b(z) \tag{8}$$

Since $\{(x_1, 0, x_3, \ldots, x_k) | x_1, x_3, \ldots, x_k \in GF(2)\}$ forms a $(k-1)$-dimensional linear subspace, $f(x_1, 0, x_3, \ldots, x_k)$ is the restriction of f to this $(k-1)$-dimensional linear subspace. Due to the assumption about f, $f(x_1, 0, x_3, \ldots, x_k)$ has a non-zero linear structure, denoted by γ, $\gamma \in V_{k-1}$. From (8), $f'(u) = f(x_1, 0, x_3, \ldots, x_n) = x_1(b(z) \oplus p(z)) \oplus b(z)$, where $u \in V_{k-1}$ and $u = (x_1, x_3, x_4, \ldots x_k)$.

There exist two cases of γ.

Case 1: $\gamma = (0, \mu)$ where $\mu \in V_{k-2}$. Since $\gamma \neq 0$, μ is non-zero. It is easy to see $f'(u) \oplus f'(u \oplus \gamma) = x_1(b(z) \oplus b(z \oplus \mu) \oplus p(z) \oplus p(z \oplus \mu)) \oplus b(z) \oplus b(z \oplus \mu)$.

Since $f'(u) \oplus f'(u \oplus \gamma)$ is a constant, $b(z) \oplus b(z \oplus \mu) \oplus p(z) \oplus p(z \oplus \mu) = 0$ and $b(z) \oplus b(z \oplus \mu) = c'$, where c' is constant. On the other hand, since $b(z)$ is bent and $\mu \neq 0$, $b(z) \oplus b(z \oplus \mu)$ is balanced and hence it is not constant. This is a contradiction. This proves that Case 1 cannot take place.

Case 2: $\gamma = (1, \nu)$ where $\nu \in V_{k-2}$ and ν is not necessarily non-zero. It is easy to see $f'(u) \oplus f'(u \oplus \gamma) = x_1(b(z) \oplus b(z \oplus \nu) \oplus p(z) \oplus p(z \oplus \nu)) \oplus b(z) \oplus p(z \oplus \nu)$.

Since $f'(u) \oplus f'(u \oplus \gamma)$ is a constant, $b(z) \oplus b(z \oplus \nu) \oplus p(z) \oplus p(z \oplus \nu) = 0$ and $b(z) \oplus p(z \oplus \nu) = c''$, where c'' is constant, and hence $b(z \oplus \nu) \oplus p(z) = c''$. From (8),

$$f(x) = x_1 x_2 \oplus x_1(b(z) \oplus b(z \oplus \nu) \oplus c'') \oplus x_2 \oplus b(z) \tag{9}$$

We now turn to the restriction of f to another $(k-1)$-dimensional linear subspace. Write $U^* = \{(x_3 \ldots, x_k) | x_3, \ldots, x_k \in GF(2)\}$ and $U_* = \{(x_1, x_2) | x_1, x_2 \in GF(2)\}$. Hence U^* is a $(k-2)$-dimensional linear subspace and U_* is a 2-dimensional linear subspace, and $V_k = (U_*, U^*)$, where $(X, Y) = \{(\alpha, \beta) | \alpha \in X, \beta \in Y\}$.

Let Λ denote an arbitrary $(k-3)$-dimensional linear subspace in U^*. Hence (U_*, Λ) is a $(k-1)$-dimensional linear subspace.

Let $f''(y)$ denote the restriction of f to (U_*, Λ), where $y \in (U_*, \Lambda)$. Hence y can be expressed as $y = (x_1, x_2, v)$ with $v = (v_1, \ldots, v_{k-2}) \in \Lambda$, where $v_1, \ldots, v_{k-2} \in GF(2)$ but not arbitrary because Λ is a proper subset of V_{k-2}.

¿From (9), $f''(y)$ can be expressed as $f''(y) = x_1 x_2 \oplus x_1(b'(v) \oplus b''(v) \oplus a) \oplus x_2 \oplus b'(v)$, where $b'(v)$ denotes the restriction of $b(z)$ to Λ and $b''(v)$ denotes the restriction of $b(z \oplus \nu)$ to Λ.

From the assumption about f, f'' has a non-zero linear structure γ', $\gamma' \in (U_*, \Lambda)$. Write $\gamma' = (a_1, a_2, \tau)$ where $\tau \in \Lambda$. Since $\gamma' = (a_1, a_2, \tau)$ is a non-zero linear structure of f'', it is easy to verify $a_1 = a_2 = 0$. This proves $\gamma' = (0, 0, \tau)$. Since γ' is non-zero, $\tau \neq 0$.

Hence $f''(y) \oplus f''(y \oplus \gamma') = x_1(b'(v) \oplus b'(v \oplus \tau) \oplus b''(v) \oplus b''(v \oplus \tau)) \oplus b'(v) \oplus b'(v \oplus \tau)$. Since $f''(y) \oplus f''(y \oplus \gamma')$ is constant, $b'(v) \oplus b'(v \oplus \tau) \oplus b''(v) \oplus b''(v \oplus \tau) = 0$ and $b'(v) \oplus b'(v \oplus \tau)$ is constant. Hence τ is a non-zero linear structure of $b'(v)$. This proves that for any $(n-3)$-dimensional linear subspace Λ, the restriction of $b(z)$ to Λ, i.e., $b'(v)$, has a non-zero linear structure. On the other hand, since $b(z)$ is a bent function on V_{k-2}, due to the induction assumption, $b(z)$ is quadratic. Hence $b(z) \oplus b(z \oplus \nu)$ must be affine. From (9), we have proved $f(x) = x_1 x_2 \oplus x_1(b(z) \oplus b(z \oplus \nu) \oplus a) \oplus x_2 \oplus b(z)$ is quadratic when $n = k$. □

Due to the low algebraic degree, quadratic functions are not cryptographically desirable, although some of them are highly nonlinear.

The following is an equivalent statement of Lemma 9.

Theorem 9. *Let f be a bent function on V_n. Then f is non-quadratic if and only if there exists an $(n-1)$-dimensional linear subspace W such that the restriction of f to W, f_W, has no non-zero linear structure.*

Theorem 9 is an interesting characterization of non-quadratic bent functions.

7   New Constructions of Cryptographic Functions

The relationships among a bent function on V_n and complementary $(n-2)$th-order plateaued functions on V_{n-1} are helpful to design cryptographic functions from bent functions. In fact, from Theorem 6, any bent function on V_n can be "split" into complementary $(n-2)$th-order plateaued functions on V_{n-1}.

We prefer non-quadratic bent functions as they are useful to obtain complementary plateaued functions that have no non-zero linear structures.

Let f be a non-quadratic bent function on V_n. By using Theorem 9, we can find an $(n-1)$-dimensional subspace W such that the restriction of f to W, f_W, has no non-zero linear structure. For any vector $\alpha \in V_n$ with $\alpha \notin W$, we have $(\alpha \oplus W) \cap W = \emptyset$ and $V_n = W \cup (\alpha \oplus W)$. From Proposition 9, the restriction of f to $\alpha \oplus W$, $f_{\alpha \oplus W}$, and f_W are complementary $(n-2)$th-order plateaued functions on V_{n-1}. Due to (i) of Proposition 6, $f_{\alpha \oplus W}$ has no non-zero linear structure. Due to (ii) of Proposition 6, one and only one of f_W and $f_{\alpha \oplus W}$ is balanced. From Propositions 4, we can see that both f_W and $f_{\alpha \oplus W}$ are highly nonlinear.

Furthermore, by using Theorem 2, we can use a nonsingular linear transformation on the variables to transform the balanced f_W or $f_{\alpha \oplus W}$ into another $(n-2)$th-order plateaued function g on V_{n-1}. The resultant function is a 1st-order correlation immune function. Obviously g is still balanced and highly nonlinear, and it does not have non-zero linear structure.

We note that there is a more straightforward method to construct a balanced, highly nonlinear function on any odd dimensional linear space, by "concatenating" known bent functions. For example, let f be a bent function on V_k, we can set $g(x_1, \ldots, x_{k+1}) = x_1 \oplus f(x_2, \ldots, x_{k+1})$. Then g is a balanced, highly nonlinear function on V_{k+1}, where $k+1$ is odd. Let η and ξ denote the sequences of g and f respectively. It is easy to see $\eta = (\xi, -\xi)$ and hence η is a concatenations of ξ and $-\xi$. We call this method *concatenating* bent functions. A major problem of this method is that f contains a non-zero linear structure $(1, 0, \ldots, 0)$.

In contrast, the method of "splitting" a bent function we discussed earlier allows us to obtain functions that do not have non-zero linear structure.

8   Conclusions

We have identified relationships between bent functions and complementary plateaued functions, and discovered a new characteristic property of bent functions. Furthermore we have proved a necessary and sufficient condition of non-quadratic bent functions. Based on the new results on bent functions, we have proposed a new method for constructing balanced, highly nonlinear and correlation immune functions that have no non-zero linear structures.

Acknowledgement

The second author was supported by a Queen Elizabeth II Fellowship (227 23 1002).

References

1. P. Camion, C. Carlet, P. Charpin, and N. Sendrier. On correlation-immune functions. In *Advances in Cryptology - CRYPTO'91*, volume 576, Lecture Notes in Computer Science, pages 87–100. Springer-Verlag, Berlin, Heidelberg, New York, 1991.

2. Claude Carlet. Partially-bent functions. *Designs, Codes and Cryptography*, 3:135–145, 1993.

3. Xiao Guo-Zhen and J. L. Massey. A spectral characterization of correlation-immune combining functions. *IEEE Transactions on Information Theory*, 34(3):569–571, 1988.

4. F. J. MacWilliams and N. J. A. Sloane. *The Theory of Error-Correcting Codes*. North-Holland, Amsterdam, New York, Oxford, 1978.

5. W. Meier and O. Staffelbach. Nonlinearity criteria for cryptographic functions. In *Advances in Cryptology - EUROCRYPT'89*, volume 434, Lecture Notes in Computer Science, pages 549–562. Springer-Verlag, Berlin, Heidelberg, New York, 1990.

6. B. Preneel, W. V. Leekwijck, L. V. Linden, R. Govaerts, and J. Vandewalle. Propagation characteristics of boolean functions. In *Advances in Cryptology - EUROCRYPT'90*, volume 437, Lecture Notes in Computer Science, pages 155–165. Springer-Verlag, Berlin, Heidelberg, New York, 1991.

7. O. S. Rothaus. On "bent" functions. *Journal of Combinatorial Theory*, Ser. A, 20:300–305, 1976.

8. J. Wang. The linear kernel of boolean functions and partially-bent functions. *System Science and Mathematical Science*, 10:6–11, 1997.

9. A. F. Webster and S. E. Tavares. On the design of S-boxes. In *Advances in Cryptology - CRYPTO'85*, volume 219, Lecture Notes in Computer Science, pages 523–534. Springer-Verlag, Berlin, Heidelberg, New York, 1986.

10. X. M. Zhang and Y. Zheng. Characterizing the structures of cryptographic functions satisfying the propagation criterion for almost all vectors. *Design, Codes and Cryptography*, 7(1/2):111–134, 1996. special issue dedicated to Gus Simmons.

11. X. M. Zhang, Y. Zheng, and Hideki Imai. Duality of boolean functions and its cryptographic significance. In *Advances in Cryptology - ICICS'97*, volume 1334, Lecture Notes in Computer Science, pages 159–169. Springer-Verlag, Berlin, Heidelberg, New York, 1997.

12. Y. Zheng and X. M. Zhang. Plateaued functions. In *Advances in Cryptology - ICICS'99*, volume 1726, Lecture Notes in Computer Science, pages 284–300. Springer-Verlag, Berlin, Heidelberg, New York, 1999.

A Technique for Boosting the Security of Cryptographic Systems with One-Way Hash Functions

Takeshi Koshiba

Telecommunications Advancement Organization of Japan
1-1-32 Shin'urashima, Kanagawa-ku, Yokohama 221-0031, Japan
koshiba@acm.org

Abstract. In this paper, we show a practical solution to the problems where one-wayness of hash functions does not guarantee cryptographic systems to be secure enough. We strengthen the notion of one-wayness of hash functions and construct strongly one-way hash functions from any one-way hash function or any one-way function.

1 Introduction

Hash functions are important primitives for many cryptographic applications. Especially, (one-way) hash functions within digital signature schemes are essential ingredients. In general, basic schemes of any digital signature systems are existentially forgeable and the one-wayness of hash functions contribute to guarantee the security of existentially unforgeability. For example, in [9], Pointcheval and Sterns showed that the (modified) ElGamal signature scheme is existentially unforgeable against an adaptively chosen message attack, if hash functions are truly random in the sense of the random oracle model. Also, in [3], Bellare and Rogaway discussed the security of the RSA signature scheme in the random oracle model. In a sense, random oracle model is rather theoretical than practical.

There are two approaches to constructing a secure cryptographic system in a practical model. One is constructing a secure cryptographic system under strong assumptions and then weakening the assumptions. The other is constructing a secure cryptographic system (in a weak sense of security) in a practical model and then strengthening the sense of security.

In this paper, we show a technique for boosting security according to the later approach. One-wayness of hash functions can be a good property that guarantees a cryptographic system to be secure while it also can be insufficient to guarantee another cryptographic system to be secure.

Intuitively, one-way functions f are easy to compute, but hard to reverse. That is, given x it is easy to compute $f(x)$, but given $f(x)$ it is hard to compute x' such that $f(x) = f(x')$. Some cryptographic applications require that given $f(x)$ it be also hard to compute x' such that $f(x)$ is close to $f(x')$. We say that functions of the above property are *neighbor-free*. In general, one-wayness

JooSeok Song (Ed.): ICISC'99, LNCS 1787, pp. 76–81, 2000.

of functions does not ensure the neighbor-freeness. We note that the notion of neighbor-freeness is different from non-malleability [5] and sibling intractability [12]. If a digital signature scheme which uses one-way hash functions does not have this property, then there is a possibility that the signature scheme is existentially forgeable. In this paper, we show a way to construct one-way neighbor-free hash functions using any one-way hash functions. Combining the results in [11] and our results, one-way neighbor-free hash functions are also constructible from any one-way functions.

2   Preliminaries

A polynomial-time computable function $f : \{0,1\}^* \rightarrow \{0,1\}^*$ is *one-way* if for any probabilistic polynomial-time algorithm A, any polynomial p, and sufficiently large k,

$$\Pr[\, f(A(f(x), 1^k)) = f(x)\,] < 1/p(k),$$

where the probability is taken over all x's of length k and the internal coin tosses of A, with the uniform probability distribution. A polynomial-time computable function $h : \{0,1\}^* \rightarrow \{0,1\}^*$ is a *single one-way hash function* if it satisfies the following:

(a) For some function ℓ, $h(\{0,1\}^k) \subseteq \{0,1\}^{\ell(k)}$ and $\ell(k) < k$ for sufficiently large k.

(b) For any probabilistic polynomial-time algorithm A, any polynomial p, and sufficiently large k,

$$\Pr[\, A(x, 1^k) = x', \ x \neq x', \ h(x) = h(x')\,] < 1/p(k),$$

where the probability is taken over all x's of length k and the internal coin tosses of A, with the uniform probability distribution.

A family $H = \bigcup_{k>0} H_k$, where $H_k = \{h \mid h : \{0,1\}^k \rightarrow \{0,1\}^{\ell(k)}\}$, is a family of one-way hash functions if for any probabilistic polynomial-time algorithm A, for any polynomial p, and sufficiently large k,

$$\Pr[\, A(h, x, 1^k) = x', \ x \neq x', \ h(x) = h(x')\,] < 1/p(k),$$

where the probability is taken over all x's of length k and all h's in H_k and the internal coin tosses of A, with the uniform probability distribution.

A neighborhood function N is a *polynomial-size neighborhood* if there exists some polynomial q such that $|N(x)| \leq q(|x|)$ for all sufficiently long x. We also denote by N_k a polynomial-size neighborhood with the input of length k. We may consider that N is a length-wise family $\{N_k\}$ of polynomial-size neighborhoods.

We show some examples of polynomial-size neighborhood. In the following examples, we assume that k is the length of inputs to N_k for the sake of simplicity.

Example 1. Let c be a constant integer. Define $N_k(y) = \{y' \mid d_H(y, y') \leq c\}$, where d_H denotes the Hamming distance. Note that $|N_k(y)|$ is $O(k^c)$.

In the above example, the distance does not depend on the parameter k. We next show an example where the distance depends on the parameter k.

Example 2. For a string y, We denote by $suf_k(y)$ the string w such that $y = vw$ and v is of length $\lfloor \log k \rfloor$. Define $N_k(y) = \{y' \mid d_2(y, y') \leq \log k\}$, where d_2 is defined as follows:

$$d_2(y, y') = \begin{cases} d_H(y, y') & \text{if } suf_k(y) = suf_k(y'), \\ k & \text{otherwise.} \end{cases}$$

It follows from Wallis's Formula (or Stirling's Formula) that

$$\binom{\log k}{i} \leq k / \sqrt{\frac{\pi}{2} \log k}$$

for any i such that $0 \leq i \leq \log k$. Therefore, $|N_k(y)|$ is bounded by a polynomial in k.

In [11], Rompel showed a way to construct a family of universal one-way hash functions from any one-way function.

In this paper, we strengthen the notion of one-way hash functions and construct families of strongly one-way hash functions from any single one-way hash function or any one-way function.

Definition 1. A family $H = \bigcup_{k>0} H_k$, where $H_k = \{h \mid h : \{0, 1\}^k \to \{0, 1\}^{\ell(k)}\}$, is a family of *one-way neighbor-free hash functions* if it satisfies the following:

For any probabilistic polynomial-time algorithm A, for any polynomial-size neighborhood N, for any polynomial p, and for sufficiently large k,

$$\Pr[A(h, x, 1^k) = x', \ x \neq x', \ h(x') \in N_k(h(x))] < 1/p(k),$$

where the probability is taken over all x's of length k and all h's in H_k and the internal coin tosses of A, with the uniform probability distribution.

For a finite set S, the notation $s \in_U S$ means that the element s is randomly chosen from the set S with the uniform probability distribution.

3   Main Results

Theorem 1. *If there exists a single one-way hash function, there exists a family of one-way neighbor-free hash functions.*

Proof. Suppose that h is a single one-way hash function. Let $H = \bigcup_{k>0} H_k$, where $H_k = \{h \circ \hat{h} \mid \hat{h} \in G_k\}$ and G_k is the set of all permutations on $\{0, 1\}^k$. Construction of G_k has been discussed in [4]. Note that, on the condition that a permutation $\hat{h}$ is randomly chosen from G_k with the uniform probability distribution, even if x is chosen according to arbitrary probability distribution, the probability distribution of $\hat{h}(x)$ accords with the uniform probability distribution. We will show that H is a family of one-way neighbor-free hash functions.

Assume, on the contrary, that there exist a probabilistic polynomial-time algorithm A, a polynomial p and a polynomial-size neighborhood N whose bound polynomial is q such that for infinitely many k,

$$\Pr[A(h \circ \hat{h}, x, 1^k) = x', \ x \neq x', \ h \circ \hat{h}(x') \in N_k(h \circ \hat{h}(x)) \mid$$
$$x \in_U \{0,1\}^k, \ h \circ \hat{h} \in_U H_k] \geq 1/p(k).$$

Let P be the above probability. Then

$$P = \Pr[A(\hat{h}, x, 1^k) = x', \ \hat{h}(x) \neq \hat{h}(x'), \ h \circ \hat{h}(x') \in N_k(h \circ \hat{h}(x)) \mid$$
$$x \in_U \{0,1\}^k, \ \hat{h} \in_U G_k]$$
$$= \Pr[A'(y, 1^k) = x', \ y \neq \hat{h}(x'), \ h \circ \hat{h}(x') \in N_k(h(y)) \mid$$
$$y \in_U \{0,1\}^k, \ \hat{h} \in_U G_k]$$
$$= \Pr[A'(y, 1^k) = \hat{h}^{-1}(y'), \ y \neq y', \ h(y') \in N_k(h(y)) \mid$$
$$y \in_U \{0,1\}^k, \ \hat{h} \in_U G_k]$$
$$= \Pr[A''(y, 1^k) = y', \ y \neq y', \ h(y') \in N_k(h(y)) \mid y \in_U \{0,1\}^k].$$

For any $y, y' \in \{0,1\}^k$, the binary event "$h(y') \in N_k(h(y))$" (or, equivalently, "$h(y) \in N_k(h(y'))$") can be partitioned into $E_1(y, y'), E_2(y, y'), \ldots, E_{q(k)}(y, y')$, where $E_i(y, y')$ is the event which occurs if y is different from y' in some fixed bit positions according to i. Note that for any i if $E_i(y, y')$ and $E_i(y', y'')$ hold then $y = y''$. That is,

$$P = \sum_{i=1}^{q(k)} \Pr[A''(y, 1^k) = y', \ y \neq y', \ E_i(y, y') \mid y \in_U \{0,1\}^k].$$

It then follows from the assumption that there exists j such that

$$\Pr[A''(y, 1^k) = y', \ y \neq y', \ E_j(y, y') \mid y \in_U \{0,1\}^k] \geq (p(k)q(k))^{-1}.$$

We set $j = j(k)$ without loss of generality.

Now, we construct Algorithm B to find a sibling of x using Algorithm A. Algorithm B, given x, picks h' at random and gives x and $h \circ h'$ to Algorithm A. Algorithm B then lets x' be the output of A, picks h'' at random, and gives x' and $h \circ h''$ to Algorithm A again. Finally Algorithm B lets x'' be the second output of A and gives x'' as its output.

Then, the probability that $B(x) = x''$ and $h(x) = h(x'')$ can be estimated as follows.

$$\Pr[B(x, 1^k) = x'', \ h(x) = h(x'') \mid x \in_U \{0,1\}^k]$$
$$\geq \Pr\left[\begin{matrix} A(\hat{h}, x, 1^k) = x', x \neq x', E_{j(k)}(\hat{h}(x), \hat{h}(x')), \\ A(\tilde{h}, x', 1^k) = x'', \ x' \neq x'', \ E_{j(k)}(\tilde{h}(x'), \tilde{h}(x'')) \end{matrix} \middle| \ x \in_U \{0,1\}^k, \hat{h}, \tilde{h} \in_U G_k\right]$$

$$= \Pr\left[\begin{matrix} A''(y, 1^k) = y', y \neq y', E_{j(k)}(y, y'), \\ A''(z, 1^k) = z', z \neq z', E_{j(k)}(z, z') \end{matrix}\middle| y, z \in_U \{0, 1\}^k\right]$$

$$= \left(\Pr[A''(y, 1^k) = y', \ y \neq y', \ E_{j(k)}(y, y') \mid y \in_U \{0, 1\}^k]\right)^2$$

$$\geq (p(k)q(k))^{-2} \tag{1}$$

We next estimate the probability that $B(x) = x$.

$$\Pr\left[\begin{matrix} A(\hat{h}, x, 1^k) = x', \ x \neq x', \ E_{j(k)}(\hat{h}(x), \hat{h}(x')), \\ A(\tilde{h}, x', 1^k) = x, \ x' \neq x, \ E_{j(k)}(\tilde{h}(x'), \tilde{h}(x)) \end{matrix}\middle| x \in_U \{0, 1\}^k, \ \hat{h}, \tilde{h} \in_U G_k\right]$$

$$= \Pr\left[\begin{matrix} A''(y, 1^k) = y', \ y \neq y', \ E_{j(k)}(y, y'), \\ A''(z, 1^k) = x, \ z \neq x, \ E_{j(k)}(z, x) \end{matrix}\middle| x, y, z \in_U \{0, 1\}^k\right]$$

$$= \Pr\left[\begin{matrix} A''(y, 1^k) = y', \ y \neq y', \ E_{j(k)}(y, y'), \\ A''(z, 1^k) = z', \ z \neq z', \ E_{j(k)}(z, z'), \\ x = z' \end{matrix}\middle| x, y, z \in_U \{0, 1\}^k\right]$$

$$= \Pr[A''(y, 1^k) = y', \ y \neq y', \ E_{j(k)}(y, y') \mid y \in_U \{0, 1\}^k]$$

$$\cdot \sum_{x' \in \{0,1\}^k} (\Pr[A''(z, 1^k) = z', z \neq z', E_{j(k)}(z, z'), x = z' \mid z \in_U \{0, 1\}^k, x = x']$$

$$\cdot \Pr[x = x' \mid x \in_U \{0, 1\}^k])$$

$$= \Pr[A''(y, 1^k) = y', \ y \neq y', \ E_{j(k)}(y, y') \mid y \in_U \{0, 1\}^k]$$

$$\cdot \frac{1}{2^k} \sum_{x' \in \{0,1\}^k} \Pr[A''(z, 1^k) = z', \ z \neq z', \ E_{j(k)}(z, z'), \ x' = z' \mid z \in_U \{0, 1\}^k]$$

$$= \frac{1}{2^k} \left(\Pr[A''(y, 1^k) = y', \ y \neq y', \ E_{j(k)}(y, y') \mid y \in_U \{0, 1\}^k]\right)^2 \tag{2}$$

Combining (1) and (2), we have

$$\Pr[B(x, 1^k) = x'', \ h(x) = h(x''), \ x \neq x'' \mid x \in_U \{0, 1\}^k] \geq (p(k)q(k))^{-2}(1 - 1/2^k).$$

This contradicts that h is a single one-way hash function.   □

The above discussion is applicable to the case of a family of universal one-way hash functions instead of any single one-way hash function.

Theorem 2. *If there exists a family of universal one-way hash functions, there exists a family of one-way neighbor-free hash functions.*

Since Rompel showed a way to construct a family of universal one-way hash functions from any one-way function [11], we obtain the following.

Corollary 1. *If there exists a one-way function, there exists a family of one-way neighbor-free hash functions.*

4   Conclusion

We showed a way to construct a family of one-way neighbor-free hash functions using a single one-way hash function and a family of permutations. Since one-way neighbor-freeness is stronger notion than one-wayness, we can say that some one-way hash functions within cryptographic systems can be replaced with one-way neighbor-free hash function families and then the modified systems achieve the stronger security.

References

1. R. Anderson. The classification of hash functions. In *Proceedings of the 4th IMA Conference on Cryptography and Coding*, pp. 83–94, 1995.
2. M. Bellare and P. Rogaway. Random oracles are practical: A paradigm for designing efficient protocols. In *Proceedings of the 1st ACM Conference on Computer and Communications Security*, pp. 62–73, 1993.
3. M. Bellare and P. Rogaway. The exact security of digital signatures – how to sign with RSA and Rabin. In *Lecture Notes in Computer Science (EUROCRYPT'96)*, Vol. 1070, pp. 399–416. Springer-Verlag, 1996.
4. J. L. Carter and M. N. Wegman. Universal classes of hash functions. *Journal of Computer and System Sciences*, 18(2):143–154, 1979.
5. D. Dolev, C. Dwork, and M. Naor. Non-malleable cryptography. In *Proceedings of the 23rd Annual ACM Symposium on Theory of Computing*, pp. 542–552. ACM Press, 1991.
6. T. ElGamal. A public key cryptosystem and a signature scheme based on discrete logarithms. *IEEE Transactions on Information Theory*, IT-31(4):469–472, 1985.
7. S. Goldwasser, S. Micali, and R. L. Rivest. A digital signature scheme secure against adaptive chosen-message attacks. *SIAM Journal on Computing*, 17(2):281–308, 1988.
8. M. Naor and M. Yung. Universal one-way hash functions and their cryptographic applications. In *Proceedings of the 21st Annual ACM Symposium on Theory of Computing*, pp. 33–43. ACM Press, 1989.
9. D. Pointcheval and J. Stern. Security proofs for signature schemes. In *Lecture Notes in Computer Science (EUROCRYPT'96)*, Vol. 1070, pp. 387–398. Springer-Verlag, 1996.
10. R. L. Rivest, A. Shamir, and L. Adleman. A method for obtaining digital signature and public key cryptosystems. *Communications of the ACM*, 21(2):120–126, 1978.
11. J. Rompel. One-way functions are necessary and sufficient for secure signatures. In *Proceedings of the 22nd Annual ACM Symposium on Theory of Computing*, pp. 387–394. ACM Press, 1990.
12. Y. Zheng, T. Hardjono, and J. Pieprzyk. Sibling intractable function families and their applications. In *Lecture Notes in Computer Science (ASIACRYPT'91)*, Vol. 739, pp. 124–138. Springer-Verlag, 1993.

Over $\mathbf{F}_p$ vs. over $\mathbf{F}_{2^n}$ and on Pentium vs. on Alpha in Software Implementation of Hyperelliptic Curve Cryptosystems

Yasuyuki Sakai[1] and Kouichi Sakurai[2*]

[1] Mitsubishi Electric Corporation
5-1-1 Ofuna, Kamakura, Kanagawa 247-8501, Japan
ysakai@iss.isl.melco.co.jp
[2] Kyushu University
6-10-1 Hakozaki, Higashi-ku, Fukuoka 812-8581, Japan
sakurai@csce.kyushu-u.ac.jp

Abstract. We consider the performance of hyperelliptic curve cryptosystems over the fields $\mathbf{F}_p$ vs. $\mathbf{F}_{2^n}$. We analyze the complexity of the group law of the Jacobians $\mathbf{J}_C(\mathbf{F}_p)$ and $\mathbf{J}_C(\mathbf{F}_{2^n})$ and compare their performance taking into consideration the effectiveness of the word size (32-bit or 64-bit) of the applied CPU (Alpha and Pentium) on the arithmetic of the definition field. Our experimental results show that $\mathbf{J}_C(\mathbf{F}_{2^n})$ is faster than $\mathbf{J}_C(\mathbf{F}_p)$ on an Alpha, whereas $\mathbf{J}_C(\mathbf{F}_p)$ is faster than $\mathbf{J}_C(\mathbf{F}_{2^n})$ on a Pentium. Moreover, we investigate the algorithm of the Jacobian and the definition-field arithmetic to clarify our results from a practical point of view, with theoretical analysis.
Keywords: Hyperelliptic curve cryptosystem, Jacobian, Efficient implementation, Lagrange reduction

1 Introduction

We implemented the group law of the Jacobians of hyperelliptic curves in software. In particular, we present here a practical comparison between the performance of hyperelliptic curve cryptosystems over $\mathbf{F}_p$ and $\mathbf{F}_{2^n}$.

1.1 ECC and HECC

Elliptic curve cryptosystems (ECCs) [Ko87,Mi85] are now being used extensively in industry [CER,RSA]. There has been much work done in recent years on their implementation [BP98,CMO98,GP97,So97,WBV96,WMPW98], and the cryptanalysis of ECC's is still being explored [FMR98,GLV98,MOV93,SA97], [SE98,SM97,WZ98]. As a natural generalization of ECCs, Koblitz [Ko88,Ko89] proposed hyperelliptic curve cryptosystems (HECCs) induced from the group law of Jacobians [CA87] defined over finite fields as a source of finite abelian groups. These Jacobian varieties seem to be a rich source of finite abelian groups for which, so far as is known, the discrete log problem is intractable [Ko89]. While

---

* Supported by Scientific Research Grant, Ministry of Education, Japan, No.11558033

attacks on HECCs have been explored [ADH94,En99,FR94,Ru97], the design of HECCs vulnerable to these attacks and their implementations have also been considered [Sm99,SS98,SSI98].

1.2   Previous Work

Win, Mister, Preneel and Wiener [WMPW98] implemented elliptic curve cryptosystems in software and compared their performance over the fields F_p and F_{2^n}. Sakai and Sakurai [SS98] implemented the group law in the Jacobian of a hyperelliptic curve. However, they only considered curves over F_{2^n}. Smart [Sm99] reported the performance of the group law in the Jacobian of curves of arbitrary genus over both F_{2^n} and F_p. As for theoretical analysis, Enge [En98] managed to determine the complexity of the group law of Jacobians and the efficiency of HECCs. In particular, Enge gives the average bit complexity of the arithmetic in hyperelliptic Jacobians.

We should note that Enge's examination [En98] assumes that the complexity of field operations is either constant, or grows with $\log q$ or $\log^2 q$. Indeed, these assumptions are theoretically reasonable. However, the complexity of field operations can also depend on the word-size (32 or 64) of the applied CPU (Pentium or Alpha). In fact, Smart [Sm99] remarks that he chose the values of p and n such that p and 2^n would be less than 2^{32} so as to ensure that the basic arithmetic over F_p and F_{2^n} could all be fitted into single words on a computer. In [Sm99], however, Smart presented experimental timing results only for HECDSA(F_{2^n}) and gave no specific data on HECDSA(F_p), although he did give results for the case when $g = 1$, i.e. ECDSA(F_p) and ECDSA(F_{2^n}).

Thus, there are currently no published reports on the performance of hyperelliptic curve cryptosystems over F_p vs. F_{2^n} based on experimental data. In the case of elliptic curves, the number of field operations for doubling a point in $E(F_p)$ is almost the same as that of $E(F_{2^n})$, but in a hyperelliptic doubling the number of field operations in $J_C(F_{2^n})$, where C has the form $v^2 + v = f(u)$, is much smaller than in $J_C(F_p)$. However, if field operations in F_p are relatively efficient compared to those in F_{2^n}, then an HECC using $J_C(F_p)$ might be faster than one that uses $J_C(F_{2^n})$. In fact, in [WMPW98], a multiplication in F_p is faster than in F_{2^n} in software implementations. Thus, the question "Which is faster, $J_C(F_{2^n})$ or $J_C(F_p)$?" is never trivial. This is the motivation for our comparison, which considers practical aspects and a real implementation.

1.3   Our Contribution

We implemented hyperelliptic curve cryptosystems over both F_p and F_{2^n}, and compared their performance. In particular, our implementation takes into account the advantage of the single words (32-bit or 64-bit) of our applied CPUs (an Alpha or a Pentium) on the arithmetic of the definition field, since modern microprocessors are designed to calculate results in units of data known as *words*.

The Arithmetic for Definition Fields. We first consider the efficiency of the arithmetic for finite field operations, [1] before implementing our hyperelliptic cryptosystems. Our HECCs were implemented on two typical CPUs with different word size, one a Pentium II (300MHz), which has 32-bit word size, and the other an Alpha 21164A (600MHz), which has 64-bit word size.

On the Performance of the Jacobians. The most significant part of computing the group law of Jacobians is finding a particular divisor's unique reduced divisor. Several algorithms along with their improvements [CA87,EN98,KO89], [PS98,SM99] are known for such a reduction step.

When performing an addition, our implemented algorithm uses *Lagrange reduction*, which was developed by Paulus and Stein [PS98], and was generalized by Enge [EN98] to fields of arbitrary characteristic. Enge [EN98] never actually implemented his findings, and Sakai-Sakurai [SS98] did not make use of Lagrange reduction in their implementation. Only Smart [SM99] adopted Lagrange reduction in his algorithm. However, no experimental data over $\mathbf{F}_p$ is available from his study [SM99].

Thus, the results of our work comprise the first reported experimental data from implementation of Lagrange reduction over $\mathbf{F}_p$ and $\mathbf{F}_{2^n}$, taking into account the word size of the CPU.

Our conclusion is this: In a typical implementation, for a field multiplication in $\mathbf{J}_C(\mathbf{F}_p)$, with $\log_2 p = 60$, of a genus 3 curve on the Alpha, we need to divide a double word integer by a single word integer. This operation is very costly on an Alpha, because integer division does not exist as hardware opcode. Therefore, this instruction is performed via a software subroutine [DEC]. In the case of $\mathbf{J}_C(\mathbf{F}_p)$, with $\log_2 p = 29$, of a genus 6 curve, an element of the field can be represented as a half-size integer on an Alpha. Although a software subroutine is still needed for division, the field multiplication is inexpensive compared to the genus 3 curve. In fact, our implementation of field operations shows that a field multiplication in $\mathbf{F}_p$, with $\log_2 p = 29$, is 4 times faster than in $\mathbf{F}_p$, with $\log_2 p = 60$. On the other hand, on the Pentium, a field multiplication in $\mathbf{F}_p$, with $\log_2 p = 15$, is only 2 times faster than in $\mathbf{F}_p$, with $\log_2 p = 30$.

2   A Hyperelliptic Curve and Its Jacobian

This section gives a brief description of Jacobians and their discrete logarithm problem. See [KO98] for more details.

2.1   Hyperelliptic Curve

Let $\mathbf{F}$ be a finite field and let $\bar{\mathbf{F}}$ be its algebraic closure. A hyperelliptic curve C of genus g over $\mathbf{F}$ is an equation of the form $C : v^2 + h(u)v = f(u)$ in $\mathbf{F}[u,v]$, where $h(u) \in \mathbf{F}[u]$ is a polynomial of degree at most g, $f(u) \in \mathbf{F}[u]$ is a monic

---

[1] In [BP98], it is stated that a multiplication in $\mathbf{F}_{2^n}$ takes $cn\frac{n}{w}$ steps.

polynomial of degree $2g + 1$, and there are no solutions $(u, v) \in \bar{\mathbf{F}} \times \bar{\mathbf{F}}$ that simultaneously satisfy the equation $v^2 + h(u)v = f(u)$ and the partial derivative equations $2v + h(u) = 0$ and $h'(u)v - f'(u) = 0$. Thus, a hyperelliptic curve does not have any singular points.

A divisor D on C is a finite formal sum of $\bar{\mathbf{F}}$-points, $D = \sum m_i P_i$, $m_i \in \mathbf{Z}$. We define the degree of D to be $\deg(D) = \sum m_i$. If $\mathbf{K}$ is an algebraic extension of $\mathbf{F}$, we say that D is defined over $\mathbf{K}$ if for every automorphism σ of $\bar{\mathbf{F}}$ that fixes $\mathbf{K}$ one has $\sum m_i P_i^\sigma = D$, where P^σ denotes the point obtained by applying σ to the coordinates of P (and $\infty^\sigma = \infty$). Let $\mathbf{D}$ denote the additive group of divisors defined over $\mathbf{K}$ (where $\mathbf{K}$ is fixed), and let $\mathbf{D}^0$ denote the subgroup consisting of divisors of degree 0. The principal divisors form a subgroup $\mathbf{P}$ of $\mathbf{D}^0$. $\mathbf{J}(\mathbf{K}) = \mathbf{D}^0/\mathbf{P}$ is called the "*Jacobian*" of the curve C.

2.2   Discrete Logarithm

The discrete logarithm problem on $\mathbf{J}_C(\mathbf{K})$ is the problem, given two divisors $D_1, D_2 \in \mathbf{J}_C(\mathbf{K})$, of determining an integer m such that $D_2 = mD_1$, if such an m exists.

As with the elliptic curve discrete logarithm problem, no general subexponential algorithms are known for the hyperelliptic curve discrete logarithm problem, except for some special cases that are easily avoided [MOV93,SA97,SE98,SM97]. Only exponential attacks, such as the baby-step giant-step method, Pollard's ρ method or the Pohlig-Hellman method, can be applied. See Appendix A, for a list of conditions necessary for the Jacobian to be secure.

3   Curve Generation

In this section, we give examples of Jacobians that have a almost prime order divisible by a large prime of size $\approx 2^{160}$. All of our Jacobians are designed to be secure against known attacks which are discussed in [ADH94,FR94,GLV98], [MOV93,PH78]. The Jacobians given in this section will be implemented and their efficiencies will be discussed in the later sections.

3.1   Over a Characteristic Two Field $\mathbf{F}_{2^n}$

Beth and Schaefer [BS91] used the zeta-function of an elliptic curve to construct elliptic cryptosystems. Analogously, Koblitz [Ko88,Ko89,Ko98] used the zeta-function of a hyperelliptic curve to construct Jacobians of hyperelliptic curves defined over finite fields. One technical difficulty in our computation is that the zeta-function has a complicated form with large degree on general hyperelliptic curves. Therefore, unlike the previous cases [BS91,Ko88,Ko89,Ko98] it is not easy to compute its exact solutions. However, it is known that the order of a Jacobian can be computed without evaluating the solution of its zeta-function [St93, Chapter V]. Therefore, we apply this algorithm to rectify our problem. See Appendix B.

By this algorithm, we have succeeded in finding Jacobians that have a almost prime order. For example, Jacobians that have a order divisible by a prime of size $\approx 2^{160}$ are shown in Table 1. [2]

Remark. Recently, Gaudry [GA99] presented a method for speeding up the discrete log computations on the Jacobian of HECs. The method is asymptotically faster than the Pollard ρ method if the genus is greater than 4. Moreover, Duursma, Gaudry and Morain [DGM99] gave a way to gain a speed-up by a factor $\sqrt{m}$ if there exist an automorphism of order m. This method is a generalization of the parallel collision search [GLV98,WZ98]. Therefore, curves with genus greater than 4 or automorphisms of large order should be used carefully in cryptosystems.

In practical point of view, we have to take into account these two attacks. Gaudry's attack has a complexity $O(N^{2/g})$, where N denotes the order of the group (suppose that N is almost prime). The attack is better than the Pollard ρ method, which has a complexity $O(N^{1/2})$, if g is greater than 4. Therefore, the attack could be applicable to the genus 5 and 6 curves given in this paper. Moreover, Duursma-Gaudry-Morain attack is applicable to the genus 6 curve in Table 1, which has an automorphism of order 4×29 [GA99]. Also for the other curves given in this paper, the attack might be applicable if an automorphism exists. However, our results are independent from the specific structure of such a curve with large automorphisms, and any technique of our implementation is valid for the curves without automorphisms of large order.

Table 1. Jacobians over $\mathbf{F}_{2^n}$ of genus 3,4,5 and 6 curves of the form $v^2 + v = f(u)$

genus	$\mathbf{F}_q$	$f(u)$	$\log_2 \sharp J_C(\mathbf{F}_q)$	$\log_2 p$, where $p \mid \sharp J_C(\mathbf{F}_q))$
3	$\mathbf{F}_{2^{59}}$	u^7	178	165
4	$\mathbf{F}_{2^{41}}$	$u^9 + u^7 + u^3 + 1$	164	161
5	$\mathbf{F}_{2^{31}}$	$u^{11} + u^5 + 1$	155	151
6	$\mathbf{F}_{2^{29}}$	$u^{13} + u^{11} + u^7 + u^3 + 1$	174	170

3.2   Over a Prime Field $\mathbf{F}_p$

Unlike elliptic curves, it is still difficult, in general, to compute the order of hyperelliptic Jacobians. However, in [BK97,Ko98], Koblitz and Buhler suggest using curves over some finite prime field $\mathbf{F}_p$ of the form $v^2 + v = u^n$, where $n = 2g + 1$ is an odd prime and $p \equiv 1 \bmod n$. Its Jacobian is a quotient of the Jacobian of the Fermat curve $X^n + Y^n = 1$. The order of such curves can be determined by computing a Jacobi sum of a certain character. It is possible to determine a Jacobi sum in polynomial time using the LLL-algorithm [LLL82].

---

[2] In [SS98], Jacobians that have a almost prime order of approximately 2^{160}, over small characteristic finite fields $\mathbf{F}_{2^n}$, $\mathbf{F}_{3^n}$, $\mathbf{F}_{5^n}$ and $\mathbf{F}_{7^n}$, have been tabulated.

See Appendix B. The details of the procedure to determine the $\mathbf{J}_C(\mathbf{F}_p)$ of the curve can be found in [BK97,Ko98].

The following Jacobians over a prime finite field $\mathbf{F}_p$ have a order divisible by a prime of size $\approx 2^{160}$. The Jacobian of the genus 6 curve should be used carefully in cryptosystems, because Duursma-Gaudry-Morain attack may be applicable.

genus 2 curve

$v^2 + v = u^5/\mathbf{F}_p$, $p = 967140655691703339766 0861$ $(\log_2 p = 84)$

$\sharp \mathbf{J}_C(\mathbf{F}_p) = 935361047892979186521140380951541036303 76284213875$

$\quad = 5^3 \cdot 7482888383143833492169123047612328290043010273711$

$\quad \log_2(\text{largest prime factor}) = 160$

genus 3 curve

$v^2 + v = u^7/\mathbf{F}_p$, $p = 631153760340591307$ $(\log_2 p = 60)$

$\sharp \mathbf{J}_C(\mathbf{F}_p) = 251423300188980936808533376314868064530443303970434811$

$\quad = 7^3 \cdot 7330125369941135183922255869238135992141204197 38877$

$\quad \log_2(\text{largest prime factor}) = 169$

genus 6 curve

$v^2 + v = u^{13}/\mathbf{F}_p$, $p = 269049691$ $(\log_2 p = 29)$

$\sharp \mathbf{J}_C(\mathbf{F}_p) = 3793166228009222337827412027254783306566277889 04081$

$\quad = 13^3 \cdot 157 \cdot 109969478588614536261880329785398830094491 2689$

$\quad \log_2(\text{largest prime factor}) = 150$

4   Arithmetic for Group Operations in a Jacobian

This section gives a brief description of the algorithm for adding two points and doubling a point on a Jacobian. Addition is accomplished by two procedures. First, we compute the *composition* step, then we compute the *reduction* step. The details of this algorithm are given in, for example, [CA87,EN98,Ko98,PS98].

4.1   Composition Step

Let $D_1 = \text{div}(a_1, b_1)$ and $D_2 = \text{div}(a_2, b_2)$ be two reduced divisors, both defined over $\mathbf{F}_q$. Then the following algorithm [3] finds a semireduced divisor $D = \text{div}(a', b')$, such that $D \sim D_1 + D_2$ [EN98].

Composition

1. Perform two extended Euclidean computations to compute
$$d = \gcd(a_1, a_2, b_1 + b_2 + h) = s_1 a_1 + s_2 a_2 + s_3(b_1 + b_2 + h)$$
2. Set $a' = a_1 a_2/d^2$ and
3. $b' = b_1 + (s_1 a_1(b_2 - b_1) + s_3(f - b_1{}^2 - b_1 h))/d \pmod{a}$

To compute d, s_1, s_2, and s_3, two extended gcd's should be performed. If a_1 and a_2 have no common factor, the composition algorithm is even simpler. Note that

---

[3] This algorithm is more efficient than the algorithm in [Ko98].

the case $\gcd(a_1, a_2) = 1$ is extremely likely if the definition field is large and a_1 and a_2 are the coordinates of two randomly chosen elements of the Jacobian.

When $a_1 = a_2$ and $b_1 = b_2$, i.e., doubling an element, we can take $s_2 = 0$. Moreover, in the case of char $\mathbf{F}_q = 2$ and $h(u) = 1$, we can take $d_1 = 1$, $s_1 = s_2 = 0$, $s_3 = 1$. Therefore, a doubling in $\mathbf{J}_C(\mathbf{F}_{2^n})$ of the curves of the form $v^2 + v = f(u)$ can be simplified as follows.

Composition : Doubling in $\mathbf{J}_C(\mathbf{F}_{2^n})$ of the curves $v^2 + v = f(u)$
1. Set $a' = a_1^2$ and
2. $b' = b_1^2 + f \pmod{a}$

Thus, in $\mathbf{J}_C(\mathbf{F}_{2^n})$ with $C : v^2 + v = f(u)$, doubling is much simpler than in $\mathbf{J}_C(\mathbf{F}_p)$.

4.2   Reduction Step

To complete the addition, we must find a unique reduced divisor $D = \mathrm{div}(a, b)$. There are three known algorithms for such a reduction step: *Gauss reduction*, *Cantor reduction* and *Lagrange reduction*[EN98]. The reduction algorithm shown in this subsection was given by Paulus and Stein for hyperelliptic curves over a field of odd characteristic [PS98]. A generalized version for arbitrary characteristic was given by Enge in [EN98]. It is called Lagrange reduction (Paulus and Stein's algorithm can be traced back to Lagrange).

Let $a_0 = a'$ and $b_0 = b'$, and compute a sequence (a_k, b_k) for $k = 1, \cdots, t$ by the algorithm, where $t \geq 0$ is the smallest index such that deg $a_t \leq g$.

Lagrange reduction
1. $a_1 = (f - b_0^2 - b_0 h)/a_0$
2. $-b_0 - h = q_1 a_1 + b_1$ with deg $b_1 <$ deg a_1
3. For $k \geq 2$:
4. $\quad a_k = a_{k-2} + q_{k-1}(b_{k-1} - b_{k-2})$
5. $\quad -b_{k-1} - h = q_k a_k + b_k$ with deg $b_k <$ deg a_k

---

Note that in the worst case, a' may have degree $2g$, i.e. deg $a_1 = $ deg $a_2 = g$. In fact, this is the typical case. Therefore, the iteration step of the above algorithm may execute quite a few times.

In most cases, this reduction algorithm is more efficient than Gauss reduction and Cantor reduction. The most costly steps in the Gauss reduction procedure are the computations of the a_k's, each involving one multiplication and one division of rather high degree polynomials. In the original reduction algorithm, each step is independent of the previous one. On the other hand, in the Lagrange reduction, as soon as one reduction step been carried out, the formula for a_k can be rewritten, using information from the previous step. In the later sections, we will implement the algorithm for an adding and a doubling in a Jacobian, using the algorithm described in this section. [4]

---

[4] In [SS98], hyperelliptic cryptosystems over fields of characteristic 2 were implemented and their efficiency was discussed. However, this implementation made no use of Lagrange reduction.

5   Arithmetic for a Finite Field and Its Efficiency

Before implementing our hyperelliptic cryptosystems, we considered the efficiency of finite field operations. Our implementation was on two different CPUs, a Pentium II (300MHz), which has 32-bit word size, and an Alpha 21164A (600MHz), which has 64-bit word size. In the later sections, we will analyze the performance of hyperelliptic curve cryptosystems based on these efficiencies.

5.1   Representation of Field Elements

Representation in $\mathbf{F}_p$. In our implementation of hyperelliptic cryptosystems, for $\mathbf{F}_p$, we represented the elements as numbers in the range, $[0, p-1]$, where each residue class is represented by its member in that range. This is clearly the most canonical way. Another possibility would be to use Montgomery residues. In our cryptosystems, the size of a field element was relatively small, and so could be represented as a single word integer on our computers. An analogous representation using Montgomery residues would not have been efficient, since we would have needed to compute an extra transformation. Therefore, we chose the more natural representation.

Representation of $\mathbf{F}_{2^n}$. For $\mathbf{F}_{2^n}$, several methods of representation are known. Two such methods are the *standard basis* representation and the *optimal normal basis* representation. A third representation lists elements of the field as polynomials over a subfield of the form $\mathbf{F}_{2^r}$, where r is a divisor of n. However, in order to make $\sharp \mathbf{J}_C(\mathbf{F}_{2^n})$ divisible by a large prime, we chose n itself to be prime. Thus, this third method was not applicable. The optimal normal basis representation enables efficient implementation in hardware. However, in our experience, this method is inefficient in software, compared to the standard basis representation. Therefore, we implemented our cryptosystems using standard basis representation. These implementations can be made more efficient if an irreducible polynomial with low Hamming weight and few terms of high degree are chosen, such as a trinomial or a pentanomial.

5.2   Field Multiplication and Inversion

In this subsection, we analyze the performance of field operations. In our settings, as we have already pointed out, an element of the definition field can be represented as a single word integer on the computers that we used. In such a case, we do not require a multi-precision integer library. This fact contributes to faster implementations.

We conducted an experiment to examine the performance of field multiplications and field inversions in $\mathbf{F}_q$, where q went up to 160 bits. We focused on the case where a field element could be represented as a single word integer on the CPUs, because we are interested in the hyperelliptic settings that can be implemented using only single word integers on CPUs. Many fast implementations of elliptic curve cryptosystems have been developed and their efficiencies analyzed for fields of size approximately 2^{160}. However, for fields of small size, where elements can be represented as a single word integer on a CPU, the performance of the field operations has not been reported in open literature.

Tables 2 and 3 show the performance of our implementation on the Alpha and the Pentium, respectively. From our implementation, we can observe that:

Over F_p

Result 1 A field multiplication has a time complexity of $c\lceil \frac{m}{w} \rceil$, where $m = \log_2 p$, w is the processor's word size and c is some constant. Moreover, if $\frac{m}{w} \leq \frac{1}{2}$, a multiplication can be much faster than the case $\frac{1}{2} < \frac{m}{w} \leq 1$. On the Alpha, a field multiplication takes 0.16 μsec when $\log_2 p < 32$, 0.61 μsec when $32 \leq \log p \leq 64$. On the Pentium, a field multiplication takes 0.15 μsec when $\log_2 p < 16$, 0.28 μsec when $16 \leq \log_2 p \leq 32$.

Result 2 The speed of a field inversion grows linearly in $\log p$. On the Alpha, a field inversion takes $0.1 \log_2 p$ μsec. On the Pentium, a field inversion takes $0.12 \log_2 p$ μsec.

Over F_{2^n}

Result 3 The speed of a field multiplication grows linearly in $n = \log_2 2^n$. On the Alpha, a field multiplication takes $0.015n$ μsec. On the Pentium, a field multiplication takes $0.04n$ μsec.

Result 4 The speed of a field inversion grows linearly in n. On the Alpha, a field inversion takes $0.2n$ μsec. On the Pentium, a field inversion takes $0.6n$ μsec.

We should note that in F_p, field multiplications for fields whose size is smaller than half the CPU's word length are faster than those for fields whose size is larger than half the CPU's word length. The reason is that the field multiplications can be computed with the instruction $(a * b) \bmod p$. Now, $(a * b)$ can be computed in advance and is at most double the larger of a and b. Therefore, if a and b are larger than the CPU's word size, then $(a * b)$ would be larger than the word size. On a Pentium, we can perform this computation with the double word instruction, $((\text{long long})a * (\text{long long})b)\%p$ in C language. As an Alpha does not have such a double word instruction, we must use some special technique or assembly language. On both the Pentium and the Alpha, such double word computation is costly compared to single word computation.

6   Enge's Analysis on the Average Number of Field Operations

In this section, we summarize the computational cost of the algorithm for the group law in Jacobians, which has been analyzed by Enge [EN98]. To compare the efficiency of $J_C(F_p)$ and $J_C(F_{2^n})$, it is important to analyze the average number of field multiplications and inversions. Enge analyzed the average

Table 2. Timings of field operations on an Alpha 21164A (600MHz) in μsec

word size	$\log_2 q$	multiplication		inversion	
		$\mathbf{F}_p$	$\mathbf{F}_{2^n}$	$\mathbf{F}_p$	$\mathbf{F}_{2^n}$
1/2	5	0.16	0.055	0.33	0.50
	10	0.16	0.16	0.83	1.66
	15	0.16	0.22	1.16	2.33
	20	0.16	0.27	1.26	3.33
	25	0.16	0.38	1.99	4.33
	30	0.16	0.44	2.11	5.50
1	35	0.61	0.55	2.33	6.66
	40	0.61	0.55	2.49	7.83
	45	0.61	0.61	4.83	9.33
	50	0.61	0.66	5.99	10.8
	55	0.61	0.66	6.83	12.4
	60	0.61	0.72	7.83	14.3
2	80	2.77	56	71	833
	100	2.77	67	101	1266
	120	2.77	80	124	1733
3	140	4.22	104	149	2300
	160	4.22	121	178	2800

complexity of the extended Euclidean algorithm on polynomials with heuristic assumptions. [5]

Several cases need to be distinguished. In this paper, we will concentrate on the following two: $\mathbf{J}_C(\mathbf{F}_p)$ and $\mathbf{J}_C(\mathbf{F}_{2^n})$ with $h = 1$. Table 4 shows the average number of field multiplications and inversions for composing two distinct reduced divisors. The proof of correctness can be found in [EN98].

We assume that the composition algorithm has yielded a random semire-duced divisor $\mathrm{div}(a', b')$ of degree $2g$, a and b are almost uniformly distributed over all polynomials of degree $2g$ and $2g - 1$, respectively. Table 5 shows the average number of field multiplications and inversions for the reduction step using Lagrange reduction. The proof of correctness can be found in [EN98]. We describe the case that the curves have genus $g \geq 3$, because we will focus on the case that the operations in the Jacobian can be implemented with a single word size integer on the CPUs.

According to the above tables, the number of field operations required for a full addition and doubling step for a genus $g \geq 3$ curve can be estimated as given in Table 6.

---

[5] **Heuristic:**[EN98] Let the hyperelliptic curve C of fixed genus g be chosen randomly according to a uniform distribution on the defining pairs of suitable polynomials (h, f). If $\mathrm{div}(a, b)$ is a uniformly selected element of $\mathbf{J}_C(\mathbf{F}_q)$, then we can assume that a varies uniformly over all polynomials of degree g and b over all polynomials of degree $g - 1$. Likewise, if $\mathrm{div}(a_1, b_1)$ and $\mathrm{div}(a_2, b_2)$ are two uniformly selected points of $\mathbf{J}_C(\mathbf{F}_q)$, then we can assume that a_1 and a_2 vary uniformly and independently over all polynomials of degree g and b_1 and b_2 over all polynomials of degree $g - 1$.

Table 3. Timings of field operations on a Pentium II (300MHz) in μsec

word size	$\log_2 q$	multiplication		inversion	
		$\mathbf{F}_p$	$\mathbf{F}_{2^n}$	$\mathbf{F}_p$	$\mathbf{F}_{2^n}$
1/2	5	0.15	0.20	0.89	2.20
	10	0.15	0.41	1.44	5.60
	15	0.15	0.62	1.87	9.00
1	20	0.28	0.82	2.33	11.7
	25	0.28	1.03	2.37	15.9
	30	0.28	1.23	3.18	20.1
2	40	2.13	60	40	573
	60	2.13	88	73	1146
3	80	3.62	128	100	1866
4	100	4.46	186	146	2813
	120	4.46	206	180	3933
5	140	6.26	253	220	5280
	160	6.26	300	253	6893

Table 4. Average number of field operations in composition step for $\mathbf{J}_C(\mathbf{F}_p)$ and $\mathbf{J}_C(\mathbf{F}_{2^n})$ with $h = 1$

	Addition		Doubling	
	mul.	inv.	mul.	inv.
$\mathbf{J}_C(\mathbf{F}_p)$	$8g^2 + 5g - 2$	$g + 2$	$\frac{15}{2}g^2 + \frac{15}{2}g$	$g + 1$
$\mathbf{J}_C(\mathbf{F}_{2^n})$	$7g^2 + 7g$	$g + 1$	$4g + 2$	1

7   Our Implementation and Comparisons of the Group Operations in the Jacobians

In this section, we will show our implementation of the group operations: an addition, a doubling and a scalar multiplication in the Jacobians. Moreover, we will compare their respective efficiencies.

7.1   The Average Number of Field Operations in a Scalar Multiplication

We will now describe the expected speed of adding two elements and doubling one element in a Jacobian. Using the average number of field operations in

Table 5. Average number of field operations for Lagrange reduction step for genus larger than or equal to 3

	multiplications	inversions
$\mathbf{J}_C(\mathbf{F}_p)$, g even	$9g^2 - 2g - 1$	$\frac{1}{2}g + 1$
$\mathbf{J}_C(\mathbf{F}_p)$, g odd	$9g^2 - g$	$\frac{1}{2}g + \frac{3}{2}$
$\mathbf{J}_C(\mathbf{F}_{2^n})$, g even	$7g^2 - 3g - 1$	$\frac{1}{2}g + 1$
$\mathbf{J}_C(\mathbf{F}_{2^n})$, g odd	$7g^2 - 2g$	$\frac{1}{2}g + \frac{3}{2}$

Table 6. Average number of field operations for a full addition and doubling

	Addition		Doubling	
	multiplications	inversions	multiplications	inversions
$\mathbf{J}_C(\mathbf{F}_p)$, g even	$17g^2 + 3g - 3$	$\frac{3}{2}g + 3$	$\frac{33}{2}g^2 + \frac{11}{2}g - 1$	$\frac{3}{2}g + 2$
$\mathbf{J}_C(\mathbf{F}_p)$, g odd	$17g^2 + 4g - 2$	$\frac{3}{2}g + \frac{7}{2}$	$\frac{33}{2}g^2 + \frac{13}{2}g$	$\frac{3}{2}g + \frac{5}{2}$
$\mathbf{J}_C(\mathbf{F}_{2^n})$, g even	$14g^2 + 4g - 1$	$\frac{3}{2}g + 2$	$7g^2 + g + 1$	$\frac{1}{2}g + 2$
$\mathbf{J}_C(\mathbf{F}_{2^n})$, g odd	$14g^2 + 5g$	$\frac{3}{2}g + \frac{5}{2}$	$7g^2 + 2g + 2$	$\frac{1}{2}g + \frac{5}{2}$

terms of the genus g shown in Table 6, we can evaluate the speed of operations in Jacobians of fixed genus curves.

Enge, in [EN98], analyzed the average bit complexity of an addition and a doubling. He examined three basic situations, in which the complexity of field operations is either constant or grows with $\log q$ or $\log^2 q$. In these settings, a comparison of the efficiency of $\mathbf{J}_C(\mathbf{F}_p)$ versus $\mathbf{J}_C(\mathbf{F}_{2^n})$ was made. He concluded that $\mathbf{J}_C(\mathbf{F}_{2^n})$ is faster than $\mathbf{J}_C(\mathbf{F}_p)$ for the simple reason that a doubling in $\mathbf{J}_C(\mathbf{F}_p)$ is costly compared to a doubling in $\mathbf{J}_C(\mathbf{F}_{2^n})$.

However, in real computations, the speed of an adding and a doubling in a Jacobian depends on the real speed of field operations. Thus, although the number of field operations in an addition and a doubling for $\mathbf{J}_C(\mathbf{F}_p)$ is larger than in $\mathbf{J}_C(\mathbf{F}_{2^n})$, if a field multiplication and a field inversion in $\mathbf{F}_p$ are faster than in $\mathbf{F}_{2^n}$, then $\mathbf{J}_C(\mathbf{F}_p)$ may be faster than $\mathbf{J}_C(\mathbf{F}_{2^n})$.

To compare the efficiencies, we introduce here the symbols M and I, denoting the real speed of a field multiplication and the real speed of a field inversion, respectively. Table 7 shows the expected speed of group operations in a Jacobian. The subscripts p and 2^n denote the symbols over $\mathbf{F}_p$ and $\mathbf{F}_{2^n}$, respectively. Note that a Jacobian has order approximately q^g. Therefore, a curve of large genus has small M and I. For a scalar multiplication, a randomly chosen element is multiplied by a 160-bit integer using a simple *"binary method"* [6].

7.2   The Average Number of Field Operations in Our Implementation

Table 8 shows the average number of field multiplications and inversions for an addition and a doubling in a Jacobian based on our implementation. The number of field multiplications grows on the order of $O(g^2)$. In a doubling, the number of multiplications for $\mathbf{J}_C(\mathbf{F}_{2^n})$ is much larger than for $\mathbf{J}_C(\mathbf{F}_p)$. These facts support Enge's analysis. The reason that the numbers in Table 8 differ slightly from Enge's analysis is that in our implementation, as soon as a reduced divisor has

---

[6] There are a number of other optimization techniques available that were not used in our implementation. A *signed binary window* method would be more useful for making an elliptic curve and hyperelliptic curve exponentiation fast. However, we need an additional computer memory for a window method. Therefore, the simple binary method is still used for a practical application such as a smart card, which has restricted hardware resources.

Table 7. Expected speed, where M and I denote the speed of a field multiplication and a field inversion, respectively

g	$\mathbf{J}_C(\mathbf{F}_q)$	Addition	Doubling	160-bit scalar multiplication
3	$\mathbf{J}_C(\mathbf{F}_p)$	$163M_p + 8I_p$	$168M_p + 7I_p$	$39920M_p + 1760I_p$
	$\mathbf{J}_C(\mathbf{F}_{2^n})$	$137M_{2^n} + 7I_{2^n}$	$44M_{2^n} + 5I_{2^n}$	$18000M_{2^n} + 1360I_{2^n}$
4	$\mathbf{J}_C(\mathbf{F}_p)$	$281M_p + 9I_p$	$285M_p + 8I_p$	$68080M_p + 2000I_p$
	$\mathbf{J}_C(\mathbf{F}_{2^n})$	$239M_{2^n} + 8I_{2^n}$	$117M_{2^n} + 4I_{2^n}$	$37840M_{2^n} + 1280I_{2^n}$
5	$\mathbf{J}_C(\mathbf{F}_p)$	$443M_p + 9.5I_p$	$445M_p + 8.5I_p$	$106640M_p + 2120I_p$
	$\mathbf{J}_C(\mathbf{F}_{2^n})$	$375M_{2^n} + 10I_{2^n}$	$187M_{2^n} + 5I_{2^n}$	$59920M_{2^n} + 1600I_{2^n}$
6	$\mathbf{J}_C(\mathbf{F}_p)$	$627M_p + 12I_p$	$628M_p + 11I_p$	$150640M_p + 2720I_p$
	$\mathbf{J}_C(\mathbf{F}_{2^n})$	$527M_{2^n} + 11I_{2^n}$	$259M_{2^n} + 5I_{2^n}$	$83600M_{2^n} + 1680I_{2^n}$

been obtained, we divide all coefficients of a (a polynomial of $\text{div}(a,b)$) by its leading coefficient, so that a becomes monic.

Table 8. Average number of field operations in an addition and a doubling with Lagrange reduction

g	$\mathbf{F}_q$	$f(u)$	Addition		Doubling	
			mull.	inv.	mull.	inv.
3	$\mathbf{F}_p, (\log_2 p = 60)$	u^7	402	7	379	7
3	$\mathbf{F}_{2^{59}}$	u^7	370	7	202	3
4	$\mathbf{F}_{2^{41}}$	$u^9 + u^7 + u^3 + 1$	631	8	341	3
5	$\mathbf{F}_{2^{31}}$	$u^{11} + u^5 + 1$	1032	10	587	4
6	$\mathbf{F}_p, (\log_2 p = 29)$	u^{13}	1591	11	1513	11
6	$\mathbf{F}_{2^{29}}$	$u^{13} + u^{11} + u^7 + u^3 + 1$	1475	11	835	4

7.3 Expected Speed

In standard discrete logarithm-based cryptographic protocols, such as DSA and ElGamal signature and verification variants, multiplications by large integers are performed. Therefore, the total performance for cryptographic protocols can be evaluated by the speed of scalar multiplications in a Jacobian.

Let $\sharp mult$ and $\sharp inv$ denote the number of field multiplications and inversions in a scalar multiplication, respectively. The efficiency of a scalar multiplication S can be formulated as follows.

$$S = \sharp mult \cdot M + \sharp inv \cdot I$$

It should be remarked that $\sharp mult$ and $\sharp inv$ depend not only on the size of the field but also on the field's characteristic, as we have seen in the previous section. Table 9 shows the efficiency of the Jacobians. In the table, $\sharp mult$ and $\sharp inv$ are followed by Enge's analysis on the average number. From this table, we can observe a few facts. Note that some of these facts are contrary to Enge's analysis. The details will be discussed in a later subsection.

Table 9. Expected Performance

Platform	g	F_q	word size	$M(\mu sec)$	$I(\mu sec)$	$\sharp mult$	$\sharp inv$	$S(\mu sec)$
Alpha	3	$F_p, \log_2 p = 60$	1	0.61	7.83	$4.0 \cdot 10^4$	$1.8 \cdot 10^3$	$3.84 \cdot 10^4$
		$F_{2^{59}}$	1	0.72	14.3	$1.8 \cdot 10^4$	$1.4 \cdot 10^3$	$3.29 \cdot 10^4$
	6	$F_p, \log_2 p = 29$	1/2	0.16	2.11	$15.1 \cdot 10^4$	$2.7 \cdot 10^3$	$2.98 \cdot 10^4$
		$F_{2^{29}}$	1/2	0.44	5.50	$8.3 \cdot 10^4$	$1.7 \cdot 10^3$	$4.59 \cdot 10^4$
Pentium	6	$F_p, \log_2 p = 29$	1	0.28	3.18	$15.1 \cdot 10^4$	$2.7 \cdot 10^3$	$5.09 \cdot 10^4$
		$F_{2^{29}}$	1/2	1.23	20.1	$8.3 \cdot 10^4$	$1.7 \cdot 10^3$	$13.6 \cdot 10^4$

7.4   Timings in $J_C(F_p)$ and $J_C(F_{2^n})$

Next, we show the timings of an addition, a doubling and a scalar multiplication in $J_C(F_p)$ and $J_C(F_{2^n})$ in our implementation. For a scalar multiplication, a randomly chosen element in the Jacobian was multiplied by a large integer. In standard discrete logarithm-based cryptographic protocols, such as DSA and ElGamal signature and verification variants, multiplications by large integers are performed. In such a case, the integer has the size of the order of the subgroup of a finite abelian group (in our case, $J_C(F_q)$). The abelian groups given in this paper have subgroups of order approximately 2^{160}, though these orders are different from each other. In our implementation, we multiplied a randomly chosen element from each of the Jacobians by a 160-bit integer, to compare performance.

The timings for hyperelliptic Jacobians, given in Tables 10 and 11, were obtained from implementations on a Pentium II (300MHz) and on an Alpha 21164A (600MHz). We used a simple "*binary method*" for scalar multiplication. Programs were written mainly in C language and compiled by gcc 2.8.1 or DEC C V5.8 with maximal optimization. Only in the implementation of multiplications in $J_C(F_p)$, with $\log_2 p = 60$, of the genus 3 curve on the Alpha, did we use assembly language, since we could not compute a 128-bit integer directly with C instructions.

Table 10. Timings of $J_C(F_p)$ (Alpha 21164A (600MHz), Pentium II (300MHz))

g	F_q	$f(u)$	Add.(msec.)		Dbl.(msec.)		Scalar(msec.)	
			Alpha	Pentium	Alpha	Pentium	Alpha	Pentium
3	$F_p, (\log_2 p = 60)$	u^7	0.39	—	0.38	—	98	—
6	$F_p, (\log_2 p = 29)$	u^{13}	0.28	0.83	0.26	0.80	66	189

Table 11. Timings of $J_C(F_{2^n})$ (Alpha 21164A(600MHz), Pentium II(300MHz))

g	F_q	$f(u)$	Add.(msec.)		Dbl.(msec.)		Scalar(msec.)	
			Alpha	Pentium	Alpha	Pentium	Alpha	Pentium
3	$F_{2^{59}}$	u^7	0.30	—	0.09	—	40	—
4	$F_{2^{41}}$	$u^9 + u^7 + u^3 + 1$	0.30	—	0.10	—	43	—
5	$F_{2^{31}}$	$u^{11} + u^5 + 1$	0.34	1.40	0.10	0.48	46	182
6	$F_{2^{29}}$	$u^{13} + u^{11} + u^7 + u^3 + 1$	0.47	1.76	0.13	0.56	61	227

7.5   Efficiency of a Jacobian and Comparison

From Table 10 and 11, we can observe the following facts:

Comparison on Genus

Fact 1: For the case $J_C(F_p)$ on the Alpha, we can see that even though the genus is larger, the scalar multiplication for the genus 6 curve $J_C(F_p)$, with $\log_2 p = 29$, is faster than that of the genus 3 curve $J_C(F_p)$, with $\log_2 p = 60$.

Fact 2: For the case $J_C(F_{2^n})$, the timing of the scalar multiplication grows as the genus grows, both on the Alpha and on the Pentium.

Comparison on the Characteristic of the Field

Fact 3: On the Alpha, for genus 3 curves, $J_C(F_{2^{59}})$ is much faster than $J_C(F_p)$, with $\log_2 p = 60$.

Fact 4: On the Alpha, for genus 3 curves, $J_C(F_{2^{29}})$ is slightly faster than $J_C(F_p)$, with $\log_2 p = 29$.

Fact 5: On the Pentium, for genus 6 curves, $J_C(F_p)$, with $\log_2 p = 29$, is slightly faster than $J_C(F_{2^{29}})$.

It should be noted that Fact 1 and 5 differ from Enge's conclusion [EN98], which was based on his theoretical analysis. These efficiencies can be evaluated based on the following points.

1. **The number of field multiplications and inversions in a scalar multiplication:**
   The number depends on the genus of the curve and on the characteristic of the definition field F_q.
2. **The efficiency of field operations:**
   The efficiency depends on the size of the fields $\log_2 q$, the word size of the fields on the CPU and on the properties of the CPU's architecture.

The properties of a CPU's architecture can affect the performance of field operations. In the Alpha 21164A processor, integer division does not exist as hardware opcode. [7] For the field multiplication of the genus 3 curve $\mathbf{J}_C(\mathbf{F}_p)$, with $\log_2 p = 60$, on an Alpha, we need to divide a double word integer by a single word integer in our implementation. This operation is costly since this instruction is done via a software subroutine [DEC]. For the genus 6 curve $\mathbf{J}_C(\mathbf{F}_p)$, with $\log_2 p = 29$, an element of the field can be represented as a half-size integer on an Alpha. Although, it still needs a software subroutine for division, a field multiplication is inexpensive compared to the genus 3 curve. In fact, our implementation of field operations, as shown in Table 2, shows that a field multiplication in $\mathbf{F}_p$, where $\log_2 p = 29$, is 4 times faster than in $\mathbf{F}_p$, where $\log_2 p = 60$. On the other hand, on a Pentium, a field multiplication in $\mathbf{F}_p$, where $\log_2 p = 15$, is only 2 times faster than in $\mathbf{F}_p$, where $\log_2 p = 30$.

References

[ADH94] L.M. ADLEMAN, J. DEMARRAIS and M. HUANG, "A Subexponential Algorithm for Discrete Logarithm over the Rational Subgroup of the Jacobians of Large Genus Hyperelliptic Curves over Finite Fields", *Algorithmic Number Theory I*, LNCS, **877** (1998), Springer-Verlag, 28–40.

[BK97] J. BUHLER and N. KOBLITZ, "An Application of Lattice Basis Reduction to Jacobi Sums and Hyperelliptic Cryptosystems", *Preprint*, (1997)

[BP98] D.V. BAILEY and C. Paar "Optimal Extension Fields for Fast Arithmetic in Public-Key Algorithms", *Advances in Cryptology – CRYPTO'98*, LNCS, **1462** (1998), Springer-Verlag, 472–485.

[BS91] T. BETH and F. SCAEFER, "Non supersingular elliptic curves for public key cryptosystems", *Advances in Cryptology – EUROCRYPT'91*, LNCS, **547** (1991), Springer-Verlag, 316–327.

[CA87] D.G. CANTOR, "Computing in the Jacobian of a Hyperelliptic Curve", *Math. Comp.*, **48**, No.177 (1987), 95–101.

[CER] http://www.certicom.com/

[CMO98] H. COHEN, A. MIYAJI and T. ONO, "Efficient Elliptic Curve Exponentiation Using Mixed Coordinates", *Advances in Cryptology – ASIACRYPT'98*, LNCS, **1514** (1998), Springer-Verlag, 51–65.

[CTT94] J. CHAO, K. TANAKA, and S. TSUJII, "Design of elliptic curves with controllable lower boundary of extension degree for reduction attacks", *Advances in Cryptology – CRYPTO'94*, LNCS, **839** (1994), Springer-Verlag, 50–55.

[DGM99] I. DUURSMA, P. GAUDRY and F. MORAIN, "Speeding up the discrete log computation on curves with automorphisms", *LIX Research Report LIX/ RR/99/03* (1999) to appear in Asiacrypt'99.

[DEC] "Alpha Architecture Handbook version 4", Compaq, http://ftp.digital.com

[EN98] A. ENGE, "The extended Euclidean algorithm on polynomials, and the efficiency of hyperelliptic cryptosystems", *Preprint*, (1998)

[EN99] A. ENGE, "Computing Discrete Logarithms in High-Genus Hyperelliptic Jacobians in Provably Subexponential Time," Tech. Report from Center for Applied Cryptographic Research at the University of Waterloo, CORR 99-04 (1999).

---

[7] **Division on an Alpha 21164A:** [DEC] Integer division does not exist as hardware opcode. Division by a constant can always be done via UMULH (the UMULH instruction can be used to generate the upper 64 bits of the 128-bit result when an overflow occurs.) of another appropriate constant, followed by a right shift. General quadword (128-bit) division by true variables can be performed via a subroutine.

[FMR98] G. FREY, M. MÜLLER,and H.G. RÜCK, "The Tate Pairing and the Discrete Logarithm Applied to Elliptic Curve Cryptosystems", *Preprints der Arbeitsgruppe Zahlentheorie*(1998),
http://www.exp-math.uni-essen.de/zahlentheorie/preprints/Index.html

[FR94] G. FREY and H.G. RÜCK, "A Remark Concerning m-Divisibility and the Discrete Logarithm in the Divisor Class Group of Curves", *Math. Comp*, **62**, No.206 (1994), 865–874.

[GA99] P. GAUDRY, "A variant of the Adleman-DeMarrais-Huang algorithm and its application to small genera", *LIX Research Report LIX/RR/99/04*, (1999) presented in Mathmatics of Public Key Cryptography'99

[GP97] J. GUAJARDO and C. PAAR, "Efficient Algorithms for Elliptic Curve Cryptosystems", *Advances in Cryptology - CRYPTO'97*, LNCS, **1294** (1997), Springer-Verlag, 342–356.

[GLV98] R. GALLANT, R. LAMBERT and S. VANSTONE, "Improving the parallelized Pollard lambda search on binary anomalous curves", http://grouper.ieee.org/groups/1363/contrib.html, (April,1998)

[Ko87] N. KOBLITZ, "Elliptic curve cryptosystems", *Mathematics of Computation*, **48** (1987), 203–209.

[Ko88] N. KOBLITZ, "A Family of Jacobians Suitable for Discrete Log Cryptosystems", *Advances in Cryptology - CRYPTO'88*, LNCS, **403** (1988), Springer-Verlag, 94–99.

[Ko89] N. KOBLITZ, "Hyperelliptic Cryptosystems", *J. Cryptology*, **1** (1989), Springer-Verlag, 139–150.

[Ko98] N. KOBLITZ, "Algebraic Aspects of Cryptography", Springer -Verlag, (1998)

[LLL82] A. LENSTRA, H. LENSTRA and L. LOVASZ, "Factoring polynomials with rational coefficients", *Math. Ann.*, **261**, (1982), 515–534.

[MI85] V. MILLER, "Uses of elliptic curves in cryptography", *Advances in Cryptology - CRYPTO'85*, LNCS, **218** (1986), Springer-Verlag, 417–426.

[MOV93] A.J. MENEZES, T. OKAMOTO and S.A. VANSTONE, "Reducing elliptic curve logarithm to logarithm in a finite field", *IEEE Trans. on IT*, **39**, (1993), 1639–1646.

[PH78] S.C. POHLIG and M.E. HELLMAN, "An improved algorithm for computing logarithms over $GF(p)$ and its cryptographic significance ", *IEEE Trans. on IT*, **24**, (1978), 106–110.

[PS98] S. PAULUS and A. STEIN, "Comparing real and imaginary arithmetics for divisor class groups of hyperelliptic curves", *Algorithmic Number Theory III*, LNCS, **1423** (1998), Springer-Verlag, 80–94.

[RU97] H.G. RÜCK, "On the discrete logarithms in the divisor class group of curves", To appear in *Math. Comp.* (1997)

[RSA] http://www.rsa.com

[SA97] T. SATOH and K. ARAKI, "Fermat Quotients and the Polynomial Time Discrete Log Algorithm for Anomalous Elliptic Curves", *Comm. Math. Univ. Sancti. Pauli*, **47**, (1998), 81–92

[SE98] I.A. SEMAEV, "Evaluation of discrete logarithms in a group of p-torsion points of an elliptic curve in characteristic p", *Math. Comp.*, **76** (1998), 353–356.

[SM97] N.P. SMART, "The Discrete Logarithm Problem on Elliptic Curves of Trace One", *Preprint* (1997): To appear in J.Cryptology.

[SM99] N.P. SMART, "On the performance of hyperelliptic cryptosystems", *Advances in Cryptology - EUROCRYPT'99*, LNCS, **1592** (1999), Springer-Verlag, 165–175.

[SO97] J.A. SOLINAS, "An Improved Algorithm for Arithmetic on a Family of Elliptic Curves", *Advances in Cryptology - CRYPTO'97*, LNCS, **1294** (1997), Springer-Verlag, 357–371.

[SS98] Y. SAKAI and K. SAKURAI, "Design of Hyperelliptic Cryptosystems in Small Characteristic and a Software Implementation over $\mathbf{F}_{2^n}$", *Advances in Cryptology - ASIACRYPT'98*, LNCS, **1514** (1998), Springer-Verlag, 80–94.

[SSI98] Y. SAKAI, K. SAKURAI and H. ISHIZUKA, "Secure hyperelliptic cryptosystems and their performance", *Public Key Cryptography, PKC'98*, LNCS, **1431** (1998), Springer-Verlag, 164–181.

[ST93] H. STICHTENOTH, "Algebraic Function Fields and Codes", Springer-Verlag, (1993)

[WBV96] E. DE WIN, A. BOSSELAERS, and S. VANDENBERGHE, "A Fast Software Implementation for Arithmetic Operations in GF(2^n)" *Advances in Cryptology - ASIACRYPT'96*, LNCS, **1163** (1996), Springer-Verlag, 65–76.

[WMPW98] E. DE WIN, S. MISTER, B. PRENEEL and M. WIENER, "On the Performance of Signature Schemes Based on Elliptic Curves", *Algorithmic Number Theory III*, LNCS, **1423** (1998), Springer-Verlag, 252–266.

[WZ98] M. WIENER and R. ZUCCHERATO, "Faster Attacks on Elliptic Curve Cryptosystems", http://grouper.ieee.org/groups/1363/contrib.html, (1998)

A   The Security of DLP in a Jacobian

We have to choose Jacobians to satisfy the following four conditions to resist all known attacks.

C1 : General Algorithms
A condition **C1** is needed to resist the Pohlig-Hellman method [PH78]. This algorithm has a running time that is proportional to the square root of the largest prime factor of $\sharp \mathbf{J}_C(\mathbf{F}_q)$. Therefore, we need to choose curves such that $\sharp \mathbf{J}_C(\mathbf{F}_q)$ has a large prime factor.

C2 : Imbedding into a Small Finite Field
A condition **C2** is needed to resist Frey and Rück's generalization [FR94] of the MOV-attack [MOV93] using the Tate pairing. This method reduces the logarithm problem over $\mathbf{J}_C(\mathbf{F}_q)$ to an equivalent logarithm problem in the multiplicative group $\mathbf{F}_{q^k}^*$ of an extension field $\mathbf{F}_{q^k}$. Ways to avoid the MOV-attack have been discussed in [BS91,CTT94]. We take a similar approach by choosing curves such that the induced Jacobian $\mathbf{J}_C(\mathbf{F}_q)$ cannot be imbedded via the Tate pairing into $\mathbf{F}_{q^k}^*$ with small extension degree k.

C3 : Large Genus Hyperelliptic Curves
A condition **C3** is needed to resist the Adleman-DeMarrais-Huang method [ADH94]. They found a sub-exponential algorithm for discrete logarithms over the rational subgroup of the Jacobians of large genus hyperelliptic curves over finite fields. It is a heuristic algorithm under certain assumptions. Therefore, we need to choose curves such that the genus is not overly large.

C4 : Additive Embedding Attack
The condition **C4** is to resist Rück's generalization [RU97] of the Semaev-Smart-Satoh-Araki attack [SE98,SM97,SA97] on elliptic cryptosystems with Frobenius trace one. The method uses an additive version of the Tate pairing to solve the discrete logarithm of a Jacobian over a finite field of characteristic p and has a running time of $O(n^2 \log p)$ for a Jacobian with cyclic group structure of order p^n.

B   Our Order Counting Method

B.1   Over a Characteristic Two Finite Field

Throughout this subsection, F denotes an algebraic function field of genus g whose constant field is the finite field $\mathbf{F}_q$ and $\mathbf{P}$ denotes the set of places of F/K. The definition, the theorem and the corollary shown below are given in the article [ST93].

Definition 1. *[ST93] The polynomial $L(t) := (1 - t)(1 - qt)Z(t)$ is called the L-polynomial of the function field $F/\mathbf{F}_q$, where $Z(t)$ denotes the zeta-function of $F/\mathbf{F}_q$.*

Theorem 1. *[ST93]*

(a) $L(t) \in \mathbf{Z}[t]$ and $\deg L(t) = 2g$
(b) $L(t) = q^g t^{2g} L(1/qt)$
(c) $L(1) = h$, the class number of $F/\mathbf{F}_q$
(d) We write $L(t) = \sum_{i=0}^{2g} a_i t^i$. Then the following holds:
*   (1) $a_0 = 1$ and $a_{2g} = q^g$.*
*   (2) $a_{2g-i} = q^{g-i} a_i$ for $0 \le i \le g$.*
*   (3) $a_1 = N - (q + 1)$ where N is the number of places $P \in \mathbf{P}_F$ of degree one.*
(e) $L(t)$ factors in $\mathbf{C}[t]$ in the form $L(t) = \prod_{i=1}^{2g}(1 - \alpha_i t)$. The complex numbers $\alpha_1, \cdots, \alpha_{2g}$ are algebraic integers, and they can be arranged in such a way that $\alpha_i \alpha_{g+i} = q$ holds for $i = 1, \cdots, g$.
(f) If $L_r(t) := (1 - t)(1 - q^r t)Z_r(t)$ denotes the L-polynomial of the constant field extension $F_r = F\mathbf{F}_{q^r}$, then $L_r(t) = \prod_{i=1}^{2g}(1 - \alpha_i t)$

Corollary 1. *[ST93] Let $S_r := N_r - (q^r + 1)$. Then we have:*
$a_0 = 1$, and $ia_i = S_i a_0 + S_{i-1} a_1 + \cdots + S_1 a_{i-1}$, for $i = 1, \cdots, g$.

We can determine the order of Jacobians by the Theorem and the Corollary in the following algorithm. It should be noted that it is easy to count $N_1, \cdots, N_g$ if $\mathbf{F}_q$ is small.

B.2   Over a Prime Finite Field

Let $n = 2g + 1$ be an odd prime, and let $p \equiv 1 \pmod{n}$. The order of a Jacobian over $\mathbf{F}_p$ of the curve of the form $v^2 + v = u^n$ can be found as follows [BK97,Ko98]:

Order Counting

Input: Hyperelliptic curve $C : v^2 + h(u)v = f(u)$ over $\mathbf{F}_q$ and extension degree n
Output: The order $\sharp J_C(\mathbf{F}_{q^n})$

---

1. Determine $N_r = \sharp J_C(\mathbf{F}_{q^r})$, for $r = 1, \cdots, g$
   by counting the number of rational points of C over $\mathbf{F}_{q^r}$
2. Determine the coefficients of $L_{\mathbf{F}_q}(t) = \sum_{i=0}^{2g} a_i t^i$ in the following:

   $a_0 = 1$
   for $1 \leq i \leq g$: $a_i = (\sum_{k=1}^{i}(N_k - (q^k + 1))a_{i-k})/i$
   for $g+1 \leq i \leq 2g$: $a_i = q^{i-g}a_{2g-i}$
3. Compute $L_{\mathbf{F}_{q^n}}(1) = \prod_{k=1}^{n} L_{\mathbf{F}_q}(\zeta^k)$, where ζ runs over the n-th root of unity
4. Return $\sharp J_C(\mathbf{F}_{q^n}) = L_{\mathbf{F}_{q^n}}(1)$

Let $\zeta = e^{2\pi i/n}$, and let $\alpha \in \mathbf{F}_p$ be a fixed non-nth-power. There is a unique multiplicative map χ on $\mathbf{F}^*_p$ such that $\chi(\alpha) = \zeta$. We extend this character χ to $\mathbf{F}_p$ by setting $\chi(0) = 0$. The Jacobi sum of the character χ with itself is defined as follows:

$$J(\chi, \chi) = \sum_{y \in \mathbf{F}_p} \chi(y)\chi(1 - y)$$

For $1 \leq i \leq n - 1$ let σ_i be the automorphism of the field $\mathbf{Q}(\zeta)$ such that $\sigma_i(\zeta) = \zeta^i$. Then the number of points on the curve of the form $v^2 + v = u^n$, including the point at infinity, is equal to

$$M = p + 1 + \sum_{i=1}^{n-1} \sigma_i(J(\chi, \chi))$$

The number N of points on the Jacobian of the curve is equal to

$$N = \prod_{i=1}^{n-1} \sigma_i(J(\chi, \chi) + 1) = \mathbf{N}(J(\chi, \chi) + 1)$$

where $\mathbf{N}$ denotes the norm of an algebraic number. Using the LLL-algorithm [LLL82], it is possible to determine $J(\chi, \chi)$ in polynomial time.

Speeding Up Elliptic Scalar Multiplication with Precomputation

Chae Hoon Lim and Hyo Sun Hwang

Information and Communications Research Center, Future Systems, Inc.
372-2, Yang Jae-Dong, Seo Cho-Gu, Seoul, 137-130, Korea
{chlim,hyosun}@future.co.kr

Abstract. It is often required in many elliptic curve cryptosystems to compute kG for a fixed point G and a random integer k. In this paper we present improved algorithms for such elliptic scalar multiplication. Implementation results on Pentium II and Alpha 21164 microprocessors are also provided to demonstrate the presented improvements in actual implementations.

1 Introduction

Let E be an elliptic curve defined over a finite field F ($F = \mathrm{GF}(2^n)$ or $\mathrm{GF}(p^n)$ for a prime p). Let G be a point of prime order in E. Elliptic scalar multiplication is to compute kG for random k. The performance of elliptic curve cryptosystems mainly depends on how efficiently this scalar multiplication can be performed. If G is random, the signed window algorithm is the most preferred algorithm for general scalar multiplication (e.g., see [6,7]). If E is defined over a small subfield such as $\mathrm{GF}(2^r)$ with $r|n$ or $\mathrm{GF}(p)$ for $n > 1$, then general scalar multiplication can be performed much faster using Frobenius expansion [10,15,18,16,4,9].

On the other hand, it is often required in elliptic curve cryptosystems to compute kG for a fixed point G. Since G is now fixed, we can substantially speed up the computation of kG using a precomputed table. Several methods have been developed for fast exponentiation using precomputation over a generic group [3,17,11], which can thus be applied equally well to the elliptic curve group. Among them, the Lim-Lee algorithm (LL algorithm, for short) is known to provide higher efficiency and flexibility in time-storage tradeoffs.

In this paper we investigate further improvements of the LL algorithm for elliptic scalar multiplication. Note that field inversion is most expensive among field operations required for elliptic curve arithmetic in most interesting fields. So, we tried to reduce the number of field inversions, at the cost of more field multiplications, utilizing the parallelizability of the LL algorithm and the simultaneous inversion technique [5, Algorithm 10.3.4]. Obviously, the amount of improvement that can be achieved with the resulting algorithm, Algorithm LL-SA, depends on the cost ratio of field inversion to multiplication. Our implementations on Pentium II and Alpha 21164 show that Algorithm LL-SA achieves about 20% speed-up over Algorithm LL in most interesting fields. Further improvement

JooSeok Song (Ed.): ICISC'99, LNCS 1787, pp. 102–119, 2000.

can be obtained by computing many scalar multiples in parallel. This simultaneous scalar multiplication algorithm, Algorithm LL-SM, may be useful for heavy loaded security servers, which often need to process hundreds of transactions (requiring scalar multiplications) at a time. We also show that these algorithms can be used to speed up general scalar multiplication using Frobenius expansion.

This paper is organized as follows. In section 2 we briefly summarize elliptic curve arithmetic (with some improvements) in $GF(2^n)$ and $GF(p^n)$. We then present improvements of the LL algorithm using simultaneous elliptic addition (Algorithm LL-SA) and simultaneous elliptic scalar multiplication (Algorithm LL-SM) in sections 3 and 4, respectively. Section 5 deals with application of LL algorithms to speed up general scalar multiplication using Frobenius expansion in $GF(p^n)$. Finally we present our implementation results in section 6 and conclude in section 7.

2 Elliptic Curve Arithmetic in Finite Fields

2.1 Affine Coordinates

A non-supersingular elliptic curve defined over a finite field F is a set of points (x, y) given by the cubic equation

$$y^2 + xy = x^3 + ax^2 + b \ (a, b \in F, \ b \neq 0) \qquad \text{if char}(F) = 2,$$
$$y^2 = x^3 + ax + b \ (a, b \in F, \ 4a^2 + 27b^3 \neq 0) \quad \text{if char}(F) > 3,$$

together with a 'point at infinity'. Addition/doubling formulas in this affine representation are summarized in Table 1.

field	operation	λ	addition formulas
$GF(2^n)$	addition (A_e) $(x_0 \neq x_1)$	$\lambda = \frac{y_1 + y_0}{x_1 + x_0}$	$x_2 = \lambda^2 + \lambda + x_0 + x_1 + a$ $y_2 = \lambda(x_0 + x_2) + x_2 + y_0$
	doubling (D_e) $(x_0 = x_1)$	$\lambda = x_0 + \frac{y_0}{x_0}$	$x_2 = \lambda^2 + \lambda + a$ $y_2 = \lambda(x_0 + x_2) + x_2 + y_0$
$GF(p^n)$	addition (A_e) $(x_0 \neq x_1)$	$\lambda = \frac{y_1 - y_0}{x_1 - x_0}$	$x_2 = \lambda^2 - (x_0 + x_1)$ $y_2 = \lambda(x_0 - x_2) - y_0$
	doubling (D_e) $(x_0 = x_1)$	$\lambda = \frac{3x_0^2 + a}{2y_0}$	$x_2 = \lambda^2 - 2x_0$ $y_2 = \lambda(x_0 - x_2) - y_0$

Table 1. Addition formulas in affine coordinates: $(x_2, y_2) = (x_0, y_0) + (x_1, y_1)$

2.2 Projective Coordinates

There is another representation of points, the so-called (weighted) projective representation, which eliminates the expensive field inversion at the cost of more field multiplications.

GF(2^n). For conversions between affine and projective coordinates, we used the transformation in [14]: $x = \frac{X}{Z}$, $y = \frac{Y}{Z^2}$. To the best of our knowledge, this is the best known conversion rule for GF(2^n). The resulting formulas for elliptic addition and doubling are given below.[1]

- Addition formula: $(X_2, Y_2, Z_2) = (X_0, Y_0, Z_0) + (X_1, Y_1, 1)$

$$\begin{aligned} A &= X_0 + X_1 Z_0 \\ B &= Y_0 + Y_1 Z_0^2, \ C = AZ_0 \end{aligned} \implies \begin{aligned} Z_2 &= C^2, \ X_2 = B^2 + A^2(C + aZ_0^2) + BC \\ Y_2 &= (BC + Z_2)(X_2 + X_1 Z_2) + (X_1 + Y_1)Z_2^2 \end{aligned}$$

- Doubling formula: $(X_2, Y_2, Z_2) = 2(X_0, Y_0, Z_0)$

$$Z_2 = X_0^2 Z_0^2, \quad X_2 = X_0^4 + bZ_0^4, \quad Y_2 = bZ_0^4(X_2 + Z_2) + X_2(aZ_2 + Y_0^2).$$

The addition formula requires 9 (8 general, 1 constant) multiplications and 5 squarings, while the doubling formula requires 5 (3 general, 2 constant) multiplications and 5 squarings. Note that the above addition formula requires one less multiplications than the formula given in [14]. If $a = 0$, then we can further reduce one multiplication in each formula.

GF(p^n). The addition/doubling formulas described here are essentially the same as those of the IEEE P1363 Draft [20]. The coordinates conversion is done by $x = \frac{X}{Z^2}$, $y = \frac{Y}{2Z^3}$.[2] So the affine coordinate (x, y) should be mapped to the projective coordinate $(X, Y, Z) = (x, 2y, 1)$. The resulting formulas for elliptic addition/doubling are described below, where we only consider the special case of $Z_1 = 1$ as before.

- Addition formula: $(X_2, Y_2, Z_2) = (X_0, Y_0, Z_0) + (X_1, Y_1, 1)$

$$\begin{aligned} A &= X_0 + X_1 Z_0^2, \ B = X_0 - X_1 Z_0^2 \\ C &= Y_0 + Y_1 Z_0^3, \ D = Y_0 - Y_1 Z_0^3, \ E = 2B \end{aligned} \implies \begin{aligned} Z_2 &= Z_0 E, \ X_2 = D^2 - AE^2 \\ Y_2 &= D(AE^2 - 2X_2) - E^2 BC \end{aligned}$$

- Doubling formula: $(X_2, Y_2, Z_2) = 2(X_0, Y_0, Z_0)$

$$\begin{aligned} A &= 3X_0^2 + aZ_0^4 \\ B &= 2X_0 Y_0^2, \ C = Y_0^4 \end{aligned} \implies \begin{aligned} Z_2 &= Y_0 Z_0, \ X_2 = A^2 - B \\ Y_2 &= A(B - 2X_2) - C \end{aligned}$$

The above formulas show that elliptic addition requires 8 multiplications and 3 squarings, while elliptic doubling requires 4 (3 general, 1 constant) multiplications and 6 squarings. If $a = -3$, then the variable A in doubling can be computed by $A = 3(X_0 + Z_0^2)(X_0 - Z_0^2)$, so one can save 2 squarings in this case.

---

[1] Here we only describe the special case of $Z_1 = 1$ for elliptic addition, which corresponds to the case where precomputation is done in affine coordinates in double-and-add algorithms for scalar multiplication. This special case gives better performances in almost all cases.

[2] The factor 2 in y is included to eliminate the modular division by 2 appearing in the addition formula when using $y = \frac{Y}{Z^3}$ (see A.10.5 in [20]). This also reduces the number of field additions/subtractions required in the doubling formula. Note that the addition/subtraction time in GF(p^n) is not negligible (see Sect.6).

2.3   Performance and Preferred Coordinates

In Table 2 we summarized the number of field operations for elliptic curve arithmetic in affine and projective coordinates. Here the capital letters I, M, S and A denote field operations of inversion, multiplication, squaring and addition, respectively. We assumed fixed values for constant a for performance reason: $a = 0$ for $GF(2^n)$ and $a = -3$ for $GF(p^n)$. It should be noted that these special values for constant a do not place much restriction in the choice of elliptic curves, since the proportion of elliptic curves that can be rescaled to have the above values for constant a is approximately $1/2$ for $GF(2^n)$ and $1/2$ or $1/4$, depending on the residue of $p \bmod 4$, for $GF(p^n)$ (see Appendix A in [20]).

field	coordinates	doubling (D_e)	addition (A_e)
$GF(2^n)$	Affine	$1I + 2M + 1S + 5A$	$1I + 2M + 1S + 7A$
$(a = 0)$	Proj.$(Z_1 = 1)$	$4M + 5S + 3A$	$8M + 5S + 8A$
$GF(p^n)$	Affine	$1I + 2M + 2S + 7A$	$1I + 2M + 1S + 6A$
$(a = -3)$	Proj.$(Z_1 = 1)$	$4M + 4S + 9A$	$8M + 3S + 9A$

Table 2. The number of field operations for elliptic addition/doubling

To simplify performance comparisons, we will use the following assumptions on speed ratios between field operations throughout this paper: $1S = 0.15M$, constant multiplication $= 0.5M$ for $GF(2^n)$ and $1S = 0.8M, 1A = 0.15M$ for $GF(p^n)$ (addition times in $GF(2^n)$ neglected). Of course, these ratios may vary from implementation to implementation, but our optimized implementations on P6 and Alpha microprocessors (see Sect.6) show that in most interesting fields the above assumptions are reasonable enough for theoretical comparison of computational complexity.

The cost ratio of field inversion to multiplication (I/M) is a key factor in determining a preferred coordinate system. So, let us find the I/M value at the break-even point between affine representation and projective representation (with $Z_1 = 1$). For this, suppose that $r = N_{D_e}/N_{A_e}$ (i.e., r elliptic doublings are required for each elliptic addition in a scalar multiplication algorithm). For example, we have $r = 6$ for the signed window algorithm with window size 4 and $r < 1$ for the LL algorithm. From Table 2 and the assumptions on speed ratios between field operations, we can obtain the following relations at the break-even point:

$$I/M = \begin{cases} 2.60 + \frac{4.00}{r+1} & \text{for } GF(2^n), \\ 3.90 + \frac{4.15}{r+1} & \text{for } GF(p^n). \end{cases}$$

Thus, for large r, it is almost always preferable to do elliptic scalar multiplication in projective coordinates. However, in the LL algorithm, we have $0.2 < r < 0.7$ for most interesting parameters, so $4.95 < I/M < 5.93$ for $GF(2^n)$ and $6.34 < I/M < 7.36$ for $GF(p^n)$. Thus, as we will see later, affine coordinates may yield better performances than projective coordinates in the case of $GF(p^n)$.

3   The Improved LL Algorithm for Scalar Multiplication

3.1   The Original LL Algorithm

We briefly describe the Lim-Lee algorithm for elliptic scalar multiplication kG for a fixed point G and analyze its performance. First, the multiplier k of l bits is divided into hv subblocks of b bits as follows (see Figure 1):

$$k = \sum_{u=0}^{l-1} 2^u e_u = \sum_{i=0}^{h-1} \left(\sum_{j=0}^{v-1} k_{i,j} 2^{bj} \right) 2^{ia}, \text{ where}$$

$$a = \lceil \frac{l}{h} \rceil, \quad b = \lceil \frac{a}{v} \rceil, \quad k_{i,j} = \sum_{t=0}^{b-1} 2^t e_{ia+jb+t}.$$

$k_{0,v-1}$	$\cdots$	$k_{0,1}$	$k_{0,0}$
$k_{1,v-1}$	$\cdots$	$k_{1,1}$	$k_{1,0}$
$\vdots$	$\cdots$	$\vdots$	$\vdots$
$k_{h-1,v-1}$	$\cdots$	$k_{h-1,0}$	$k_{h-1,0}$

$$|\leftarrow b \rightarrow|$$
$$|\leftarrow \qquad a \qquad \rightarrow|$$

Fig. 1. Partition of an l-bit multiplier k for the LL Algorithm

In the (off-line) precomputation stage, we compute and store the point $GG[I][j]$ as follows:

$$G_{i,j} = 2^{ia+jb}G \quad \text{for } 0 \le i < h \text{ and } 0 \le j < v,$$

$$GG[I][j] = \sum_{i=0}^{h-1} e_i G_{i,j} \quad \text{for } 0 \le j < v, \quad \text{where } I = \sum_{i=0}^{h-1} 2^i e_i. \tag{1}$$

Using these precomputed values, we can express kG as

$$kG = \sum_{j=0}^{v-1}\sum_{i=0}^{h-1} k_{i,j} G_{i,j} = \sum_{t=0}^{b-1} 2^t \left(\sum_{j=0}^{v-1}\sum_{i=0}^{h-1} e_{ia+jb+t} G_{i,j} \right),$$

$$= \sum_{t=0}^{b-1} 2^t \left(\sum_{j=0}^{v-1} GG[I_{j,t}][j] \right), \quad \text{where } I_{j,t} = \sum_{i=0}^{h-1} 2^i e_{ia+jb+t}. \tag{2}$$

Algorithm LL
$T := \sum_{j=0}^{v-1} GG[I_{j,b-1}][j]$;
for $t := b - 2$ **to** 0 **step** -1
$\quad T := 2T$;
$\quad T := T + \sum_{j=0}^{v-1} GG[I_{j,t}][j]$;
return T;

Note that $I_{j,t}$ corresponds to the t-th bit column of the j-th block column in Figure 1. Now, we can compute kG for each new value of k using equation (2) as shown in Algorithm LL.

Let us count the number of additions/doublings required by Algorithm LL. Obviously, we only need $(b-1)$ doublings. For the number of additions required, we note that the number of $GG[I_{j,t}][j]$ to be added is at most a. Therefore, we can see that the total cost for the worst case is given by

$$C_{LLw}(l, h, v) = (a - 1)A_e + (b - 1)D_e.$$

Let q be the probability of a bit being zero (so the probability of $I_{j,t}$ being zero is q^h). Then we can easily derive the expected number of additions/doublings as

$$C_{LLa}(l, h, v) = (a - q^h(a + (ah - l)(q^{-1} - 1)) - 1)A_e + (b - 1)D_e. \quad (3)$$

For random k, we may assume that $q = 1/2$. In this cse, equation (6) becomes

$$C_{LLa}(l, h, v) = \left(a - 1 - \frac{a + (ah - l)}{2^h}\right) A_e + (b - 1)D_e. \quad (4)$$

It is also easy to see that Algorithm LL requires the storage for $(2^h - 1)v$ precomputed points and that the cost for precomputation is given by

$$C_{LLp}(l, h, v) = v(2^h - h - 1)A_e + b(hv - 1)D_e.$$

Table 3 shows the average number of field inversions and multiplications, $(N_I + N_M)$, given by $C_{LLa}(160, h, v)$ for some selected parameters h and v, where we used the assumptions in Sect.2.3 to compute the equivalent number of field multiplications required for elliptic addition/doubling. In the case of projective coordinates, we also included the cost for coordinates conversion back to affine coordinates. The last two columns of Table 3 show the I/M ratios at the break-even point between computations in affine and projective coordinates. The ratios range from 5 to 6 in $GF(2^n)$ and from 6.5 to 8 in $GF(p^n)$. Our implementations on P6 and Alpha (see Sect.6) show that actual I/M ratios are larger than 10 for elliptic curves in $GF(2^n)$ and in $GF(p^n)$ with small n, so projective coordinates are preferred for these cases.

config.	storage	Affine		$(N_I + N_M)$	Proj.	I/M at B.E.P.	
$h \times v$	$(2^h - 1)v$	$GF(2^n)$	$GF(p^n)$	$GF(2^n)$	$GF(p^n)$	$GF(2^n)$	$GF(p^n)$
2×2	6	98.0+210.7	98.0+399.6	1+684.1	1+1031	4.88	6.51
2×4	12	78.0+167.7	78.0+306.6	1+599.1	1+859.8	5.60	7.18
3×2	14	72.0+154.8	72.0+291.1	1+515.1	1+766.9	5.08	6.70
3×4	28	59.0+126.8	59.0+230.6	1+459.9	1+655.8	5.74	7.33
4×2	30	55.5+119.3	55.5+223.4	1+402.3	1+595.4	5.19	6.83
4×4	60	45.5+ 97.8	45.5+176.9	1+359.8	1+509.9	5.89	7.48
5×2	62	45.0+ 96.8	45.0+180.8	1+328.4	1+484.9	5.26	6.91
5×4	124	37.0+ 79.5	37.0+143.5	1+294.4	1+416.5	5.97	7.58
6×2	126	38.5+ 82.9	38.5+155.0	1+280.9	1+415.4	5.27	6.94
6×4	252	31.5+ 67.8	31.5+122.4	1+251.2	1+355.6	6.00	7.63
7×2	254	32.8+ 70.5	32.8+131.9	1+239.8	1+354.4	5.32	7.00
7×4	508	26.8+ 57.6	26.8+104.0	1+214.3	1+303.1	6.07	7.72
8×2	510	27.9+ 60.0	27.9+111.9	1+206.0	1+303.4	5.42	7.11
8×4	1020	22.9+ 49.3	22.9+ 88.6	1+184.7	1+260.6	6.18	7.85

Table 3. Average performance of Algorithm LL for computing kG with $|k| = 160$

3.2   The Improved LL Algorithm

Computation of multiple inverses modulo the same modulus can be substantially speeded up using Montgomery's trick to parallel inversion [5, Algorithm 10.3.4]. For example, to compute inverses of A and B modulo p, we first compute $C = (AB)^{-1} \bmod p$ and then $A^{-1} = CB \bmod p$ and $B^{-1} = CA \bmod p$. In general, this simultaneous inversion algorithm requires 1 inversion and $3(t-1)$ multiplications mod p for t-simultaneous inversion. Therefore, from Table 2, we can see that t-simultaneous elliptic addition in $GF(p^n)$ requires the computational cost of $(I - 3M) + t(5M + S + 6A)$. Similarly, t-simultaneous elliptic addition in $GF(2^n)$ requires $(I - 3M) + t(5M + S + 7A)$. This technique thus enables us to replace one field inversion by about 3 field multiplications for large t. The resulting cost savings are substantial, since field inversion costs more than 3 field multiplications in most interesting fields.

Now, let us consider how to achieve a maximal improvement of Algorithm LL using the simultaneous addition algorithm. First note that in Algorithm LL we may precompute and store the following b points ahead of time:

$$GGG[t] = \sum_{j=0}^{v-1} GG[I_{j,t}][j] \quad \text{for } 0 \le t \le b - 1. \tag{5}$$

Then, we can just add one point $GGG[t]$ to T in the t-th iteration of the for-loop (see Algorithm LL-SA). Thanks to the high degree of parallelism existing in equation (5), we can take much advantage of simultaneous inversion in this on-line precomputation stage.

Algorithm LL-SA
for $t := 0$ **to** $b - 1$ **step 1**
$\quad GGG[t] := \sum_{j=0}^{v-1} GG[I_{j,t}][j];$
$T := GGG[b - 1];$
for $t := b - 2$ **to** 0 **step -1**
$\quad T := 2T;$
$\quad T := T + GGG[t];$
return $T;$

A naive way to evaluate equation (5) is to iterate b-simultaneous elliptic addition $v - 1$ times (so, $v - 1$ inversions required). However, the number of inversions required can be further reduced by performing b-simultaneous elliptic additions in parallel (based on the binary tree structure). E.g., if $v = 4$, we do the computation as follows:

1. $GGG[t] = GG[I_{0,t}][0] + GG[I_{1,t}][1]$ for $0 \leq t \leq b - 1$ and
   $TTT[t] = GG[I_{2,t}][2] + GG[I_{3,t}][3]$ for $0 \leq t \leq b - 1$.
2. $GGG[t] = GGG[t] + TTT[t]$ for $0 \leq t \leq b - 1$.

This way we can reduce the number of inversions from $v - 1$ to $\lceil \log_2 v \rceil$. This method of course increases the requirement for temporary storage from b to $\lfloor \frac{v}{2} \rfloor b$.

Suppose that c elliptic additions are required for the on-line precomputation of $GGG[t]$'s. This requires field operations given by

$$C_{sa}(c) = \begin{cases} \lceil \log_2 v \rceil (I - 3M) + c(5M + S + 7A) & \text{for } GF(2^n), \\ \lceil \log_2 v \rceil (I - 3M) + c(5M + S + 6A) & \text{for } GF(p^n). \end{cases} \quad (6)$$

Since c elliptic additions in Algorithm LL are now replaced by $C_{sa}(c)$ in Algorithm LL-SA, we can obtain the cost of Algorithm LL-SA as

$$C_{LL\text{-}SA}(l, h, v) = C_{LL}(l, h, v) - \Delta C(c), \text{ where } \Delta C(c) = cA_e - C_{sa}(c). \quad (7)$$

Thus we only need to find the average and worst case values of c, c_a and c_w.

In the worst case, we need $a - b$ additions in the precomputation stage, so we have $c_w = a - b$. Considering the probability of $GG[I][j]$ being 'point at infinity', we can find the expected value c_a as

$$c_a = a - b - q^h(a + (ah - l)(q^{-1} - 1)) + \delta, \text{ where}$$
$$\delta = q^{hv}(b + (ah - l)(q^{-1} - 1) + (bv - a)(q^{-h} - 1)). \quad (8)$$

As before, assuming that $q = 1/2$, we can simplify equation (8) to

$$c_a = a - b - \frac{a + (ah - l)}{2^h} + \delta, \text{ where } \delta = \frac{b + (ah - l) + (bv - a)(2^h - 1)}{2^{hv}}. \quad (9)$$

The cost advantage ΔC of Algorithm LL-SA over Algorithm LL can be expressed in terms of field operations as follows.

$$\text{Affine}: \quad \Delta C(c) = (c - \lceil \log_2 v \rceil)(I - 3M),$$
$$\text{Proj.}: \quad \Delta C(c) = \begin{cases} 3(c + \lceil \log_2 v \rceil)M + 4cS + \ cA - \lceil \log_2 v \rceil I & \text{for GF}(2^n), \\ 3(c + \lceil \log_2 v \rceil)M + 2cS + 3cA - \lceil \log_2 v \rceil I & \text{for GF}(p^n). \end{cases} \quad (10)$$

We evaluated the average performance of Algorithm LL-SA using equation (10) and Table 3. The result is shown in Table 4. Obviously, the amount of improvement of Algorithm LL-SA over Algorithm LL depends on the I/M ratio and becomes larger as the I/M ratio increases, since the improvement comes from the replacement of field inversions in Algorithm LL with about 3 field multiplications in Algorithm LL-SA. The I/M ratios at the break-even point shown in the last two columns can be used to determine which coordinates are preferred for the implementation of Algorithm LL-SA. From Tables 3 and 4 and the measured I/M ratios (Table 8 in Sect.6), we can see that the amount of improvement can be about 10 to 25% for projective coordinates and about 5 to 40% for affine coordinates.

config.	storage	Affine	$(N_I + N_M)$		Proj.	I/M at B.E.P.	
$h \times v$	(temp.)	GF(2^n)	GF(p^n)	GF(2^n)	GF(p^n)	GF(2^n)	GF(p^n)
2×2	6(40)	76.5+275.2	76.5+464.1	2+600.1	2+914.2	4.36	6.04
2×4	12(40)	39.7+282.6	39.7+421.6	3+448.0	3+650.2	4.51	6.23
3×2	14(27)	52.1+214.5	52.1+350.8	2+436.9	2+658.3	4.44	6.14
3×4	28(28)	27.8+220.6	27.8+324.4	3+334.3	3+481.9	4.59	6.36
4×2	30(20)	38.7+169.8	38.7+273.8	2+335.1	2+502.5	4.51	6.23
4×4	60(20)	20.0+174.4	20.0+253.5	3+254.6	3+364.9	4.73	6.56
5×2	62(16)	30.9+139.1	30.9+223.1	2+270.9	2+405.5	4.57	6.31
5×4	124(16)	16.0+142.6	16.0+206.6	3+205.5	3+294.2	4.85	6.75
6×2	126(14)	26.7+118.4	26.7+190.5	2+231.7	2+347.5	4.59	6.36
6×4	252(14)	13.9+120.7	13.9+175.3	3+174.6	3+250.5	4.93	6.88
7×2	254(12)	22.7+100.8	22.7+162.1	2+196.9	2+295.5	4.63	6.43
7×4	508(12)	11.9+102.3	11.9+148.6	3+147.5	3+211.9	5.06	7.09
8×2	510(10)	19.0+ 86.8	19.0+138.7	2+167.2	2+250.2	4.73	6.57
8×4	1020(10)	10.0+ 88.1	10.0+127.4	3+125.0	3+179.3	5.28	7.41

Table 4. Average performance of Algorithm LL-SA for computing kG with $|k| = 160$

4   Simultaneous Scalar Multiplication

A central security server often needs to handle thousands of transactions, e.g., involving Diffie-Hellman key exchanges or digital signatures, at a peak time. For such a heavy loaded application, further speed up can be obtained by computing many scalar multiples simultaneously.

Suppose that we want to evaluate t scalar multiplications, k_iG ($0 \leq i < t$, $|k_i| = l$), at a time. We can then perform t instances of Algorithm LL-SA simultaneously, one for each k_iG. Let us call this algorithm as Algorithm LL-SM. Then the simultaneous inversion technique can be applied even to double-and-add parts of concurrent Algorithm LL-SA instances. We thus perform all elliptic curve arithmetic in affine coordinates, but the number of field inversions required for t scalar multiplications is reduced to about $\lceil \log_2 v \rceil + 2b - 2$.

From the analysis of Sect.3, we can easily see that the average performance of Algorithm LL-SM (assuming that $q = 1/2$) is given by

$$C_{LL-SMa}(t, l, h, v) = (\lceil \log_2 v \rceil + 2b - \delta - 2)(I - 3M) +$$
$$t\left(a - 1 - \frac{a + (ah - l)}{2^h}\right)\tilde{A}_e + t(b-1)\tilde{D}_e, \quad (11)$$

where δ is the same as before (equation (9)) and $\tilde{A}_e$ and $\tilde{D}_e$ are given by

$$\tilde{A}_e = \begin{cases} 5M + 1S + 7A & \text{for GF}(2^n), \\ 5M + 1S + 6A & \text{for GF}(p^n). \end{cases}$$
$$\tilde{D}_e = \begin{cases} 5M + 1S + 5A & \text{for GF}(2^n), \\ 5M + 2S + 7A & \text{for GF}(p^n). \end{cases} \quad (12)$$

Therefore, for large t, the cost of Algorithm LL-SM per scalar multiplication, i.e., $C_{LL-SMa}(t, l, h, v)/t$, is almost the same as the cost of Algorithm LL in affine coordinates with field inversion replaced by 3 field multiplications. This would be the best achievable performance per scalar multiplication, as far as field inversion is more expensive than 3 field multiplications. For example, we tabulated in Table 5 the cost of Algorithm LL-SM per scalar multiplication for $t = 100$. As can be seen from the table, the number of field inversions required per scalar multiplication is less than 1 for large t. We can also see that Algorithm LL-SM improves over Algorithm LL-SA by more than 15% under the reasonable assumption of I/M ratios (see Table 8 in Sect.6).

5   Speeding Up Scalar Multiplication Using ϕ-Expansion

We can view an elliptic curve E defined over $\text{GF}(p)$ as an elliptic curve defined over $\text{GF}(p^n)$. For such a subfield curve we can achieve a much higher performance using Frobenius expansion [9]. Let $P = (x, y)$ be a $\text{GF}(p^n)$-point on E. The Frobenius map ϕ is defined as $\phi : (x, y) \rightarrow (x^p, y^p)$ and satisfies the equation

$$\phi^2 - t\phi + p = 0 \text{ and } \phi^n = 1, \quad -2\sqrt{p} \leq t \leq 2\sqrt{p}. \quad (13)$$

This map ϕ can be evaluated only using $2(n-1)$ multiplications mod p (see [9]).

configuration		storage		$(N_I + N_M)/t$	
$h \times v$	(a,b)	perm/temp	c_a	GF(2^n)	GF(p^n)
2×2	$(80,40)$	6/4000	22.5	0.77+502.4	0.77+691.4
2×4	$(80,20)$	12/4000	40.3	0.40+400.5	0.40+539.5
3×2	$(54,27)$	14/2700	20.9	0.52+369.2	0.52+505.5
3×4	$(54,14)$	28/2800	33.2	0.28+303.0	0.28+406.8
4×2	$(40,20)$	30/2000	17.8	0.39+284.7	0.39+388.7
4×4	$(40,10)$	60/2000	27.5	0.20+233.7	0.20+312.8
5×2	$(32,16)$	62/1600	15.1	0.31+230.8	0.31+314.8
5×4	$(32, 8)$	124/1600	23.0	0.16+190.1	0.16+254.1
6×2	$(27,14)$	126/1400	12.9	0.27+197.7	0.27+269.8
6×4	$(27, 7)$	252/1400	19.6	0.14+162.0	0.14+216.6
7×2	$(23,12)$	254/1200	11.1	0.23+168.3	0.23+229.6
7×4	$(23, 6)$	508/1200	16.9	0.12+137.7	0.12+184.0
8×2	$(20,10)$	510/1000	9.9	0.19+143.2	0.19+195.1
8×4	$(20, 5)$	1020/1000	14.9	0.10+117.7	0.10+157.1

Table 5. Average performance of Algorithm LL-SM for $t = 100$ and $|k_i| = 160$

To compute kP, we first express the multiplier k using equation (13) as

$$k = \sum_{i=0}^{n-1} k_i \phi^i, \quad \text{where } |k_i| < \frac{p}{2}, \tag{14}$$

precompute n points $P_i = \phi^i(P)$ for $0 \le i < n$ and then compute kP as

$$kP = \sum_{i=0}^{n-1} k_i P_i. \tag{15}$$

Note that the bit-length of k_i's in equation (14) can always be made one bit less than the bit-length m of p ($|p| = m$) and that the negative signs can be absorbed into the precomputed points P_i's. So, we may assume that the coefficients k_i's in equation (15) are always positive integers of bit-length $m - 1$.

The main source of efficiency in this scalar multiplication using base-ϕ expansion is that the intermediate points P_i's can be evaluated almost free, only using $2(n-1)^2$ subfield multiplications, and thus about $\frac{n-1}{n}|k|$ elliptic doublings can be saved, compared to general scalar multiplication. The cost we have to pay for this improvement is a small amount of on-line precomputation (i.e., base-ϕ expansion of k and $(n-1)$ evaluations of ϕ), which costs less than a few elliptic additions.

The right-hand side of equation (15) can be efficiently evaluated using the signed binary algorithm with optimal signed encoding of k_i's [18] (this is actually the same as Type-II expansion in [9]). Since an optimal signed encoding of a t-bit integer can produce an integer of bit-length at most $t + 1$ and probability of a bit being zero $2/3$, this computation can be done in $(m - 1)$ elliptic doublings and $(\frac{nm}{3} - 1)$ elliptic additions on average.

$n = 7$

k_1	k_0
k_3	k_2
k_5	k_4
	k_6

$n = 11$

k_3	k_2	k_1	k_0
k_7	k_6	k_5	k_4
	k_{10}	k_9	k_8

$n = 13$

k_4	k_3	k_2	k_1	k_0
k_9	k_8	k_7	k_6	k_5
		k_{12}	k_{11}	k_{10}

Fig. 2. Arrangements of k_i's for base-ϕ scalar multiplication using Algorithm LL

Further speed-up can be achieved using Algorithms LL/LL-SA, as can be expected from equation (15). Figure 2 shows some possible (actually best on average) arrangements of k_i's for using Algorithms LL/LL-SA. Here we only consider three field extensions of degree 7, 11 and 13, since they are most interesting in practice. Unlike Algorithms LL/LL-SA in Sect.3, we now have to do the precomputation required on-line. It is easy to see that the costs of on-line precomputation for the configurations shown in Figure 2 are given by $C_{LLp}(7m, 4, 2) = 15A_e$, $C_{LLp}(11m, 3, 4) = 13A_e$, and $C_{LLp}(13m, 3, 5) = 14A_e$, respectively. Since it is preferable to do the precomputation in affine coordinates, we can obtain some speed-up with simultaneous inversion. In this case, the costs are given by $C_{LLp}(7m, 4, 2) = 2I + 94.5M$, $C_{LLp}(11m, 3, 4) = 2I + 81.1M$, and $C_{LLp}(13m, 3, 5) = 2I + 87.8M$, respectively.

Since the average Hamming weight of k_i's can also be reduced to approximately $\frac{1}{3}$ with some clever weight minimization strategy, we can obtain average performances for the evaluation of equation (15) using Algorithms LL/LL-SA by substituting $a = vm, b = m, l = nm$ and $q = 2/3$ in equations (3) and (8):

$$C_{LLa}(nm, h, v) = C_{LLp}(nm, h, v) + \left(\left(v - \left(\frac{2}{3} \right)^h \left(v + \frac{hv - n}{2} \right) \right) m - 1 \right) A_e + (m - 1)D_e,$$

$$C_{LL\text{-}SAa}(nm, h, v) = C_{LLa}(nm, h, v) - \Delta C(c_a), \text{ where}$$

$$c_a = \left(v - 1 - \left(\frac{2}{3} \right)^h \left(v + \frac{hv - n}{2} \right) + \left(\frac{2}{3} \right)^{hv} \left(1 + \frac{hv - n}{2} \right) \right) m.$$

Table 6 shows the number of elliptic additions and field inversions required for three methods of evaluating equation (15), where the computational costs for base-ϕ expansion and ϕ evaluations are not included.

Finally, it is worth noting that though we can obtain much higher efficiencies using Frobenius expansion with subfield curves, we should be careful for their security consequences. The structure allowing faster implementations may also allow faster attacks (e.g., see [19]). $\#E/\text{GF}(p^m)$ (the order of $E/\text{GF}(p^m)$) divides $\#E/\text{GF}(p^n)$ if m divides n, so $\#E/\text{GF}(p^n)$ contains at least small prime factors of size$\#E/\text{GF}(p)$. Since we have to use a prime order subgroup for ECC, this

coord.	n	$m = \|p\|$	signed binary	Algorithm LL	Algorithm LL-SA
Affine	7	28	83.0+307.1	70.2+372.4	55.4+416.8
	11	16	68.0+251.6	58.7+305.0	33.8+379.6
	13	14	67.9+251.2	59.1+311.5	30.9+396.0
Proj.	7	28	1.0+978.2	3.0+812.1	4.0+729.2
	11	16	1.0+802.0	3.0+701.9	5.0+560.3
	13	14	1.0+800.8	3.0+720.2	6.0+553.8

Table 6. Average performances ($N_I + N_M$) of three algorithms for base-ϕ scalar multiplication

may increase the order of subfield curves more than necessary. Furthermore, the small prime factors in $\#E/\mathrm{GF}(p^n)$ may considerably weaken the resulting cryptosystem in many applications if proper precautions are not taken (by the small order subgroup attack in [12]).

6   Implementation and Discussion

We have implemented Algorithms LL, LL-SA and LL-SM on two different architectures: Pentium II/266MHz (32-bit μP; Windows 98, MSVC 5.0 with in-line assembly) and Alpha 21164/533MHz (64-bit μP; Linux, GCC 2.95 with in-line assembly). Table 7 summarizes the parameters used for field constructions. The three field parameters with degree of n^* ($n = 7, 11, 13$) were included for use in building subfield curves. The figures of the 'order' column in Table 7 denote the largest possible prime orders (in bits) in $E/\mathrm{GF}(p^n)$. See [13] for details on selection criteria of field parameters and timings for field/EC arithmetic.

field	n	order	p	irred. poly.
$\mathrm{GF}(2^{162})$	162	162		$x^{162} + x^{27} + 1$
$\mathrm{GF}(p^n)$	13^*	168	$2^{14} - 3$	$x^{13} - 2$
	12	168	$2^{14} - 3$	$x^{12} - 2$
	11^*	160	$2^{16} - 437$	$x^{11} - 2$
	10	160	$2^{16} - 165$	$x^{10} - 2$
	7^*	168	$2^{28} - 57$	$x^7 - 2$
	6	168	$2^{28} - 165$	$x^6 - 2$
	5	160	$2^{32} - 5$	$x^5 - 2$
	3	171	$2^{57} - 13$	$x^3 - 2$
	2	178	$2^{89} - 1$	$x^2 - 3$
	1	160	$p = 2^{160} - 2933$	

Table 7. Field constructions for elliptic curves

For better understanding of this presentation, we provided Table 8 summarizing various speed ratios between field operations and elliptic doubling to addition.[3] From Table 8, we can see that our assumptions given in Sect.2.3 are quite reasonable at least on P6 and Alpha family microprocessors, i.e., $A/M \approx 0.15, S/M \approx 0.8$ in $GF(p^n)$ and $S/M \approx 0.15$ in $GF(2^n)$. Also note that I/M ranges from 5 to 7 for most fields (except for $GF(p)$, $GF(p^2)$ and $GF(2^n)$).

μP	Pentium II/266MHz				Alpha 21164/533MHz			
Field	A/M	S/M	I/M	D_e/A_e	A/M	S/M	I/M	D_e/A_e
$GF(2^{162})$	0.03	0.13	14.0	0.47	0.05	0.16	10.5	0.48
$GF(p^{13})$	0.17	0.73	5.54	0.72	0.15	0.59	4.99	0.66
$GF(p^{12})$	0.18	0.74	6.63	0.73	0.15	0.61	6.11	0.64
$GF(p^{11})$	0.12	0.77	6.46	0.71	0.14	0.62	6.39	0.69
$GF(p^{10})$	0.14	0.78	6.04	0.77	0.18	0.66	5.98	0.70
$GF(p^7)$	0.11	0.79	6.05	0.73	0.14	0.87	6.17	0.72
$GF(p^6)$	0.13	0.82	6.41	0.73	0.18	0.87	5.79	0.70
$GF(p^5)$	0.18	0.82	5.89	0.74	0.12	0.81	4.60	0.76
$GF(p^3)$	0.18	0.80	7.59	0.75	0.15	0.89	6.88	0.74
$GF(p^2)$	0.16	0.86	19.9	0.73	0.13	0.90	10.4	0.75
$GF(p)$	0.15	0.88	42.7	0.74	0.12	0.85	31.7	0.75

Table 8. Speed ratios of field and elliptic curve operations

Timings for Algorithms LL/LL-SA/LL-SM on Pentium II/266MHz are given in Table 10, and timings on Alpha 21164/533MHz are given in Table 11. Here are some observations on the implementation results:

- As expected from the analysis in Sect.3 (compare the I/M ratios in Tables 3 and 4 with those in Table 8), Algorithms LL and LL-SA yield better performances in projective coordinates than in affine coordinates for $GF(2^n)$ and $GF(p^n)$ with $n \leq 3$.
- Algorithm LL-SA improves over Algorithm LL by about 10 to 25% in either coordinates, with some exceptions in $GF(p)$, $GF(p^2)$ and $GF(2^n)$. The exceptions in these fields are actually expected from the speed ratio of I/M in Table 8 (i.e., much higher values of I/M).
- Compared to individual scalar multiplication using Algorithm LL-SA in preferred coordinates, simultaneous scalar multiplication using Algorithm LL-

---

[3] The figures in Table 8 are different from the figures in Tables 9-11 in [13]. At the time of writing [13], we didn't implement the field inversion method using exponentiation from [2] (Algorithm BP, for short). Though the multiplicative complexity of Algorithm BP seems higher than that of Algorithm IM in [13], our actual implementations show that Algorithm BP runs about 20 to 30% faster than Algorithm IM due to smaller overheads in other simple operations and loop controls.

SM can significantly reduce the time per scalar multiplication (up to 40% for GF(p)).

- Algorithms LL/LL-SA/LL-SM can achieve 2 to 10 times speedup over the ordinary signed window algorithm for scalar multiplication.

Timings for elliptic scalar multiplication using Frobenius expansion are given in Table 9. We can see that Algorithm LL achieves about 15 to 20% improvement over the signed binary algorithm and that Algorithm LL-SA again improves over Algorithm LL by 5 to 15%.

	algorithm		w/o Frob.		binary		Alg. LL		Alg. LL-SA	
	field	$\|k\|$	A	P	A	P	A	P	A	P
Pentium	GF(p^{13})	178	4.19	3.48	1.66	1.86	1.43	1.58	1.34	1.36
II	GF(p^{11})	160	5.17	4.07	2.03	2.11	1.78	1.81	1.62	1.58
266MHz	GF(p^{7})	168	3.03	2.42	1.40	1.44	1.24	1.24	1.19	1.16
Alpha	GF(p^{13})	178	3.19	2.51	1.27	1.41	1.08	1.17	1.03	1.04
21164	GF(p^{11})	160	3.15	2.23	1.27	1.24	1.09	1.05	1.01	0.95
533MHz	GF(p^{7})	168	1.75	1.32	0.79	0.77	0.69	0.67	0.66	0.63

Table 9. Timings for scalar multiplication using Frobenius expansion (in msec, A: affine, P: projective)

7   Conclusion

Simultaneous inversion is a simple but powerful technique to speed up elliptic curve arithmetic with high degree of parallelism. Lim-Lee's algorithm for elliptic scalar multiplication for a fixed point (Algorithm LL) allows a very high degree of parallelism and thus can be substantially speeded up using the simultaneous inversion technique. This paper investigated such improvement on Lim-Lee's algorithm. More specifically, we presented and analyzed improved Lim-Lee's algorithms using simultaneous inversion: Algorithm LL-SA for computing a single scalar multiple and Algorithm LL-SM for computing many scalar multiples at a time. Implementation results of these algorithms on Pentium II and Alpha 21164 microprocessors were also provided to demonstrate practical performance improvement. We also showed that the presented algorithms can be used to speed up general elliptic scalar multiplication using Frobenius expansion.

References

1. D.V.Bailey and C.Paar, Optimal extension field for fast arithmetic in public key algorithms, *Advances in Cryptology-CRYPTO'98*, LNCS 1462, Springer-Verlag, 1998, pp.472-485.
2. D.V.Bailey and C.Paar, Inversion in Optimal Extension Fields, presented at *The Mathematics of Public-Key Cryptography*, Jun. 1999 (see also Elliptic curve cryptosystems over large characteristic extension fields by the same authors, preprint, 1999).
3. E.F.Brickell, D.M.Gordon, K.S.McCurley and D.Wilson, Fast exponentiation with precomputation, *Advances in Cryptology-EUROCRYPT'92*, LNCS 658, Springer-Verlag, 1993, pp.200-207.
4. J.H.Cheon, S.M.Park, S.W.Park and D.H.Kim, Two efficient algorithms for arithmetic of elliptic curves using Frobenius map, *Public Key Cryptography*, LNCS 1431, S-V, 1999, pp.195-202.
5. H.Cohen, *A course in computational number theory*, Graduate Texts in Math. 138, Springer-Verlag, 1993, Third corrected printing, 1996.
6. H.Cohen, A.Miyaji and T.Ono, Efficient elliptic curve exponentiation, *Information and Communications Security*, LNCS 1334, Springer-Verlag, 1997, pp.282-290.
7. H.Cohen, A.Miyaji and T.Ono, Efficient elliptic curve exponentiation using mixed coordinates, *Advances in Cryptology-ASIACRYPT'98*, LNCS 1514, Springer-Verlag, 1998, pp.50-65.
8. J.Guajardo and C.Paar, Efficient algorithms for elliptic curve cryptosystems, *Advances in Cryptology-CRYPTO'97*, LNCS 1294, Springer-Verlag, 1997, pp.342-356.
9. T.Kobayashi, H.Morita, K.Kobayashi and F.Hoshino, Fast elliptic curve algorithm combining Frobenius map and table reference to adapt to higher characteristic, *Advances in Cryptology-EUROCRYPT'99*, LNCS 1592, Springer-Verlag, 1999, pp.176-189.
10. N.Koblitz, CM curves with good cryptographic properties, *Advances in Cryptology-CRYPTO'91*, LNCS 576, Springer-Verlag, 1992, pp.279-287.
11. C.H.Lim and P.J.Lee, More flexible exponentiation with precomputation, *Advances in Cryptology-CRYPTO'94*, LNCS 839, Springer-Verlag, 1994, pp.95-107.
12. C.H.Lim and P.J.Lee, A key recovery attack on discrete log-based schemes using a prime order subgroup, *Advances in Cryptology-CRYPTO'97*, LNCS 1294, Springer-Verlag, 1997, pp.249-263.
13. C.H.Lim and H.S.Hwang, Fast implementation of elliptic curve arithmetic in $GF(p^n)$, *Public Key Cryptography*, LNCS 1751, Springer-Verlag, 1999.
14. J.Lopez and R.Dahab, Improved algorithms for elliptic curve arithmetic in $GF(2^n)$, *Selected Areas in Cryptography*, LNCS 1556, Springer-Verlag, 1999, pp.201-212.
15. W.Meier and O.Staffelbach, Efficient multiplication on certain non-supersingular elliptic curves, *Advances in Cryptology-CRYPTO'92*, LNCS 740, Springer-Verlag, 1993, pp.333-344.
16. V.Muller, Fast multiplication on elliptic curves over small fields of characteristic two, *J. of Cryptology*, vol.11, no.4, 1998, pp.219-234.
17. P.de Rooij, Efficient exponentiation using precomputation and vector addition chains, *Advances in Cryptology-EUROCRYPT'94*, LNCS 950, Springer-Verlag, 1995, pp.389-399.
18. J.A.Solinas, An improved algorithm for arithmetic on a family of elliptic curves, *Advances in Cryptology-CRYPTO'97*, LNCS 1294, Springer-Verlag, 1997, pp.357-371.
19. M.J.Wiener and R.J.Zuccherato, Faster attacks on elliptic curve cryptosystems, *Selected Areas in Cryptography*, LNCS 1556, Springer-Verlag, 1999, pp.190-200.
20. IEEE P1363: Standard Specifications for Public Key Cryptography, *Working Draft*, Aug. 1999.

	field	GF(p^n)										GF(2^n)
	n	1	2	3	5	6	7	10	11	12	13	162
AFFINE	Win. Alg.	8.93	4.72	3.08	2.04	2.41	2.91	4.21	5.18	3.96	3.99	12.0
	LL 2×4	3.68	1.91	1.20	0.78	0.93	1.12	1.63	2.00	1.55	1.53	5.01
	3×4	2.79	1.44	0.91	0.59	0.70	0.85	1.22	1.52	1.16	1.15	3.77
	4×4	2.14	1.11	0.70	0.45	0.54	0.65	0.94	1.17	0.90	0.88	2.91
	5×4	1.75	0.91	0.57	0.37	0.44	0.53	0.77	0.95	0.73	0.73	2.37
	6×4	1.49	0.77	0.49	0.32	0.38	0.45	0.66	0.81	0.63	0.62	2.02
	7×4	1.27	0.66	0.41	0.27	0.32	0.39	0.56	0.69	0.53	0.53	1.72
	8×4	1.09	0.56	0.35	0.23	0.28	0.33	0.48	0.59	0.46	0.45	1.47
	LL-SA 2×4	2.19	1.29	0.99	0.69	0.80	0.97	1.39	1.69	1.28	1.35	3.39
	3×4	1.57	0.94	0.73	0.51	0.59	0.72	1.03	1.26	0.95	1.01	2.47
	4×4	1.15	0.70	0.55	0.39	0.45	0.55	0.79	0.95	0.72	0.77	1.84
	5×4	0.92	0.56	0.45	0.32	0.37	0.44	0.64	0.77	0.59	0.62	1.48
	6×4	0.81	0.49	0.39	0.27	0.32	0.38	0.55	0.67	0.51	0.54	1.28
	7×4	0.69	0.42	0.33	0.23	0.27	0.33	0.47	0.57	0.44	0.46	1.10
	8×4	0.58	0.36	0.29	0.20	0.23	0.28	0.40	0.49	0.37	0.39	0.93
	LL-SM 2×4	0.65	0.66	0.78	0.59	0.68	0.83	1.21	1.43	1.12	1.26	1.75
	3×4	0.48	0.49	0.58	0.44	0.51	0.62	0.90	1.07	0.83	0.93	1.30
	4×4	0.37	0.37	0.44	0.34	0.38	0.47	0.68	0.81	0.62	0.69	1.00
	5×4	0.30	0.30	0.36	0.27	0.31	0.38	0.55	0.65	0.50	0.56	0.81
	6×4	0.26	0.26	0.31	0.24	0.27	0.33	0.47	0.56	0.43	0.48	0.69
	7×4	0.22	0.22	0.26	0.20	0.23	0.28	0.40	0.48	0.37	0.41	0.59
	8×4	0.19	0.19	0.23	0.17	0.19	0.24	0.35	0.41	0.32	0.35	0.51
PROJECTIVE	Win. Alg.	1.96	1.86	2.23	1.68	1.88	2.31	3.41	4.06	3.01	3.34	3.91
	LL 2×4	0.93	0.92	1.12	0.86	0.96	1.19	1.74	2.06	1.54	1.70	2.45
	3×4	0.72	0.70	0.86	0.66	0.74	0.90	1.33	1.58	1.18	1.31	1.89
	4×4	0.57	0.55	0.67	0.51	0.57	0.71	1.04	1.23	0.92	1.02	1.50
	5×4	0.48	0.46	0.55	0.42	0.47	0.58	0.85	1.01	0.75	0.84	1.23
	6×4	0.42	0.40	0.47	0.36	0.41	0.50	0.73	0.87	0.65	0.72	1.06
	7×4	0.36	0.34	0.41	0.31	0.35	0.43	0.63	0.75	0.56	0.62	0.92
	8×4	0.32	0.30	0.35	0.27	0.30	0.37	0.54	0.65	0.48	0.53	0.80
	LL-SA 2×4	0.84	0.78	0.91	0.69	0.77	0.95	1.38	1.63	1.23	1.36	2.01
	3×4	0.66	0.60	0.68	0.52	0.58	0.71	1.03	1.23	0.92	1.02	1.54
	4×4	0.53	0.47	0.52	0.40	0.44	0.54	0.79	0.94	0.71	0.78	1.22
	5×4	0.45	0.39	0.43	0.32	0.36	0.45	0.65	0.78	0.58	0.64	1.01
	6×4	0.41	0.34	0.37	0.28	0.32	0.39	0.56	0.67	0.51	0.56	0.89
	7×4	0.37	0.30	0.32	0.24	0.28	0.33	0.49	0.58	0.44	0.48	0.78
	8×4	0.33	0.27	0.28	0.21	0.24	0.29	0.42	0.50	0.38	0.42	0.69

Table 10. Timings (in msec) for Algorithms LL, LL-SA and LL-SM for computing kG with $|k| = 160$ on Pentium II/266MHz (timings for Algorithm LL-SM denote timings per scalar multiplication for $t = 100$)

field			GF(p^n)										GF(2^n)
		n	1	2	3	5	6	7	10	11	12	13	162
A F F I N E		Win. Alg.	4.22	1.80	1.06	1.41	1.14	1.63	2.61	3.14	3.08	3.05	2.82
	L L	2 × 4	1.75	0.72	0.42	0.53	0.45	0.62	1.00	1.24	1.12	1.16	1.18
		3 × 4	1.32	0.54	0.32	0.40	0.34	0.47	0.76	0.94	0.85	0.87	0.90
		4 × 4	1.02	0.42	0.24	0.31	0.26	0.36	0.58	0.72	0.65	0.67	0.69
		5 × 4	0.83	0.34	0.20	0.25	0.21	0.30	0.48	0.59	0.54	0.55	0.56
		6 × 4	0.71	0.29	0.17	0.22	0.18	0.25	0.41	0.51	0.46	0.47	0.48
		7 × 4	0.60	0.25	0.15	0.18	0.16	0.22	0.35	0.43	0.39	0.40	0.41
		8 × 4	0.52	0.21	0.13	0.16	0.13	0.19	0.30	0.37	0.34	0.34	0.35
	L L | S A	2 × 4	1.09	0.56	0.35	0.49	0.40	0.55	0.88	1.05	0.97	1.04	0.88
		3 × 4	0.79	0.41	0.26	0.37	0.30	0.41	0.66	0.78	0.72	0.78	0.65
		4 × 4	0.58	0.31	0.20	0.28	0.23	0.31	0.50	0.59	0.55	0.59	0.49
		5 × 4	0.47	0.25	0.16	0.23	0.19	0.25	0.40	0.49	0.44	0.49	0.40
		6 × 4	0.41	0.22	0.14	0.20	0.16	0.22	0.35	0.42	0.38	0.42	0.34
		7 × 4	0.35	0.19	0.12	0.17	0.14	0.19	0.30	0.36	0.33	0.36	0.29
		8 × 4	0.30	0.16	0.11	0.15	0.12	0.16	0.26	0.31	0.28	0.31	0.25
	L L | S M	2 × 4	0.42	0.41	0.31	0.49	0.39	0.50	0.80	0.92	0.87	1.00	0.57
		3 × 4	0.31	0.31	0.23	0.36	0.29	0.37	0.58	0.68	0.65	0.74	0.43
		4 × 4	0.24	0.23	0.18	0.28	0.22	0.28	0.44	0.52	0.49	0.56	0.33
		5 × 4	0.19	0.19	0.14	0.22	0.18	0.23	0.36	0.42	0.40	0.45	0.26
		6 × 4	0.17	0.16	0.12	0.19	0.15	0.20	0.31	0.36	0.34	0.39	0.23
		7 × 4	0.14	0.14	0.11	0.17	0.13	0.17	0.27	0.31	0.30	0.33	0.19
		8 × 4	0.12	0.12	0.09	0.14	0.11	0.14	0.23	0.26	0.25	0.29	0.17
P R O J E C T I V E		Win. Alg.	1.13	1.06	0.79	1.24	0.90	1.23	2.00	2.24	2.12	2.39	1.22
	L L	2 × 4	0.56	0.57	0.41	0.63	0.47	0.63	1.02	1.18	1.10	1.28	0.77
		3 × 4	0.43	0.44	0.31	0.48	0.36	0.48	0.78	0.91	0.84	0.98	0.59
		4 × 4	0.34	0.34	0.24	0.38	0.28	0.38	0.61	0.71	0.66	0.77	0.47
		5 × 4	0.28	0.28	0.20	0.31	0.23	0.31	0.50	0.59	0.55	0.63	0.38
		6 × 4	0.24	0.24	0.17	0.27	0.20	0.27	0.43	0.50	0.47	0.54	0.33
		7 × 4	0.21	0.21	0.15	0.23	0.17	0.23	0.37	0.43	0.40	0.47	0.29
		8 × 4	0.19	0.18	0.13	0.20	0.15	0.20	0.32	0.37	0.35	0.40	0.25
	L L | S A	2 × 4	0.50	0.45	0.33	0.52	0.39	0.52	0.85	0.96	0.92	1.04	0.63
		3 × 4	0.39	0.35	0.25	0.39	0.30	0.39	0.64	0.73	0.68	0.79	0.49
		4 × 4	0.31	0.27	0.19	0.30	0.23	0.30	0.49	0.56	0.53	0.60	0.38
		5 × 4	0.26	0.22	0.16	0.25	0.19	0.25	0.40	0.47	0.43	0.50	0.32
		6 × 4	0.24	0.19	0.14	0.22	0.17	0.22	0.35	0.40	0.38	0.43	0.28
		7 × 4	0.21	0.17	0.12	0.19	0.14	0.19	0.30	0.35	0.33	0.37	0.24
		8 × 4	0.19	0.15	0.11	0.16	0.13	0.16	0.26	0.30	0.28	0.32	0.21

Table 11. Timings (in msec) for Algorithms LL, LL-SA and LL-SM for computing kG with $|k| = 160$ on Alpha 21164/533MHz (timings for Algorithm LL-SM denote timings per scalar multiplication for $t = 100$)

Why Hierarchical Key Distribution Is Appropriate for Multicast Networks

Chandana Gamage, Jussipekka Leiwo*, and Yuliang Zheng

Peninsula School of Computing and Information Technology
Monash University, McMahons Road, Frankston, VIC 3199, Australia
{chandag,skylark,yuliang}@pscit.monash.edu.au

Abstract. The design rationale for many key distribution schemes for multicast networks are based on heuristic arguments on efficiency, flexibility and scalability. In most instances the choice of key server placement in a multicast network architecture is based on intuitive cryptographic considerations. We use an analytical model of multicast group formation and network growth to look at the selection of a key distribution scheme from a network operation perspective. Thereafter, this model is used to validate the choice of hierarchical (hybrid) key distribution model as the most appropriate.

Keywords: Network security, Multicast networks, Key distribution architectures

1   Introduction

The phenomenal growth of wide area networks, in the form of ubiquitous *Internet*, have given rise to many new applications that are different from the typical one-to-one (unicast) communication model of standard network applications. Many of the new applications in information distribution and collaborative activities such as web-casting, shared white-boards, on-line auctions, etc., have a one-to-many (multicast [6, 7]) model of communications. There are two main reasons that motivate the use of multicast for highly distributed network applications:

1. The number of messages a sender needs to transmit is reduced. This is due to the fact, that a single multicast address represents a large number of individual receivers. This results in a lower processing load for the sender and also simplifies the application design.
2. The number of messages in-transit over the network is reduced. As the correct message delivery is handled by multicast-capable routers, which normally make redundant copies of a message only when transmitting on divergent network links, data meant for a group of receivers is transmitted as a single message for most part of the network. This in turn improves the overall network bandwidth utilization.

---

* Since Sept. 1999, author has been with Vrije Universiteit, Department of mathematics and computer science, De Boelelaan 1081a, 1081 HV Amsterdam, The Netherlands, leiwo@cs.vu.nl

JooSeok Song (Ed.): ICISC'99, LNCS 1787, pp. 120–131, 2000.

Therefore, multicast data transmission provides significant benefits to both the applications and the network infrastructure and consequently is an important network technology for emerging applications. The basic difference between broadcast networks and multicast networks is that in multicast, delivery is to a specifically targeted group. This group may be created based on many metrics such as affiliation to a certain institution, long-duration membership subscriptions, short-duration tickets, etc. Many of the group management functions such as join, leave or re-join that control membership of a multicast group require cryptographic techniques to ensure that integrity of the control process is not compromised by malicious users or intruders. Furthermore, the multicast application itself may require secure data transmission to and from members. As the communication model of multicasting is different from unicast communication, the attacks and threat models are also different for multicast networks and in fact more severe [2].

To provide secure group management services, standard security functions such as identification, authentication and message transmission with confidentiality and integrity are required. The basic support service for secure group management in multicast networks is session key distribution which incorporates the primary functions of member identification, authentication and session key transport. The key distribution schemes described in literature can be classified under three basic models of centralized, distributed or hierarchical as shown in figure 1. In the fully distributed scheme, although shown as a tree, a fixed root may not be physically present.

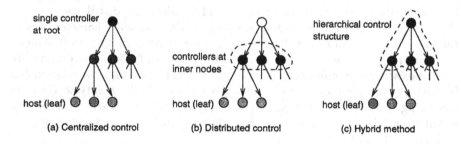

Fig. 1. Standard multicast group control methods. The fully distributed method shown in (b) requires horizontally structured coordination among participating controller nodes

Motivation. The cryptographic research literature is replete with sophisticated key distribution architectures for multicasting based on wide ranging assumptions while the networking community have adopted only a handful of techniques in proposed or experimental secure multicasting schemes. The work presented in this paper was motivated by the inadequate consideration given to network-centric issues when developing solutions that are grounded in cryptography.

Organization. In section 2, we overview several secure multicast schemes to see if their key distribution scheme selection is based on network considerations or cryptographic issues (or a combination of both). In section 3, we develop analytical arguments from a network perspective to validate the choice of hierarchical key distribution as the preferred framework. We make concluding remarks in section 4.

2   Related Work

In general, control and routing tree structure selection (shared trees, shortest-path trees, etc.) and protocol algorithm design for multicasting is based on expected sparseness/denseness of multicast group, efficiency in terms of number of messages, low message propagation delay, ease of recovery from message loss and low overhead in group management. For *secure* multicasting, in which the main design aspect is the key distribution scheme, designers may opt to consider underlying multicast network characteristics or mainly use cryptographic metrics such as number of rounds required for key distribution, size of security control messages and key update/change techniques. Next we briefly review previous work from literature that have taken different approaches to implementing secure multicasting.

A design for a secure key distribution architecture is presented in [14] that is overlaid on the core-based tree (CBT) multicast routing protocol [3]. The justification for the hybrid control structure of [14] in which key distribution centers (KDC) are co-located with routers is based on the favorable characteristics of the multicast protocol rather than on multicast network structure itself. Among the main reasons given for the use of CBT framework for key distribution are the pre-existing scalability properties of the routing protocol, close relationship between grouping structure and router placement and the ability to combine processing workload for router setup and key distribution. Early work on key distribution schemes based closely on underlying multicast protocol structures appeared in [1, 2, 11].

Similarly, the *Iolus* secure multicasting framework [15] is based on a distributed tree of group security intermediaries (GSI) for subtrees and an overall group security controller (GSC) for coordination of GSIs. The collection of these group security agents constitutes a hybrid key distribution architecture. However, the framework is designed to operate over many different multicast protocols including CBT and protocol independent multicasting (PIM) [7]. The distributed registration and key distribution (DiRK) technique presented in [16] is another multicast protocol independent decentralized and distributed model that simply assumes a hybrid model is better suited for large scale multicast groupings. Similar proposals appear in [9]

In contrast, the SecureRing suite of group communication protocols [12] use multicasting and a fully distributed control structure to provide membership management and message distribution under Byzantine errors but does not depend on any particular characteristics of the underlying multicast routing protocol for efficient or reliable operation. Their scheme uses cryptographic message digests and Byzantine fault detectors among other techniques to achieve efficiency and reliability. Similar cryptographic protocol based work also appear in [4, 8, 10, 13].

In summary, we can see that most key distribution schemes for secure multicasting use the hybrid model of key server placement. While this approach is intuitively reasonable, there are no analytical basis to support the model selection. In the next section, we analyze the growth and formation of multicast groups in wide area networks to provide evidence for the correctness of choosing a hybrid model.

3   Analysis of Key Distribution Agent Placement Models

We start our analysis using a regular tree structure which is more tractable than a general network topology. Consider a multicast distribution tree as shown in figure 2 with arity k and depth D where all the leaf nodes represent hosts that could be potential members of a multicast group. The inner nodes represent routers and the nodes at depth $D-1$ denotes sites (or local clusters). Therefore we have a regular network structure with total number of hosts $M = k^D$ and total number of sites $m = k^{D-1}$.

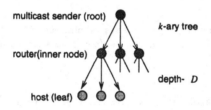

Fig. 2. The basic k-ary tree used to model the multicast distribution tree

3.1   Clustering of Hosts in the Multicast Distribution Tree

First we look at the effect on key distribution schemes due to the clustering of hosts. When we select a number of hosts to create a multicast group (say, of total size n), they could be arbitrarily distributed among several sites. While a single member multicast group will have a node from only a single site, a two member multicast group can select nodes from one or two distinct clusters. Following this argument, we can determine the best possible and worst possible clustering of hosts in sites when creating a multicast group. The plots of the two curves (equations 1 and 2) are shown in the graph of figure 3.

$$\text{Best case curve: } m = \left\lceil \frac{n}{k} \right\rceil \tag{1}$$

$$\text{Worst case curve: } m = \min\{n, k^{D-1}\} \tag{2}$$

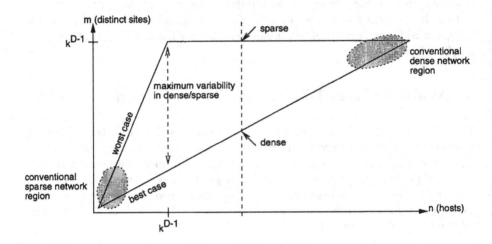

Fig. 3. The graph of number of distinct sites vs. number of hosts in the multicast group shows the allowable variability in denseness/sparseness for a multicast group of given size

We make following observations on the distribution of hosts in sites when setting up a multicast group as derived from the uniform tree structure:

1. The *conventional sparse region* was defined based on the observation that it has relatively very small number of hosts in the group and therefore even in *worst case* can only get distributed into few sites. The recommend key distribution architecture for this scenario is the centralized model. In using any other model, the multicast network will be needlessly using key distribution (sub)agents in inner nodes where most will be unused. In this region, the main issue is efficient use of security agents and not scalability.

2. The *conventional dense region* was defined based on the observation that it has relatively very large number of hosts in the group and therefore even in the *best case* can easily get distributed to nearly all the sites. In this instance, the recommended architecture is the distributed model. Any other model will create a bottleneck situation at the root affecting performance and also make it difficult for the key distribution architecture to scale with

the growth of the multicast network. In this region, the main issues concern both efficiency and scalability.

3. From a practical sense, the most interesting region is the middle area where the variability range is significant. Essentially, this means we might have either a densely populated or sparsely populated multicast network depending on the host distribution among site. Given the large range of sites (m) to which a multicast group of given size (n) can form into, it is quite impractical to discuss an average case scenario. The standard approach would be to use the hierarchical model as the key distribution agent architecture.

3.2 Total Size of the Multicast Distribution Tree

Next we look at the effect of clustering of hosts on the total size of the multicast distribution tree. For the purpose of analyzing the cost of message distribution, we assume a fixed transmission cost for any link in the multicast tree. For a multicast distribution tree represented as a uniform tree structure, the lowest total cost is obtained when hosts are densely located in the smallest possible number of sites as shown in figure 4 (a). The total size of the distribution tree L for a multicast group with n members is obtained by progressively counting the total number of links in all the full sub trees below a given level from top to bottom as shown in equation 3. The quantity ϕ_l denotes the total number of nodes counted prior to level l and the value ρ_l accounts for the link traversed when moving to the next level below to process a partially filled sub tree.

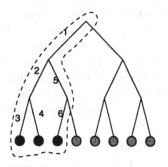

(a) Shortest path length grouping
for multicast hosts

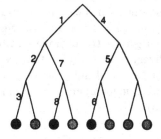

(b) Longest path length grouping
for multicast hosts

Fig. 4. The best and worst case grouping with respect to number of network links over which messages should pass are given by (a) depth-first search tree and (b) breadth-first search tree

Best case curve:

$$L(n) = \sum_{l=1}^{D} \left\{ \left\lfloor \frac{n - \phi_{l-1}}{k^{D-l}} \right\rfloor \times \left(\left(\sum_{j=1}^{D-l} k^j \right) + 1 \right) + \rho_l \right\}$$

$$\text{where } \phi_l = \begin{cases} 0 & l = 0 \\ \phi_{l-1} + \left\lfloor \frac{n - \phi_{l-1}}{k^{D-(l-1)}} \right\rfloor \times k^{D-(l-1)} & l > 0 \end{cases} \quad (3)$$

$$\text{and } \rho_l = \begin{cases} 0 & n = \phi_{l-1} \\ 1 & otherwise \end{cases}$$

The highest total cost for a multicast distribution tree occurs when the hosts are sparsely distributed among as many sites as possible as shown in figure 4 (b). The distribution is limited by the saturation value φ shown in equation 4 which the maximum number of clusters possible. Several sample plots of the two curves are shown in figure 5.

Worst case curve:

$$L(n) = \begin{cases} nD & n \le k \\ kD + \sum_{i=1}^{\varphi} \left((k^{i+1} - k^i)(D - i) \right) + (n - k^{\varphi})(D - \varphi) & k^i < n \le k^{i+1} \end{cases}$$

$$\text{where } \varphi = \left\lfloor \frac{\ln n}{\ln k} \right\rfloor$$

$$(4)$$

Previously we have discussed non-random clustering of hosts to form a multicast distribution tree in order to study the worst case and best case costs of the delivery tree. Next we look at the random formation of a multicast tree to analyze the total delivery cost for average case. When a host is selected at the leaf level of the tree to form a multicast group, at level l, a route through one of k^l links need to be selected. Therefore, the probability that a given link at level l is in the multicast delivery tree is $\frac{1}{k^l}$. Furthermore, the probability of a link being used in the delivery tree after n hosts have been selected at leaf level is $1 - \left(1 - \frac{1}{k^l}\right)^n$. If hosts are being selected at random at leaf level to form the multicast group, the average number of links at level l that will be included in the delivery tree is $k^l \left(1 - \left(1 - \frac{1}{k^l}\right)^n\right)$. Finally, assuming the link selection process to be a set of independent events, the total size of the multicast tree for a group with n members can be expressed as equation 5 (this result appears in [17] also).

$$L(n) = \sum_{l=1}^{D} k^l \left(1 - \left(1 - \frac{1}{k^l} \right)^n \right) \quad (5)$$

The set of graphs in figure 6 plots the curves for best, average and worst case scenarios for the same k and D. As can be seen from the graphs, the average cost of the multicast delivery tree is closer to the worst case cost for small (and therefore sparse) groups and tends toward best case cost for large (and therefore dense) groups. This result is intuitively correct and validates the expressions

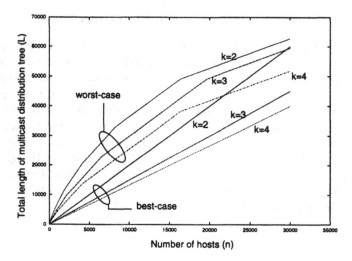

Fig. 5. The graph plots the total size of the multicast distribution tree (in terms of inter-node links) vs. number of hosts assembled in to both worst case and best case multicast groupings. The regular trees have k values 2 ($D = 15$), 3 ($D = 10$) and 4 ($D = 8$)

developed previously to analyze the structure of the multicast tree with respect to clustering of hosts. However, as can be seen from the graphs, the accuracy of the average case curve is lost as the number of hosts increase where the curve dips below the best case result.

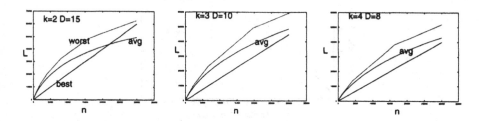

Fig. 6. Total size of the multicast distribution tree vs. number of hosts

The outcome of the foregoing analysis is that for most values of the multicast group size (n), the total size of the multicast distribution tree (L) can vary widely. This behavior again leaves the hierarchical key distribution architecture as the preferred option.

3.3   Applicability of Results to General Multicast Trees

Our analysis so far was based on uniform multicast distribution trees. However, practical multicast distribution structures normally take the shape of irregular trees. An important question at this point is, how relevant the results of an analysis based on uniform trees to real multicast networks? To answer this question, we look at the results obtained by Chuang and Sirbu [5] on the relationship between multicast distribution tree size and size of the membership for general multicast networks. According to the Chuang-Sirbu scaling law, the normalized multicast tree cost is directly proportional to the 0.8 power of the group size (shown in equation 6) for randomly selected group members. The normalized tree cost is obtained as the ratio between total multicast distribution tree length (L_m) and average unicast delivery path length (L_u).

$$\left[\frac{L_m}{L_u}\right]_{general} \propto n^{0.8} \tag{6}$$

We can compute the normalized tree cost for the uniform distribution tree with random member selection using equation 5. The average unicast tree length in this case is the tree depth D. Therefore, for the uniform multicast tree, the normalized tree cost can be given as equation 7.

$$\left[\frac{L_m}{L_u}\right]_{uniform} = \frac{L(n)}{D} = \frac{1}{D}\sum_{l=1}^{D} k^l \left(1 - \left(1 - \frac{1}{k^l}\right)^n\right) \tag{7}$$

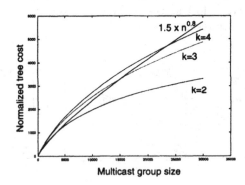

Fig. 7. The graph of normalized distribution tree cost vs. multicast group size with constant of proportionality for Chuang-Sirbu curve set at 1.5

The graph in figure 7 shows that the shape of normalized distribution tree curves for different k values of uniform trees follows that of the general curve due to Chuang-Sirbu scaling law for the range of n in which the average curve lies between best case and worst case curves of figure 5. The selection of the proportionality constant is admittedly arbitrary, but its function is simply to scale the curves with no distortion of the shape. As shown in the log-scale graph of figure 8, the value was selected for a close fit with plots for uniform trees. The implication of this matching of curves representing theoretical multicast networks to a curve of general multicast networks is, we can expect that for most group membership sizes (n), the average distribution cost (L) of *real* multicast networks also to be in the approximate middle of best case (dense) and worst case (sparse) values.

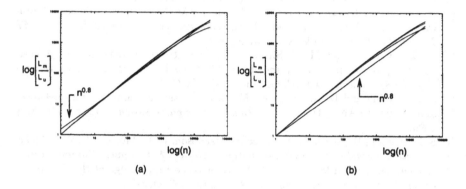

Fig. 8. The graph of normalized distribution tree cost vs. multicast group size with constant of proportionality for Chuang-Sirbu curve set at (a) 1.5 and (b) 1.0

In summary, the significance of this average total distribution cost curve of real multicast networks not being closer to sparse or dense formation of groups is that it is not meaningful to use a centralized or fully distributed control structure for key distribution. This in turn provides an analytical basis for using the hybrid control structure for key distribution.

4   Conclusion

A key distribution framework provides the backbone for any secure multicast architecture. Although the most widely used model for key distribution is the hybrid scheme, the reasons for its selection are usually heuristic arguments of flexibility and scalability. In this work we have used a different approach to validate the use of hybrid model by providing analytical arguments to exclude the use of both centralized and fully distributed control models. Although this work is based on key distribution in multicast networks, the results are applicable in other contexts such as loss recovery where a hierarchical control structure may be used.

References

[1] A. Ballardie. Scalable multicast key distribution. RFC 1949, Network Working Group, May 1996.

[2] A. Ballardie and J. Crowcroft. Multicast-specific security threats and counter-measures. In *Proceedings of the Internet Society Symposium on Network and Distributed System Security (NDSS'95)*, pages 2–16, San Diego, CA, February 1995. IEEE Computer Society Press.

[3] A. Ballardie, P. Francis, and J. Crowcroft. Core based trees (CBT): An architecture for scalable inter-domain routing. *SIGCOMM Computer Communication Review*, 23(4):85–95, October 1993. Conference Proceedings of the Communication Architectures, Protocols and Applications Conference.

[4] R. Canetti, J. Garay, G. Itkis, D. Micciancio, M. Naor, and B. Pinkas. Multicast security: A taxonomy and efficient constructions. In *Proceedings of IEEE INFOCOM'99*, New York, NY, March 1999.

[5] J. C.-I. Chuang and M. A. Sirbu. Pricing multicast communication: A cost-based approach. In *Proceedings of the 8th Annual Internet Society Conference (INET'98)*, Geneva, Switzerland, July 1998. ISOC.

[6] S. E. Deering and D. R. Cheriton. Multicast routing in datagram internetworks and extended LANs. *ACM Transactions on Computer Systems*, 8(2):85–110, May 1990.

[7] S. E. Deering, D. Estrin, D. Farinacci, V. Jacobson, C.-G. Liu, and L. Wei. An architecture for wide-area multicast routing. *SIGCOMM Computer Communication Review*, 24(4):126–135, October 1994. Conference Proceedings of the Communication Architectures, Protocols and Applications Conference.

[8] L. Gong. Enclaves: Enabling secure collaboration over the Internet. In *Proceedings of the 6th USENIX Security Symposium*, pages 149–159, San Jose, CA, July 1996. USENIX.

[9] L. Gong and N. Shacham. Multicast security and its extension to a mobile environment. *ACM-Baltzer Journal of Wireless Networks*, 1(3):281–295, October 1995.

[10] L. Gong and N. Shacham. Trade-offs in routing private multicast traffic. In *Proceedings of IEEE GLOBECOM'95*, Singapore, November 1995.

[11] H. Harney, C. Muckenhirn, and T. Rivers. Group key management protocol (GKMP) architecture. Internet Draft, 1994.

[12] K. P. Kihlstrom, L. E. Moser, and P. M. Melliar-Smith. The SecureRing protocols for securing group communication. In *Proceedings of the 31st Annual Hawaii International Conference on System Sciences (HICSS-31)*, volume 3, pages 317–326, Kona, Hawaii, January 1998. IEEE Computer Society Press.

[13] D. Malkhi, M. Merrit, and O. Rodeh. Secure reliable multicast protocols in a WAN. In *Proceedings of the 17th International Conference on Distributed Computing Systems (ICDCS'97)*, pages 87–94, Baltimore, MD, May 1997. IEEE Computer Society Press.

[14] K. Matsuura, Y. Zheng, and H. Imai. Compact and flexible resolution of CBT multicast key-distribution. In Y. Masunaga, T. Katayama, and M. Tsukamoto, editors, *Proceedings of the Second International Conference on Worldwide Computing and Its Applications (WWCA'98)*, volume 1368 of *Lecture Notes in Computer Science*, pages 190–205, Tsukuba, Japan, March 1998. Springer-Verlag.

[15] S. Mittra. Iolus: A framework for scalable secure multicasting. *SIGCOMM Computer Communication Review*, 27(4):277–288, October 1997. Conference Proceedings of the Communication Architectures, Protocols and Applications Conference.

[16] R. Oppliger and A. Albanese. Participant registration, validation, and key distribution for large-scale conferencing systems. *IEEE Communications Magazine*, 35(6):130–134, June 1997.

[17] G. Phillips, S. Shenker, and H. Tangmunarunkit. Scaling of multicast trees: Comments on the Chuang-Sirbu scaling law. *SIGCOMM Computer Communication Review*, 29(4), October 1999. Conference Proceedings of the Communication Architectures, Protocols and Applications Conference.

Secure Selection Protocols*

Kapali Viswanathan, Colin Boyd, and Ed Dawson

Information Security Research Centre
Queensland University of Technology
GPO Box 2434, Brisbane, Australia
{viswana,boyd,dawson}@fit.qut.edu.au

Abstract. The process of selection is omnipresent in the real world and modeling this process as a cryptologic protocol will enable cross use of techniques among *similar* protocol applications, which will eventually lead to better understanding and refinement of these applications. We present a proposal for a specialised selection protocol with anonymity as the security service. An area for its application is anonymous peer review, where no peer should know the identity of the reviewer.

1 Introduction

Many seemingly different protocol problems share a variety of common properties. The reason for this may be because confidentiality, integrity and identification are the basic services that cryptologic protocols employ. Complex services can be built using different building blocks. For example:

1. Signature systems employ integrity and identification services.
2. Anonymity systems employ confidentiality and identification services.
3. Blind signature systems employ services of anonymity and signature systems.

The first step towards developing secure and efficient solutions for such systems is to precisely understand their *goals*. Faster, better and improved understanding of such protocols can be achieved when analysing a *collection of similar* protocol problems that possess similar goals. In this paper we shall concentrate on a protocol problem that requires setting up of a peer review system. A feature of this system is its similarity to many other problem instances currently under investigation. We shall call this collection the set of *secure selection protocols*.

A second phenomenon that is also common is *compliance*. Research into a special class of cryptosystems called compliant cryptosystems offering services to different sets of users with logically contradicting requirements has become prominent. The existence of *safety valve* mechanisms is a fundamental property in these systems. Popular examples of such systems are escrowed encryption, fair electronic cash, electronic voting and group signature. The term (or concept) compliance has been used in the literature either implicitly or explicitly. For instance Desmedt [5] introduced society and group oriented cryptography, which implicitly identified the issue of compliance when certain functionalities were

---

* Research supported by the Australian Research Council grant A49804059

JooSeok Song (Ed.): ICISC'99, LNCS 1787, pp. 132–146, 2000.
© Springer-Verlag Berlin Heidelberg 2000

shared by sets of users. There are informal and explicit discussions of this term in the literature [7].

This paper will present our proposal for a cryptologic protocol for the design of a peer review system, which is a selection process with one-way anonymity as the security service. A proposal for transforming the system into a compliant (or fair) selection system will also be discussed.

2 Cryptographic Tools and Primitives

The important tools and primitives used are:

1. Proof of knowledge of discrete logarithm.
2. Proof of partial knowledge of discrete logarithm.
3. Electronic cash technology.

2.1 Proof of Knowledge of Discrete Logarithm

We will use the proof of knowledge introduced by Schnorr [14] in the non-interactive mode. Here the prover P has to prove the he knows the discrete logarithm of a public value u, where $u = g^v \bmod p$ and g is a publicly known generator of the group $\mathcal{Z}_p^*$. The prover performs the following function:

Begin Function PKGen
          Choose at random $k \in_R \mathcal{Z}_p$
          Compute $r = g^k$ and $c = \mathcal{H}(u,\ r)$
                             $d = cv + k \pmod{p-1}$
          Send to verifier $(c,\ d,\ r)$
End Function PKGen

The verifier performs the following function:

Begin Function PKVer
          Check $g^d \stackrel{?}{=} u^c r$
                          $c \stackrel{?}{=} \mathcal{H}(u,\ r)$
          If SUCCESS   output 1
          Else   output 0
End Function PKVer

If $\mathcal{H}$ is a cryptographically secure hash function, the verifier can be convinced that the prover knows $\log_g u \bmod p$ when the function PKVer outputs 1.

2.2 Proof of Partial Knowledge of Discrete Logarithm

Cramer et al. [3, 4] proposed a scheme to transform an interactive proof system into a proof system that will convince a verifier that the prover knows some secret, using a suitable secret sharing scheme with an appropriate access structure.

In this section we propose a modification to the witness indistinguishable variant of the Schnorr identification protocol [14] proposed in [4] to obtain a more computationally efficient protocol construct that can be used for the proof of knowledge of discrete logarithm. Our proposal transforms their interactive proof system into a non-interactive proof system and applies the screening technique used in batch verification methods [15,1] to the protocol proposed in [4]. The soundness and completeness properties of the protocol in [4] are not affected by the changes when a cryptographically secure hash function is used. This is due to the use of standard hashing technique [6] for the transformation. We also integrate the Schnorr signature scheme [14], so that the prover will provide the verifier with transcripts for the proof that also contains his/her signature.

Suppose that a set of values $\mathcal{U} = \{u_i = g^{v_i} \mid i = 1, \cdots, n\}$ are publicly known and a prover, possessing the public key y_j ($y_j = g^{x_j}$), wishes to prove to a verifier that he/she knows the discrete logarithm of *at least one* of the public values. For this to happen the verifier must allow the prover to *simulate* (or cheat) at most $n-1$ proofs. Assume that the prover knows v_j, which is the secret value corresponding to u_j for some $j \in \{1, \cdots, n\}$. The prover performs the following function:

Begin Function PPKGen

Choose at random $k_j \in_R \mathcal{Z}_p$, $\{c_l, d_l \in_R \mathcal{Z}_p \mid l \neq j\}$

Compute

$$r_j = g^{k_j} \tag{1}$$

$$\{r_l = g^{d_l} u_l^{c_l} \mid l \neq j\} \tag{2}$$

$$r = \prod_{i=i}^{n} r_i \bmod p \tag{3}$$

$$c = \mathcal{H}(u_1, \cdots, u_n, r) \tag{4}$$

$$c_j = c - \sum_{l \neq j} c_l \tag{5}$$

$$d = k_j - v_j c_j - x_j c + \sum_{l \neq j} d_l \pmod{p-1} \tag{6}$$

Send to verifier $(d, , c, \{c_i \mid i = 1, \cdots, n\})$

End Function PPKGen

The verifier performs the following function:

Begin Function PPKVer

Check

$$c \overset{?}{=} \mathcal{H}(u_1, \cdots, u_n, g^d y_j^c \prod_{i=1}^{n} u_i^{c_i}) \tag{7}$$

$$c \overset{?}{=} \sum_{i=i}^{n} c_i \tag{8}$$

If SUCCESS   output 1

Else   output 0

End Function PPKVer

If the function PPKVer outputs 1 when the transcripts from the prover are provided as inputs, then the verifier can, with a very high probability, decide that the prover knows the discrete logarithm of at least one of the n public values.

Computational Requirements: The function PPKGen requires $2n - 1$ modular exponentiations and the function PPKVer requires $n + 2$ modular exponentiations.

Analysis. We shall assume that the hash function, $\mathcal{H}$, used for the proof of partial knowledge, is cryptographically secure.

When $|\mathcal{U}| = 1$, $i = j$ and the proof is the usual proof for knowledge of discrete logarithm. So the verification equation will be of the form,

$$c \stackrel{?}{=} \mathcal{H}(u_1, \cdots, u_n, g^d y_j^c u_j^c) \tag{9}$$

which is a standard Schnorr signature with the signature inputs to the hash function of the form $g^d(y_j u_j)^c$. It is evident that this transcript can be formed without the knowledge of discrete logarithm of *both* y_j and u_j, if and only if the Schnorr signature can be forged. Thus, based on the assumption that the Schnorr signature is unforgeable, the above equation is sound, in that the prover cannot cheat the verifier.

When $y_j = 1$, equation 7 will be of the form,

$$c = \mathcal{H}(u_1, \cdots, u_n, g^d \prod_{i=1}^{n} u_i^{c_i}) \tag{10}$$

which is the verification equation for the non-interactive version of the protocol proposed in [4]. If the prover can form this equation without the knowledge of discrete logarithms for any of the u_i's then the protocol construct proposed by Cramer *et al.* [4] is flawed *or* the standard hashing technique proposed by Fiat and Shamir [6] is flawed. Thus on the assumption that *both* the techniques [6,4] are not flawed, the above equation is the sound and complete, in that the prover cannot cheat the verifier and the honest prover will generate transcripts that will be accepted by the verifier.

Observe that the Schnorr signature scheme [14] and the partial proof of knowledge protocol [4] are derivatives of the Schnorr identification protocol [14]. The Schnorr identification scheme is a three move protocol, the moves being commitment, challenge and response. Our protocol constrains the prover to use the same commitment and challenge to generate (two) different responses, namely Schnorr signature and partial proof of knowledge, that can be independently interpreted by the verifier. Thus, the proposed protocol constrains valid transcripts to contain a message tuple $(u_1, \cdots, u_n)$, commitment (r), challenge (c), and the response (d) that is interpreted as a Schnorr signature on the message tuple *and* proof of knowledge of at least one discrete logarithm in the set of public values $(u_1, \cdots, u_n)$.

2.3 Electronic Cash Technology

In our proposal, the electronic cash technology will be employed to generate anonymous tokens. This section will provide a brief summary of the fair off-line electronic cash scheme proposed by Frankel *et al.* [8,9], where a conditional anonymity service is offered when no coin is spent more than once. It employs the restrictive blind signature scheme presented by Brands [2]. The protocol specifications for the fair electronic cash system [8] are presented in Appendix A. In their proposal the anonymity of a coin can be revoked by a set of Trustees (or Ombudsman).

An e-cash system consists of the withdrawal, payment and deposit protocols.
Withdraw: In the withdrawal protocol, the client:

1. Authenticates to the mint and conveys its intention to withdraw some cash.
2. Generates a random message with a predefined structure and blinds this message.
3. Obtains a signature on the blinded message from the bank using an appropriate public key corresponding to the denomination and unblinds the signature.

This process is called restrictive blind signature because the client is not allowed to obtain the bank's blind signature on arbitrary messages, but only on messages with pre-defined structure (could contains information on identity of the client).
Payment: In the payment protocol, the client performs the following step 1 and if anonymity revocation is required it performs step 2.

1. The client anonymously contacts the merchant and proves that he/she knows the representation of the coin (restrictive blind signature on a random message by the mint). In this step, the transcripts bind the identity of the merchant to the coin.
2. The client proves to the merchant that the transcript contains the encryption of his/her identity under the public key of a trustee (typically, a distributed entity), without revealing the identity. Since the mint would expect the merchant to provide the transcript of this proof along with the transcript from the previous step before crediting the merchants account, the merchant must obtain a valid transcript from the client.

Deposit: In the deposit protocol, the merchant and the mint perform the following steps.

1. The merchant sends the transcripts of the payment protocol along with its public key or credentials.
2. The mint verifies the transcripts and, if successful, credits an appropriate amount to the merchant's account.

Trace: The mint and the trustees perform the following steps to trace the owner of a coin (received from the merchant).

1. The mint sends the deposit transcripts to the trustee.

2. The trustee decrypts the ciphertext that is present in the transcript and sends the information on identity of the owner to the mint.

Actually, there are two forms of anonymity revocation, namely coin tracing and owner tracing [8, 9]. We are only interested in the owner tracing facility, in this paper. This protocol module will be applicable only to the systems in which conditional anonymity service, as opposed to unconditional anonymity service, is offered.

3   Protocol Phases

We propose a three phase protocol schema to solve the peer review problem. The actual protocol will be presented in Section 4. The peer review problem consists of a set of participants called peers, having two roles in the system, namely reviewer and candidate to be reviewed. Since no participant should review itself, a solution to the peer review problem is a permutation of a set *with no fixed points*. The properties of the peer review protocol are:

1. The solution must define a permutation without any fixed points.
2. Every reviewer is also a candidate.
3. The solution must provide one-way anonymity service for the reviewers. That is the reviewers know the identity of the candidate, but the candidate does not know the identity of the reviewer.

The number of participants in the system must be greater than three, otherwise the peer review system cannot provide anonymity. Suppose that A, B and C are the participants, and the set of ordered pairs containing the reviewer and the candidate is $\{(A, B), (B, C), (C, A)\}$. A will know that C is its reviewer because it is reviewing B and if B is reviewing A then C has to review itself, which is not allowed. The reasoning for the case when $n = 2$ is trivial.

A challenging (and interesting) problem that is inherent in the problem statement is that when two participants collude they will be able to obtain some information that could *weaken* the anonymity of honest participants. Clearly, the information that colluding participants obtain is inversely proportional to the total number of participants in the system and directly proportional to the number of colluding participants. We believe that overcoming this problem would be difficult without weakening the security services for honest participants.

3.1   Basic Solution

A simple solution to solve the peer review problem could consist of three steps.

Step 1   Every participant wishing to participate in the protocol signs a random message and, publishes the signature and message in a publicly readable bulletin board, B_1. Let the number of signatures in B_1 be n, which is the number of participants.

Step 2 Each participant generates a random pseudonym and *anonymously* publishes its pseudonym in a publicly readable bulletin board B_2. The step completes when n pseudonyms are published. Let the set of pseudonyms be represented by PS.

Step 3 Each participant in turn chooses a pseudonym from B_2, such that it does not select the pseudonym it submitted. Let this choice be ps. The participant then generates the proof for its knowledge of the secret corresponding to one of the pseudonyms in the set $PS\backslash\{ps\}$, without revealing its pseudonym. It signs its identity, choice and the proof, and submits the signature along with the message to a bulletin board B_3. It also removes its choice from B_1, so that nobody else can make the same choice. This phase completes when n valid messages along their signatures are present in the bulletin board B_3. Anyone may check if every public key used for verifying the signatures in B_1 is also used in B_3.

Drawbacks and Solution: The protocol proposed assumes honest participants, which may not be very desirable. The protocol has the following drawbacks:

P1 Two participants, say i and j, can reveal their pseudonyms as u_i and u_j to each other, so that they can select each other.

P2 Two participants, say i and j, can generate the transcripts in Step 3 for each other, so that they can select themselves.

P3 Since v_i is only a short term secret, participant i can reveal this value to j, so that j can select twice. This would allow j to select itself.

P4 The system does not provide anonymity revocation, which may be required in common applications.

P5 An attacker can mount a denial of service attack on the system and be unidentified, because in Step 2 does not guarantee that only the participants involved in Step 1 submit *only one* pseudonym.

It seems difficult to overcome problem P1. Moreover, P1 does not *adversely* affect the goals of the protocol. But P2 and P3 do adversely affect the goals of the protocol. These problems can be solved if the participants are forced to use their long term secret values, namely private key corresponding to their certified public key, to generate the transcripts in Step 3. P4 can be solved by linking Step 2 to Step 1, so that the link can be computed if necessary. P5 can be solved by issuing *only one* anonymous token to every participant who registered in Step 1 and accepting *only one* pseudonym for every anonymous token in Step 2. Note the similarities in the solutions for P4 and P5.

It is interesting to note that problems with similar traits as P2 and P3 are observed in other protocol applications as well. Non-transferability of electronic cash [13], receipt-free electronic voting [12] and prevention of purchase of votes [10] are some examples for these traits of problem noticed in other protocol applications.

3.2   The Protocol Schema

We shall now describe a three phase protocol schema that overcomes problems P2 through P5. We assume that all participants possess certified public keys that support digital signature and authentication schemes. We also assume the existence of a token issuer TI, whose public key y_t is available to all the participants through a secure channel, and a supervisor S whose role is to act as a monitor of the system. Note that no explicit trust need be placed on S because all participants are required to generate publicly verifiable proofs. Let the system have n participants. The three phases of the schema are:

Phase 1 Participant i generates a message C_i (for commitment), signs this message using its public key, say y_i, sends the message and the signature, say D_i, to TI and obtains an anonymous token, AT_i, such that only participant i knows the ordered pair (y_i, AT_i). Note that AT_i could be a blind signature or an electronic coin that can be verified using the public key of TI, y_t. All participants must participate in this phase before proceeding to the next phase. This can be checked when n valid signature tuples, (C_i, D_i), are submitted and n tokens are withdrawn from TI.

Phase 2 Participant i (anonymously) submits AT_i to S, proves ownership of AT_i, submits its pseudonym $u_i = secret(v_i)$, where $secret$ could be a one way function, and keeps v_i as its secret. After verifying the proofs, S publishes (u_i, AT_i) in a publicly accessible directory, along with the proofs. All participants must participate in this phase before proceeding to the next phase. This can be checked when n tokens are submitted to S. Note that to create a strong link between Phase 1 and this phase, the value of u_i must be a function (or part) of the anonymous token AT_i. Or in other words, it cannot be randomly generated.

Phase 3 Participant i chooses its reviewer to be the owner of the pseudonym u_j, such that $j \neq i$, generates transcripts to prove that it knows the secret value corresponding to *one* of the $n-1$ public values in the set $\{u_l \mid l \neq j\}$ and commits to the choice by signing the choice and the transcripts of the proof. If S successfully verifies the proofs and the signature, it publishes the tuple (y_i, u_j) along with the proof and signature in a publicly accessible directory. Participant j can query the public directory to know the identity of its candidate, y_i. If n participants complete this phase and the public key used for verifying D_i was used to verify the signature for commitment to the choice then, S announces the protocol to be complete. If participant n cannot prove that it knows the secret corresponding to one of the $n-1$ public values in the set $\{u_l \mid l \neq j\}$, then u_j must be its pseudonym. This event results in a deadlock. Our preliminary analysis suggests the probability that deadlock will occur reduces as n increases and is bounded above by the value $1/2e$, where e is the base of the natural logarithm. In this case, S announces the protocol to be incomplete and all the participants must start the protocol anew from Phase 1.

Since the technology used to generate AT_i provides computational anonymity, the resulting system will provide *fair* peer review. If AT_j can be linked to y_j in Phase 1, then y_j can be linked to y_i using the information from the tuples (y_j, AT_j) and (u_j, y_i).

4   The Protocol

We shall now present our proposal to realise the functionality of each phase mentioned in Section 3.2. We shall use the electronic coin technology [8] for the anonymous token facility (Section 2.3) in Phases 1 & 2, and partial proof for knowledge of discrete logarithm (Section 2.2) in Phase 3.

System Setup. The supervisor of the system, S, selects a large prime p such that computing discrete logarithms in $\mathcal{Z}_p$ is intractable. S also selects a generator g, of the group $\mathcal{Z}_p^*$. Henceforth, all arithmetic will be computed in the congruence class modulo p, unless stated otherwise. The token issuer, TI, possesses a public key y_t of the form $y_t = g^{x_t}$, where $x_t \in_R \mathcal{Z}_p^*$ is the private key corresponding to y_t. The tuple (g, p, y_t) are published as the public parameters for the selection system. The supervisor maintains two bulletin boards with read permission for everyone and edit permission only for the supervisor. Let the two bulletin boards be labelled $\mathcal{A}$ and $\mathcal{B}$. Bulletin board $\mathcal{A}$ will contain unselected pseudonyms and bulletin board $\mathcal{B}$ will contain the selected pseudonyms.

Let there be n participants in the system, such that $n \geq 4$. The public key of participant i, $y_i (= g^{x_i} \mid x_i \in_R \mathcal{Z}_p^*)$, is published in a certified public directory with x_i as the corresponding private key. Every participant in the system possesses a certified public key.

Additional system parameters required for the electronic cash technology [8], which will be used as an anonymous token, are published.

4.1   The Proposal

We shall now present our proposal for the peer review problem, assuming *honest* participants. The next section will present modifications to this protocol that will greatly relax this assumption.

Phase 1. Participant i generates and signs a message to obtain a message-signature tuple (C_i, D_i) and, sends the tuple to TI (who verifies the signature using i's public key). The token will be a blind signature on a message by TI that can be verified using its public key y_t. We will use Brands blind signature scheme [2] in the withdrawal protocol. Participant i chooses a random value $v_i \in_R \mathcal{Z}_p^*$ and computes $u_i = g^{v_i}$. It then lets u_i be the message to be blindly signed by TI and obtains an anonymous token AT_i by executing the *Withdraw* protocol of the electronic cash technology with TI. Thus, $AT_i = \text{blindSignature}(u_i, \cdots, y_t)$. We refer to Appendix A for an exposition on the details of the withdrawal protocol.

Computational Requirements: The computational requirements will be the same as that required for the electronic cash technology. Suppose the electronic cash scheme proposed by Frankel, Tsiounis and Yung [8] is employed, then TI must perform $2n(\mathbf{dl}+1)$ modular exponentiations and each participant must perform $13(\mathbf{dl}+1)$ modular exponentiations, where $\mathbf{dl}$ denotes the number of times the dead lock situation occurs before the protocol completes.

Phase 2. The following steps are performed by individual participants and S:

Step 2.1   Participant i anonymously contacts S, presents the tuple $(AT_i,\ u_i)$ to S, engages in the *Payment* protocol with S. S checks if AT_i contains the blind signature by TI on u_i.

Step 2.2   If S successfully verified the transcripts then it publishes the tuple $(AT_i,\ u_i)$ in a public directory along with the transcripts of the *Payment* protocol.

Step 2.3 S enters u_i into $\mathcal{A}$.

All participants must complete this phase before the protocol can proceed to the next phase.

Computational Requirements: S must perform $11n(\mathbf{dl}+1)$ modular exponentiations and the participants must perform $6(\mathbf{dl}+1)$ modular exponentiations.

Phase 3. The following steps are performed by individual participants and S:

Step 3.1   Participant i authenticates to S using its public key y_i.

Step 3.2   Participant i chooses a pseudonym u_j such that $j \neq i$ from $\mathcal{A}$.

Step 3.3   Participant i presents u_j to S along with the transcripts generated using the function PPKGen (see Section 2.2) with $\{u_l \mid l \neq j\}$ and v_i as the inputs to the function and its signature on the transcript that can be verified using its public key y_i.

Step 3.4 S verifies the transcripts sent by participant i using the function PPKVer with $\{u_l \mid l \neq j\}$ as the input to the function. If it correctly verifies the transcripts and the signature on the transcripts using the public key y_i, it removes u_j from $\mathcal{A}$, adds u_j to $\mathcal{B}$ and publishes the tuple $(u_j,\ y_i)$ in a public directory.

Step 3.5   Participant j can consult with the public directory to find y_i as its candidate to be reviewed.

When participant n engages in this protocol, there will be only one entry in $\mathcal{A}$. If the last entry happens to be u_n (the pseudonym of participant n), then a deadlock is said to have occurred. In this case participant n cannot generate valid transcripts in Step 3.3, as it will not know possess the knowledge of discrete logarithm for any of the objects in the set $\{u_l \mid l \neq j\}$. Participant n then must prove that it knows the discrete logarithm of u_j using the protocol construct described in Section 2.1. If S successfully checks this then it publishes the transcripts sent by participant n along with its signature on the transcript and announces the protocol to be incomplete. In which case all participants must restart the protocol from Phase 1.

Computational Requirements: S must perform $(n^2 - n)(dl + 1)$ modular exponentiations and each participant must perform $(2n - 3)(dl + 1)$ modular exponentiations (see Section 2.2).

Anonymity Revocation: In the system an additional entity T, the trustee, will be involved to facilitate anonymity revocation. Then S, TI or any other authorised entity can engage in a *Trace* protocol (see owner tracing in [8]) with T to obtain/compute the tuple (y_i, AT_i), which can link y_i to u_i when the public information (AT_i, u_i) is used. To reduce the level of trust that must be placed on T, its functionality can be distributed.

4.2   Security Analysis

This section will present an analysis of the phases to elucidate its achievement of the desired properties.

Property 1: It achieves permutation without any fixed points. In Phase 1, when participant i authenticates to TI using its public key y_i, it receives only one AT_i. If more than one token was issued to participant i using y_i, then TI can be held responsible (all transcripts are publicly verifiable and signed by individual entities). Phase 2 allows only one pseudonym to be submitted for every AT_i. Phase 3 requires participant i to prove its knowledge for at least one pseudonym in the set of pseudonyms that does not contain its choice. In order to pass this phase, participant i cannot choose itself. Thereby, the protocol is a permutation without fixed points.

Property 2: Since every user is allowed to submit only one pseudonym and selects a different pseudonym (in the same category of pseudonyms), every reviewer is also a candidate.

Property 3: Reviewers are anonymous from the candidate and the candidate is not anonymous from the reviewer. Since every user chooses the pseudonym of its reviewer *after* authentication (using the public key, say y_i), this choice is public and the reviewer (say u_j) can know the identity of the candidate. From the publicly known tuples (AT_j, u_j) and (u_j, y_i), candidate i cannot know the identity of reviewer j, if the technology used for generating anonymous tokens does provide anonymity. Candidate i cannot obtain the tuple (y_j, u_j) by observing the protocol runs in Phase 3, if the proof system used is witness indistinguishable. Proposition 1 provides the proof for this property.

Proposition 1 *The system provides anonymity service to the reviewers.*

Proof: Assume that the participants do not collude and the electronic cash technology prevents any entity other than participant i to compute the tuple (y_i, AT_i). S, by itself or in collusion, cannot correlate between the values y_i and u_i, using the public knowledge (AT_i, u_i). Since the functions PPKVer and PPKGen are witness indistinguishable (see [4]), S, by itself or in collusion, cannot correlate the value y_i with u_i using the outputs of the function PPKGen, as computed by participant i.   □

If the proof systems used for the anonymous token technology and partial proof of knowledge protocol construct are *publicly verifiable*, then the trust level on the token issuer, TI, and the supervisor, S, can be considerably reduced. The advantage of this approach is that it does not make any assumptions on the possible inclusion of anonymity revocation mechanism. This is the advantage of abstracting anonymous token, AT_i, to provide this service. Anonymity revocation mechanisms can be built into the token technology without affecting other core functionalities of the protocol (permutation without fixed points).

5 Discussion

We proposed a protocol schema that can be employed to design a peer review system. If we look at the peer review process from a higher abstraction, it is evident that peer review is a secure selection process with one-way anonymity service, most of which is achieved only in Phase 3.

The generic selection protocol $\mathcal{P}$, is a mapping defined as $\mathcal{P} : \mathcal{S} \mapsto \mathcal{C}$ with additional constraints to achieve specific security services, where $\mathcal{S}$ is the set of selectors and $\mathcal{C}$ is the set of choices. The peer review system is a special case of the selection protocol when, $\mathcal{S}$ is the set of reviewers, $\mathcal{C}$ the set of candidates and $\mathcal{S} = \mathcal{C}$ with constraints to ensure no-fixed point and one-way anonymity for participants in $\mathcal{S}$.

There are similarities between the basic peer review process and other protocol problems, as follows:

1. **Electronic Voting:** Participants in the set of voters $\mathcal{S}$ cast a vote that *selects* an entity from the set of candidates $\mathcal{C}$. Only the participant knows the selection or the voter-vote relation is known only to the voter. To achieve this goal, confidentiality of information on selection can be translated to confidentiality of information on the identity (anonymity) of the selector (voter).
2. **Contract Bidding:** Participants in the set of bidders $\mathcal{S}$ commit to a bid that *selects* a value from the set of bid options $\mathcal{C}$. Only the participants knows the value of the bid.
3. **Conference Paper Review:** Reviewers from the set of programme committee members $\mathcal{S}$ *select* objects from the set of submitted papers $\mathcal{C}$ with constraints on the selection, like the member must not be from the same organisation from where the paper originated *etc*. The reviewer should not know the details pertaining to the authors of the paper *and* the author of the paper should not know the identity of the reviewer. This seems to be a very complex system primarily due to the number of possible constraints and the requirement for two way (mutual) anonymity.

Some of the seemingly conflicting requirements that are evident in these problems, as in the peer review problem, are:

1. authentication and anonymity;
2. confidentiality and validation of the information.

Future research will be directed towards solving these protocols by employing the method described in this paper. It is also planned to investigate the possibility for obtaining an exact expression for the probability of deadlock.

References

1. Mihir Bellare, Juan A. Garay, and Tal Rabin. Fast batch verification for modular exponentiation and digital siagntures. In *Advances in Cryptology – EURO-CRYPT'98*, number 1403 in Lecture Notes in Computer Science, pages 236–250. Springer-Verlag, 1998.
2. Stefan Brands. Untraceable off-line cash in wallet with observers. In *Advances in Cryptology – CRYPTO'93*, Lecture Notes in Computer Science, 1993.
3. Ronald Cramer, Ivan Damgård, and Berry Schoenmakers. Proofs of partial knowledge and simplified design of witness hiding protocols. In Yvo G. Desmedt, editor, *Advances in Cryptology – CRYPTO'94*, number 839 in Lecture notes in computing science, pages 174–187. Springer, 1994.
4. Ronald Cramer, Ivan Damgård, and Berry Schoenmakers. Proofs of partial knowledge and simplified design of witness hiding protocols. Technical Report CS-R9413, Computer Science/Department of Algorithmics and Architecture, P.O. Box 94079, 1090 GB Amsterdam, The Netherlands, 1994.
5. Yvo Desmedt. Society and group oriented crytpography: A new concept. In *Advances in Crytology – CRYPTO'87*, 1987.
6. A. Fiat and A. Shamir. How to prove yourself: practical solutions to identification and signature problems. In A. M. Odlyzko, editor, *Advances in Cryptology – CRYPTO'86*, Lecture Notes in Computer Science, pages 186–194. Springer-Verlag, 1986.
7. Yair Frankel and Moti Yung. Escrow encryption systems visited: Attacks, analysis and designs. In *Advances in Cryptology – CRYPTO'95*, Lecture Notes in Computer Science, 1995.
8. Yair Frankel, Yiannis Tsiounis, and Moti Yung. Indirect discourse proofs: Achieving efficient fair off-line e-cash. In *Asiacrypt'96*, Lecture Notes in Computer Science, 1996.
9. Yair Frankel, Yiannis Tsiounis, and Moti Yung. Fair off-line e-cash made easy. In Kazuo Ohta and Dingyi Pei, editors, *Advances in Cryptology – ASIACRYPT'98*, number 1514 in Lecture Notes in Computer Science, pages 257–270, 1998.
10. Valtteri Niemi and Ari Renvall. How to prevent buying of votes in computer elections. In Josef Pieprzyk and Reihanah Safavi-Naini, editors, *Advances in Cryptology – ASIACRYPT'94*, number 917 in Lecture Notes in Computer Science, pages 164–170. Springer-Verlag, 1994.
11. Hannu Nurmi, Arto Salomaa, and Lila Santean. Secret ballot elections in computer networks. *Computers & Security*, 10:553–560, 1991.
12. Tatsuaki Okamoto. Receipt-free electronic voting schemes for large scale elections. In *Security Protocol Workshop*, 1997.
13. Holger Petersen and Guillaume Poupard. Efficient scalable fair cash with off-line extortion prevention. In *Information and Communication Security, ICICS'97*, pages 463–473, November 1997.
14. C.P. Schnorr. Efficient signature generation for smart cards. *Journal of Cryptology*, 4:161–174, 1991.
15. S. Yen and C. Laih. Improved digital signature suitable for batch verification. *IEEE Transactions on Computers*, 44(7):957–959, July 1995.

Appendix

A   Fair Off-Line Cash

In this section a brief overview of the fair off-line cash scheme, proposed by Frankel Tsiounis and Yung [8,9], will be provided.

System settings The mint chooses primes p and q such that $p - 1 = \delta + k$ for a specified constant δ, and $p = \gamma q + 1$ for a small integer γ. A unique subgroup $\mathcal{G}_q$ of prime order q of the multiplicative group $\mathcal{Z}_p$ and generators g, g_1, g_2 of $\mathcal{G}_q$ are defined. The mints secret key $X_B \in_R \mathcal{Z}_q$ is created. Hash functions $\mathcal{H}, \mathcal{H}_0, \mathcal{H}_1, \cdots$, from a family of correlation-free one way hash functions are defined. The mint published $p, q, g, g_1, g_2, (\mathcal{H}, \mathcal{H}_0, \mathcal{H}_1, \cdots)$ and its public keys $h = g^{X_B}$, $h_1 = g_1^{X_B}$, $h_2 = g_2^{X_B}$. The public key of the trustee, $\mathcal{T}$, of the form $f_2 = g_2^{X_T}$ is also published, where $X_T \in_R \mathcal{Z}_q$. Note that $\mathcal{T}$ should be a distributed entity to reduce the level of trust placed on it.

The mint associates the user with the identity $I = g_1^{u_1}$, where $u_1 \in_R \mathcal{G}_q$ is generated by the user such that $g_1^{u_1} g_2 \neq 1$. The user is expected to prove the knowledge of discrete logarithm of I w.r.t. g_1. The user computes $z' = h_1^{u_1} h_2 = (Ig_2)^{X_B}$.

Function Withdraw: This protocol creates a restrictive blind signature on I, so that at the completion of the protocol the user obtains a valid signature of the mint on $(Ig_2)^s$ for a random secret value s known only to the user. The signature verification equation $sig(A, B, u_i) = (z, a, b, r)$ satisfies the following equation:

$$g^r = h^{\mathcal{H}(u_i, A, B, z, a, b)} a \text{ and } A^r = z^{\mathcal{H}(u_i, A, B, z, a, b)} b \tag{11}$$

In our proposal the anonymous token AT_i would be the tuple (u_i, A, B, z, a, b), h the public key of the token issuer TI and u_i the pseudonym of participant i whose identity in the system is I. The withdrawal protocol is of the form:

User	Mint
	$w \in_R \mathcal{Z}_q$
	$\xleftarrow{\quad a', b' \quad}$ $a' = g^w, b' = (Ig_2)^w$
$s, v_i \in_R \mathcal{Z}_q$	
$A = (Ig_2)^s$ and $u_i = g^{v_i}$	
$z = z'^s$	
$x_1, x_2, u, v \in_R \mathcal{Z}_q$	
$B_1 = g_1^{x_1}$	
$B_2 = g_2^{x_2}$	
$B = [B_1, B_2]$	
$a = (a')^u g^v$	
$b = (b')^{su} A^v$	
$c = \mathcal{H}(u_i, A, B, z, a, b)$	
$c' = c/u$	$\xrightarrow{\quad c' \quad}$ $r' = c' X_B + w \bmod q$
$r = r'u + v \bmod q$	$\xleftarrow{\quad r' \quad}$

The user verifies if $g^{r'} \stackrel{?}{=} h^{c'} a', (Ig_2)^{r'} \stackrel{?}{=} z'^{c'} b'$.

Function Payment: This protocol is performed in an anonymous channel. The following table provides the sketch for the protocol. Note that, in the protocol, the tuple (D_1, D_2) is an ElGamal encryption of the identity of the user I under the public key of $\mathcal{T}$. Also note how the identity of the shop $\mathcal{I}_S$ is bound to the coin when the challenge d is generated by the shop.

User		Shop
$A_1 = g_1^{u_1 s}$, $A_2 = g_2^s$		
$m \in_R \mathcal{Z}_q$		
$D_1 = I g^{X_T m}$, $D_2 = g_2^m$	$\xrightarrow{\ D_1, D_2\ }$	$D_2 \overset{?}{\neq} 1$
	$\xrightarrow{\ u_1, A_1, A_2, A, B, (z,a,b,r)\ }$	$A \overset{?}{=} A_1 A_2, A \overset{?}{\neq} 1$
		$sig(A, B, u_1) \overset{?}{=} (z, a, b, r)$
		$d = \mathcal{H}_1(A_1, B_1, A_2, B_2, \mathcal{I}_S, \text{date/time})$
		$s_0, s_1, s_2 \in_R \mathcal{Z}_q$
		$D' = D_1^{s_0} g_2^{s_1} D_2^{s_2}$
$V = \mathcal{H}_1((D')^s / (f_2')^{ms})$	$\xleftarrow{\ d, D', f_2'\ }$	$f_2' = f_2^{s_0} g_2^{s_2}$
$r_1 = d(u_1 s) + x_1$		
$r_2 = ds + x_2$	$\xrightarrow{\ r_1, r_2, V\ }$	$V \overset{?}{=} \mathcal{H}_1(A_1^{s_0} A_2^{s_1})$
		$g_1^{r_1} \overset{?}{=} A_1^d B_1, g_2^{r_2} \overset{?}{=} A_2^d B_2$

Function Deposit: The shop deposits the payment transcripts to the mint, which checks the transcripts using the same checking equations that the shop employed during the payment protocol. If the equations hold then the mint deposits a suitable amount into the shop's account.

Function Trace: To trace the identity of the user who engaged in the payment protocol that resulted in a particular deposit, the mint contacts $\mathcal{T}$ and presents the transcripts submitted by the shop. $\mathcal{T}$ can then decrypt the information on identity using the ElGamal ciphertext tuple (D_1, D_2). *System settings* The mint chooses primes p and q such that $p - 1 = \delta + k$ for a specified constant δ, and $p = \gamma q + 1$ for a small integer γ. A unique subgroup $\mathcal{G}_q$ of prime order q of the multiplicative group $\mathcal{Z}_p$ and generators g, g_1, g_2 of $\mathcal{G}_q$ are defined. The mints secret key $X_B \in_R \mathcal{Z}_q$ is created. Hash functions $\mathcal{H}, \mathcal{H}_0, \mathcal{H}_1, \cdots$, from a family of correlation-free one way hash functions are defined. The mint published $p, q, g, g_1, g_2, (\mathcal{H}, \mathcal{H}_0, \mathcal{H}_1, \cdots)$ and its public keys $h = g^{X_B}, h_1 = g_1^{X_B}, h_2 = g_2^{X_B}$. The public key of the trustee, $\mathcal{T}$, of the form $f_2 = g_2^{X_T}$ is also published, where $X_T \in_R \mathcal{Z}_q$. Note that $\mathcal{T}$ should be a distributed entity to reduce the level of trust placed on it.

The mint associates the user with the identity $I = g_1^{u_1}$, where $u_1 \in_R \mathcal{G}_q$ is generated by the user such that $g_1^{u_1} g_2 \neq 1$. The user is expected to prove the knowledge of discrete logarithm of I w.r.t. g_1. The user computes $z' = h_1^{u_1} h_2 = (I g_2)^{X_B}$.

The Efficient 3-Pass Password-Based Key Exchange Protocol with Low Computational Cost for Client[1]

Hyoungkyu Lee[1], Kiwook Sohn[2], Hyoungkyu Yang[3], and Dongho Won[1]

[1]Information and Communications Security Laboratory
School of Electrical and Computer Engineering, Sungkyunkwan University
300 Chunchun-dong, Jangan-gu, Suwon, Kyunggi-do, Korea.
hklee@dosan.skku.ac.kr, dhwon@dosan.skku.ac.kr
[2] Department of Coding Technology
Electronics and Telecommunications Research Institute
161 Kajong-dong, Yusong-Gu, Taejon, Korea
kiwook@etri.re.kr
[3] Division of Computer Science, Electronics, and Industrial Engineering
Kangnam University, San 6-2, Kugal-ri, Gihung-up, Yongin, Kyunggi-do, Korea
hkyang@kns.kangnam.ac.kr

Abstract. We propose the efficient password-based key exchange protocol, which resists against dictionary attack mounted by a passive or active adversary and is a 3-pass key exchange protocol, whereas existing protocols are 4-pass or more. Thus, considering network traffic, it will be able to reduce the total execution time in comparison with other several schemes. Especially, from the view point of the client's computational cost, our protocol is suitable for mobile communications. It is because we can reduce the modular exponentiation of client (or mobile) in comparison with other several password-based protocols. Besides, the proposed scheme has the characteristics of perfect forward secrecy, and resists against a known key attack. It also offers resistance against a stolen verifier attack as A-EKE, B-SPEKE, and SRP. Finally, two parties involved in protocol are able to agree on Diffie-Hellman exponential g^{xy} in the proposed scheme.

1 Introduction

Generally, a public key cryptosystem is more convenient than a secret key cryptosystem in key establishment, but it is not easy to use due to the unmemorizable secret key or the difficulty of public key infrastructure establishment. To solve these problems, Bellovin and Merrit proposed the password-based protocol, which is Encrypted Key Exchange (EKE) [1], that allows both parties to share a common key by using only password. Later, several schemes have followed [2-6]. These password-based protocols are not a limited scheme in comparison with public key

---

[1] This work was supported by KSEF (Korea Science and Engineering Foundation) under project 97-01-13-01-05

scheme using certification. Since users can accomplish their purpose (eg, key exchange) by using only passwords, they don't need to use certification and long secret key. However, these password-based protocols must be immune to password-guessing attacks such as a dictionary attack, where the dictionary means a list of probable passwords, because an adversary can use the dictionary to guess the correct password by on-line or off-line. In most cases, it is difficult to detect off-line password-guessing attacks since an adversary tries to verify guessed values using publicly available information. However, on-line password-guessing attacks are easily thwarted by counting access failures since an adversary use the guessed value whenever he logs in. Thus, our main concern is to protect passwords from off-line password-guessing attacks. The password-based key exchange protocol is divided into two branches [5]: One is a plaintext-equivalent mechanism [1,3], the other is a verifier-based mechanism by *secret public key* [2,4-6, 10]. In the verifier-based mechanism, password and verifier correspond to private and public key. Recently, in aspect of computational cost, Kwon and Song [6] proposed more efficient protocol than other several protocols such as the combined B-SPEKE and the optimized SRP [4, 5]. But, the total execution time of protocol depends on the number of protocol steps rather than computational cost. The protocol proposed in [6] is a 4-pass key exchange protocol as well as those in [4,5]. In this paper, we propose a 3-pass password-based key exchange protocol whose computational cost is similar to [6]. Also, the proposed scheme doesn't require a safe prime or a primitive root to thwart a partition attack or a subgroup confinement attack in comparison with several protocols proposed in [1-4]. Therefore, we suggest the use of large prime-order subgroup for efficiency. This paper examines security and efficiency of protocol. In section 2, we describe and analyze several protocols and standard techniques based on password. In section 3, the new key exchange scheme is proposed. In section 4, we examine the security of proposed scheme, while we examine the efficiency of proposed scheme in section 5. Finally, section 6 is conclusion.

2   Historical Review of Existing Protocols

In this section, we review several password-based key exchange protocols. In such protocols, the client performs key exchange by using only the secret password or something equivalent to it. On the other hand, server comes to perform key exchange by two mechanisms [5]: As we mentioned earlier, one is the plaintext-equivalent mechanism that both client and server can access to the same secret password or something equivalent to it. Thus, in the plaintext-equivalent mechanism, there is no difference of knowledge between the client and the server. The other is a verifier-based mechanism where the verifier has similar properties to a public key. That is, the verifier is easily computed from the password whereas obtaining the password from the verifier is computationally infeasible. In the verifier-based mechanism, the server cannot access to the secret password. The server can access to only the verifier. On the other hand, the client can access to both password and verifier. Thus, there is the difference of knowledge between the client and the server. Of course, the verifier is kept secret by the server.

As shown in Table 1, we summarize the characteristics of existing protocols. In Table 1, f() and h() represent public one-way functions stretching or hashing the secret exponent such as a password. Generally, only client knows the plain password P. In SRP and KS, server stores the salt s or s⊕P, and sends it to client.

Table 1. The Characteristics of Existing Protocols

Protocol	Client	Server	Session Key	Modular p	Authentication Mechanism
			G : Primitive Element x, y, u : Random Number		P: Password s : Salt
DH-EKE	$h(P)$	$h(P)$	g^{xy}	safe prime	plaintext
SPEKE	$h(P)$	$h(P)$	$h(P)^{2xy}$	safe prime	plaintext
A-EKE	$S(P), V(P)$	$V(P)$	g^{xy}	safe prime	verifier
B-SPEKE	$P, h(P)$	$h(P), g^P$	$h(P)^{2xy}$	safe prime	verifier
SRP	$k=h(salt,P)$	$s, g^{h(s,P)}$	g^{xy+uky}	safe prime (recommendation)	verifier
KS	$f(salt,P)$	$s\oplus P, g^{f(s,P)}$	g^{xy}	non-smooth prime	verifier

In 1992, DH-EKE is introduced by Bellovin and Merritt [1]. It is a combination of asymmetric and symmetric cryptography that allows two participants to share a common key by plaintext-equivalent mechanism. In DH-EKE, two participants encrypt their key material with symmetric cryptosystem using the shared hash value of secret password as a key. Then, they have come to share the Diffie-Hellman exponential g^{xy}. Besides, even if legitimate users pick bad passwords, a reasonable level of security is maintained. However, this scheme has several shortcomings. That is, it requires a safe prime and primitive root due to the use of symmetric cryptosystem such as DES. Also, it requires the careful choice of symmetric cryptography. ; Otherwise, it is vulnerable to information leakage or partition attack [1]. In 1996, Jablon proposed a noble scheme that is called SPEKE (Simple Password Exponential Key Exchange) [3]. In contrast with DH-EKE, SPEKE does not require a symmetric cryptography. In this scheme, Jablon proposed that the base of group used in Diffie-Hellman key exchange should be chosen as a function of password. That is, $h(P)$ is used as the base of group where $h()$ is a hash function and P is a password. However, it also requires a safe prime to thwart the subgroup confinement attack which middle person confine key materials such as $g^x \bmod p$ in Diffie-Hellman key exchange to the subgroup of small order [3]. Later, several verifier-based protocols including the extended versions of above two protocols were followed [2, 4-6]. These protocols were motivated by password file compromise such as a stolen verifier attack [5]. That is, the adversary obtaining verifier should still perform an off-line password-guessing attack using a dictionary to impersonate client. In this section, we go into more details about these protocols. DH-EKE is extended to A-EKE, while SPEKE is done to B-SPEKE [2, 4]. As shown in Table 1, A-EKE uses {S(P), V(P)} as a private/public key pair for digital signature [2]. In A-EKE, digital signature is used to prove client's knowledge of the password. Similarly, B-SPEKE authenticate the client by a second Diffie-Hellman method instead of digital signature (see section 3 in [4]). In 1998, SRP was introduced by Wu [5]. Compared with B-SPEKE, SRP has the advantage of being more flexible in that it performs the key exchange without the base chosen as a function of password. Furthermore, SRP has less execution time than that of B-SPEKE. However, SRP cannot agree on the correct Diffie-Hellman exponential g^{xy} due to the use of another random integer u for security (see section 3 in [5]). The agreed exponential of SRP is g^{xy+uky} as shown in Table 1. Recently, Kwon and Song proposed a new scheme, denoted by KS in Table 1, for the correct Diffie-Hellman exponential g^{xy} [6]. Compared with the related protocols, their protocol is simple and efficient. Unlike A-EKE and B-SPEKE, both SRP and KS use the salt. Especially, KS uses s⊕P which is exclusive-or of password and salt whereas SRP uses salt transmitted as a plaintext. However, we have to note that such use of salt still can not preclude an adversary from obtaining information on password by stolen verifier since the entropy of password is low.

3   Proposed Scheme

In this section, we introduce a new scheme reduced as a 3-pass without compromise of security and performance. The user name *Alice* and *Bob* correspond to *client* and *server*, respectively. $h()$ is the hash function, and α is the element with large prime order q in $GF(p)$. Let $f()$ be stretch function which extends the bit length of pre-image to that of secure exponent. Also, we omit *mod p* for simplicity when we describe our equation. We summarize the details of a notation in Table 2.

Table 2. The Details of a Notation in Proposed Scheme

ID	User's name or address
p	Large prime modular
q	Large prime factor of p-1
g	Primitive element in GF(p)
α	Element of order q in GF(p)
P	Password
v	Verifier stored in server's storage
x, y, r	Randomly chosen integers
e, t	Alice's transmissions for key exchange and authentication
h()	Public hash function
f()	Public stretch function
K	Session key

The key exchange and authentication are performed by verifier-based mechanism using extra Nyberg-Rueppel one-pass scheme [7]. In this paper, the terms, *extra* means conversion of a public key scheme into a password-based scheme. Thus, in Nyberg-Rueppel one-pass scheme, Bob's secret key is removed and Alice's secret key is replaced by $f(P)$. First, we start with password setup steps. To establish a password P with Bob, Alice computes $v=\alpha^{f(P)}$. Bob stores v with *ID* as Alice's verifier. For authentication and key agreement, Alice selects two random integer x and r, where $x, r \in_R \{1, 2, \ldots, q-1\}$, and computes $e= \alpha^{-r}$ and $t=r+e \cdot f(P) \bmod q$. Then, Alice sends to Bob (e, t) with *ID*, initializing the key exchange.

$$\text{Alice} \Rightarrow \text{Bob} : \text{ID}, (e, t) \tag{1}$$

Bob looks up Alice's verifier, and obtains α^x by computing $(g^t \cdot v^e \cdot e)$ from message (1). Then, he chooses a random integer, $y \in_R \{1, 2, \dots, q\text{-}1\}$, and computes $K = \alpha^{xy}$ as a session key. Bob sends α^y with $h(K, e, t)$ to Alice.

$$\text{Bob} \Rightarrow \text{Alice} : \alpha^y, h(K, e, t) \tag{2}$$

Alice can also compute $K = \alpha^{xy}$ and a corresponding hash image from message (2). Then, she checks if the hash images match each other. If it is satisfied, Alice sends $h(K, \alpha^y)$ to Bob.

$$\text{Alice} \Rightarrow \text{Bob} : h(K, \alpha^y) \tag{3}$$

Bob can also compute the hash image corresponded with Alice's message (3). If the hash images match each other, then both parties agree on the Diffie-Hellman exponential, i.e. α^{xy}. Simple description of each step is in Figure 1.

	Alice		Bob
1	$e = \alpha^{x \cdot r}$		
2	$t = r + e \cdot f(P)$	$\xrightarrow{\text{ID}, e, t}$	
3			Look up : (ID, v)
4			$\alpha^x = \gamma^t \alpha^x_?$
5			α^y
6			$K = (\alpha^x)^y$
7	$K = (\alpha^y)^x$	$\xleftarrow{\alpha^y, h(K, e, t)}$	$h(K, e, t)$
8	Verify $h(K, e, t)$		
9	$h(K, \alpha^y)$	$\xrightarrow{h(K, \alpha^y)}$	
10			Verify $h(K, \alpha^y)$

Fig. 1. The Proposed 3-Pass Password-Based Key Exchange Protocol

4   Security Analysis

A password-based protocol must not leak the information about the passwords. The leakage of such information may cause the *verifiability* that allow an adversary to guess the password. Thus, an important requirement of password-based protocol is that it must be immune to password-guessing attacks such as a dictionary attack. It can be realized by a protocol where its transmissions offer *unverifiability* for guessed passwords. Typically, password guessing attack proceeds as follows (Also, see [9]) :

- A password is stored in a computer file as the image of an unkeyed hash function. When a user logs on and enters a password, it is hashed and the image is compared to the stored value. An adversary can take the hash values of guessed passwords using dictionary which is a list of probable password. Subsequently, if an adversary compare this to the list of true encrypted passwords, he or she may find the correct password.

The proposed protocol is worked by secret public key referred to as verifier. It resists against dictionary attack mounted by passive or active adversary. Also, by Diffie-Hellman problem, the proposed scheme has the characteristics of perfect forward secrecy and is protected from a known key attack such as Denning-Sacco attack [11, 15]. The details are as follows ;

4.1   Protection from a Passive Adversary

A passive adversary can eavesdrop all transmissions over protocol and perform off-line password-guessing attack. In the above protocol, an adversary can obtain (e, t), α^{y}, and the hash values. But, an adversary cannot learn any useful information from which he can guess the correct password.

4.2   Protection from an Active Impersonator

An active impersonator can masquerade as a legitimate user. Also, he can modify the transmissions of legitimate user. For description, let A^* and B^* be impersonator with the guessed value P^* for Alice and Bob respectively. First, A^* can send to Bob (e, t) where $e=\alpha^{y\cdot r}$, $t=r+e\cdot f(P^*) \bmod q$. Then, Bob computes $\alpha^{(x+(f(P^*)-f(P))x)y}$ as a session key. Thus, A^* cannot obtain the correct Diffie-Hellman exponential α^{xy}. Also, A^* cannot perform the off-line password-guessing attack since he cannot know $y \bmod q$ by discrete logarithm problem. In a similar way, B^* is also detected easily since he cannot generate $h(\alpha^{xy}, e, t)$ of message (2).

4.3   Resistance against a Stolen Verifier Attack [5]

In the proposed scheme, Alice has to know $f(P)$ to agree on Diffie-Hellman exponential. Thus, even if an active adversary can obtain verifier v, he cannot impersonate Alice. That is, to impersonate Alice, an adversary has to try to guess the effective $f(P)$ from $\alpha^{f(P)}$. Note that Bob is always a possible enemy by his own verifier.

4.4   Resistance against a Partition Attack and a Subgroup Confinement Attack [1, 3]

To thwart a partition attack [1], the proposed scheme was designed not to leak information about the password. It is possible because our protocol is generalized in such a group that has the fixed base without other cryptographic technology. Also, a subgroup confinement attack [3] can be easily thwarted by the use of large prime-order subgroup. That is, an adversary cannot confine transmissions of each participant to the subgroup with small order. Thus, in this paper, an adversary cannot perform the password guessing attacks such as a subgroup confinement attack or a partition attack because all computations are performed over large prime-order subgroup.

4.5   Comments on Parameter

In 1978, Pohlig and Hellman introduced technique known as Pohlig-Hellman decomposition can be used to reduce the running time by decomposing the original large discrete log problem into a number of smaller such sub-problems [14]. By pohlig-Hellman decomposition, given a group GF(p) of order $p-1$, primitive element g and g^x, the running time of finding x becomes $O(q^{1/2})$ where $p=qw+1$, q is a large prime factor and w is the product of smooth factors.

Oorschot et al indicated that the basic Diffie-Hellman key exchange had a potential drawback [13]. They showed that the shared key of both participants would become g^{xyq} by middle person attack. Thus, middle person attacker could easily find the session key, $K=g^{xyq}$, by exhaustive search. As we mentioned above, this attack is referred as subgroup confinement attack. Consequently, it motivates the use of subgroups with large prime-order. The use of prime-order subgroup has the advantage in constructing prime p in comparison with that of safe prime. The prime-order subgroup requires only that q is secure exponent, whereas a safe prime requires guaranteeing a large prime divisor $q=(p-1)/2$. Thus, compared with existing password-based protocols, our protocol has the flexibility in constructing p and precludes a subgroup confinement attack. See also [12] for the prime-order subgroup.

5   Efficiency

The proposed scheme has two important advantages in light of efficiency. One is the number of protocol steps and the other is the low computational cost for client. The former is described by the number of message exchanges required between the parties involved in the protocol. The proposed scheme is reduced to three steps. Thus, compared with the related schemes [4, 5, 6], the proposed scheme can be more suitable to remotely distributed environment. It is because the total execution time of protocol depends on the number of protocol steps rather than computation time. Consequently, the proposed scheme will be able to reduce frequent disconnections or transmission errors by network traffic. In light of each participant's computational cost, it is efficient too. Typically, the computational cost of protocol can be summarized as the number of modular exponentiation computed by each participant. In our protocol, the number of modular exponentiations required to the client is two, to the server four. On the other hand, in [6], the number of modular exponentiations required to the client is four, to the server three. Considering above results, we can also apply our protocol to mobile communication environments. Generally, mobile communication environments require a low computational cost for a mobile device (see section 2 in [8]). That is, since mobile devices are made small and light to be portable, they usually come to have comparatively fewer resources and computational power than those of server or base station. Also, though the total execution time of protocol is much related with the number of protocol steps, we try to represent the total amount of execution time in parallel as shown in [6]. It is as follows: For E(client : server), $E(g^{x \cdot r} :)$, $E(: g')$ and $E(:v^c)$ between step 1 and 2, and $E(:g^y)$, $E(:(g^x)^y)$ and $E((g^y)^x:)$ between step 2 and 3. Thus, the proposed scheme requires six for three steps whereas KS proposed in [6] five for four steps. These facts are well described in Table 3. We exclude the execution time of A-EKE from Table 3 due to dependence on digital signature scheme.

Table 3. The Amount of Computational Cost

Protocols	The Number of Modular Exponentiations		Total Execution time (in Parallel)	The Number of Passes
	Client	Server		
Combined B-SPEKE	3	4	7	4
Optimized SRP	3	3	4	4
KS	4	3	5	4
Proposed scheme	2	4	6	3

6   Conclusion

Several password-based key exchange protocols have been presented ever since DH-EKE is introduced by Bellovin and Merritt. Such password-based protocols were designed to have strong authentication without depending on external infrastructure. In this paper, we described the history and properties of such password-based protocols, and proposed a new 3-pass password-based key exchange protocol which both participants could agree on the correct Diffie-Hellman exponential g^{xy}. The proposed scheme has excellent performance without compromise of security in comparison with existing protocols. As we mentioned earlier, the proposed scheme can be suitable for mobile communications due to the low computational cost of client. From the viewpoint of the number of protocol steps, we proposed more efficient 3-pass scheme than several 4-pass schemes such as [2, 4-6]. Besides, the proposed scheme requires less restriction against group parameter in comparison to the related schemes such as A-EKE or B-SPEKE.

References

1. S. BELLOVIN and M. MERRITT, 'Encrypted key exchange: password-based protocols secure against dictionary attacks', IEEE Comp. Society Symp. On Research in Security and Privacy, 1992, pp. 72-84
2. S. BELLOVIN and M. MERRIT, 'Augmented encrypted key exchange: a password-based protocol secure against dictionary attacks and password file compromise', ACM Conf. Comp. And Comm. Security, 1993, pp. 244-250
3. D. JABLON, 'Strong password-only authenticated key exchange', ACM Comput. Commun. Rev., 1996, 20, (5) pp. 5-26
4. D. JABLON, 'Extended password key exchange protocols', WETICE Workshop on Enterprise Security, 1997
5. T. WU, 'Secure remote password protocol', Internet Society Symp. Network and Distributed System Security, 1998
6. T. Kwon and J. Song, 'Secure agreement scheme for g^{xy} via password authentication', Electron. Lett., 1999, 35, pp. 892-893
7. K. Nyberg and R. Rueppel, 'A new signature scheme based on DSA giving message recovery', Proc. 1st ACM Conf. On Comput. Commun. Security, November, 1993, pp. 58-61
8. C. Boyd and A. Mathuria, 'Key Establishment Protocols for Secure Mobile Communications: A Selective Survey', Information Security and Privacy, LNCS 1498, Springer-Verlag, 1998, pp. 344-355
9. A.J. Menezes, P.C. vanOorschot and S.A. Vanstone, 'Handbook of Applied Cryptography', CRC Press, 1996, pp. 388-397.
10. L. Gong, M.A. Lomas, R. Needham, and J. Saltzer, 'Protecting poorly chosen secrets from guessing attacks', IEEE Journal on Selected Areas in Communications, Vol. 11, No. 5, June, 1993, pp. 648-656
11. M. Steiner, G. Tsudik, and M. Waidner, 'Refinement and Extension of Encrypted Key Exchange', Operating Systems Review, Vol. 29, No. 3, July, 1995, pp. 22-30
12. R.Anderson, S.Vaudenay, 'Minding your p's and q's', Advances in Cryptology-Asiacrypt'96, Springer-verlag, LNCS 1163, 1996, pp.15-25
13. P.C. vanOorschot and M.J.Wiener, 'On Diffie-Hellman Key Agreement with Short Exponents', In Advances in Cryptology- Eurocrypt'96, Springer-verlag, LNCS 1070, 1996, pp.332-343
14. S.C. Pohlig and M.E.Hellman, 'An improved algorithm for computing logarithms over GF(p) and its cryptographic significance', IEEE Trans. Inform. Theory, IT-24(1), 1978, pp.106-110.
15. D.E. Denning and G.M. Sacco, 'Timestamps in key distribution protocols', Communications of the ACM, Vol. 24, No. 8, 1981. pp. 533-536

A 2-Pass Authentication and Key Agreement Protocol for Mobile Communications

Kook-Heui Lee[1], Sang-Jae Moon[1], Won-Young Jeong[2], and Tae-Geun Kim[2]

[1] School of Electronic and Electrical Eig., Kyungpook National Univ.
1370, San-Kyuk Dong, Pook-Gu, Taegu, 702-201, Korea
lkh@palgong.knu.ac.kr, sjmoon@ee.knu.ac.kr
[2] Access Network Research Lab., Korea Telecom
17 Woomyun-Dong, Suhcho-Gu, Seoul, Korea
wyjeong@kt.co.kr, tkimm@kt.co.kr

Abstract. We present a new 2-pass authentication and key agreement protocol for mobile communications. The protocol solves the weaknesses of the PACS and Zheng's 1.5-move protocol for air-interface of mobile systems. The paper outlines the new protocol, examines its various aspects, and compares them to those of other 2-pass protocols of the PACS and the 1.5-move protocol.

1 Introduction

The demand for a variety of value-added services, such as electronic commerce, continues to grow in the mobile communications. It is likely that public key based techniques will be rapidly employed in the implementation of security services for mobile communications due to its facility of key management and variety of security services including digital signature.

To provide authentication and key establishment between two involving entities in a mobile communication system, several public key based 2-pass protocols have been presented, including the PACS protocol [1] and the 1.5-move protocol [2]. The PACS is a standard adopted by ANSI for personal communications systems. It is based on Bellcore's WACS (wireless access communications system) [3] and on Japan's PHS (personal handyphone system) [4]. The 1.5-move protocol was proposed by Zheng. Its interesting feature is that most operations for mobile user device can be done in off-line. These schemes were designed to meet certain requirements that occur specifically in a mobile communication environment. These requirements include user anonymity, shortage of radio bandwidth, and computational limitation of mobile user device.

Most public key based authentication and key agreemet (hereinafter AKA) protocols use either 2-pass or 3-pass scheme to exchange messages. The 2-pass method includes advantages such as high bandwidth efficiency and rapid connection achievement, while the 3-pass scheme can provide more variety of security services. It would be preferable if the advantages of both schemes could be combined while satisfying more requirements.

JooSeok Song (Ed.): ICISC'99, LNCS 1787, pp. 156–168, 2000.
© Springer-Verlag Berlin Heidelberg 2000

This paper presents a 2-pass public key based AKA protocol for mobile communications. The protocol provides non-repudiation of the mobile user to the base station, yet satisfies the merits of the 2-pass scheme. Various aspects of the protocol are examined and then compared with those of other existing 2-pass solutions.

2   Properties for Mobile Communications

The security features required for a protocol providing authentication and key establishment between a mobile user and base station have been identified in previous literature. Horn and Preneel [5] recommended some goals for wireless communication, while the European ASPeCT project [6] extended this to initialize a mechanism for enabling payment of the value-added service.

Since this study is primarily interested in achieving AKA between a mobile user and a base station, some of these properties associated with just payment initialization have been ignored.

Definitions of security services for the AKA are well specified in the American National Standard X.9.63-199x[7], and also described in [5, 6, 8]. For this study, the properties requiring consideration in the development of a public key based protocol for mobile communications were as follows.

S1 Entity Authentication: The assurance provided to entity U that entity V has been involved in a real-time communication with entity U.

S2 Key Authentication: For a clearer description, this property is classified into two categories; implicit key authentication and explicit key authentication.

- **Implicit key authentication:** The assurance that the corresponding key is possibly computed by only the involving entities.
- **Explicit key authentication:** The assurance that the corresponding key is possibly computed by only the involving entities, and the corresponding key is actually computed by only the involving entities.

S3 Key Agreement: A key agreement scheme is a key establishment scheme in which the keying data established is a function of the contributions provided by both entities in such a way that neither party can predetermine the value of the keying data.

S4 Assurance of Key Freshness: The assurance provided to each entity that a new establishing key is randomly fresh so that attacks based on use of information associated with compromised keys or previous data are prevented

S5 Anonymity of Mobile User: This property provides confidentiality of the mobile user's location and movement. This property is also one of the most important requirements in mobile communications

S6 Non-repudiation of User: This service guarantees undeniable evidence related to the user charge or important data. The significance of this service is increasing as a variety of value-added services are expected

An AKA protocol must satisfy certain requirements that occur specifically in a mobile communication environment, including bandwidth limitation on the air interface and computational performance limitation of the mobile user device. Accordingly, the following efficiency properties need to be fully considered when building a protocol.

R1 Minimum Number of Passes: To reduce a latency time, the number of message exchanges required between entities should be minimal

R2 Efficient Usage of Bandwidth: Due to the high cost of bandwidth on the air interface, the total number of bits transmitted should be kept as small as possible

R3 Low Computational Load: The computational load for each entity should be small. In particular, since the on-line performance of a mobile user device is limited, it is desirable to reduce the load at the mobile side by employing off-line computation.

3   The New Proposal

3.1   Protocol Description

Throughout this paper the application of an elliptic curve cryptosystems [9–11] and the use of the notation in Table 1 are assumed. For clarity of exposition, optional data are omitted.

Table 1. Notation and its meaning.

Notation	Meaning
ID_E	An identifier of the entity E
$RT(TS)$	A real time value(a time stamp)
$hash(x)$	The result of hashing to input x
x_E	A secret key of the entity E
P_E	A public key of the entity E, $P_E = x_E \cdot G$
K_{MB}	The common session key between M and B
G	A generator point of the elliptic curve
$Cert_E$	A certificate of the P_E
$Sig_E\{x\}$	A value x signed by the entity E
r_E	A random number generated by entity E
$x\|y$	Concatenation of x and y
$E_K\{x\}(D_K\{x\})$	The symmetric encryption (decryption) of x using key K
M	A mobile station(user)
B	A base station(service provider)

Figure 1 shows a flow diagram of the ordinary operation of the protocol in which flows are faithfully relayed between two entities M and B. Points on an

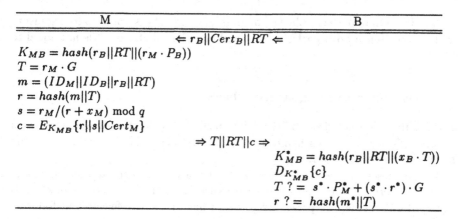

M	B

$$\Leftarrow r_B||Cert_B||RT \Leftarrow$$

$K_{MB} = hash(r_B||RT||(r_M \cdot P_B))$
$T = r_M \cdot G$
$m = (ID_M||ID_B||r_B||RT)$
$r = hash(m||T)$
$s = r_M/(r + x_M) \bmod q$
$c = E_{K_{MB}}\{r||s||Cert_M\}$

$$\Rightarrow T||RT||c \Rightarrow$$

$K_{MB}^* = hash(r_B||RT||(x_B \cdot T))$
$D_{K_{MB}^*}\{c\}$
$T \ ? = \ s^* \cdot P_M^* + (s^* \cdot r^*) \cdot G$
$r \ ? = \ hash(m^*||T)$

Fig. 1. Flow diagram of the new protocol.

elliptic curve such as $r_M \cdot P_B$ and $x_B \cdot T$ are regarded as binary strings when involved in hashing.

The procedure of the developed protocol is executed as follows.

Prerequisites: It is assumed that the set of the elliptic curve domain parameters associated with the flow have been validated. It will also be shown that the use of the elliptic curve system is the best choice to satisfy the requirements of **R2** and **R3**.

The first pass from B to M :

- B broadcasts a random number r_B, a real-time value RT, and its public key certificate $Cert_B$, where the identifier ID_B and B's public key $x_B \cdot G$ are included in the certificate $Cert_B$.

The second pass from M to B :

- Entity M executes the following actions:
  1. Extract ID_B and P_B from $Cert_B$
  2. Generate a random number r_M
  3. Compute a common session key $K_{MB} = hash(r_B||(r_M \cdot P_B))$ and temporary key $T = r_M \cdot G$
  4. Compute r and s which serve as signature
  5. Encrypt r, s, and some other data using K_{MB}
  6. Send T and RT to B together with the encrypted message
- Entity B executes following actions:
  1. B takes T and RT from the received message and compute the common session key $K_{MB}^* = hash(r_B||RT||(x_B \cdot T))$
  2. Decrypt the received message c using K_{MB}^* and extract the M's identifier and public key from $Cert_M^*$
  3. Compute $(s^* \cdot P_M^* + (s^* \cdot r^*) \cdot G)$ and compare with the received T. It aborts the protocol if the received T is invalid
  4. Compute $hash(m^*||T)$ and compare with the decrypted r^*

If s^* is equal to 0, the public key of mobile station P_M is not used for signature verification. Hence both participants should confirm that s or s^* is not equal to 0.

3.2   Assessments on Signature Scheme

In this subsection, we describe the relationship between the presented protocol and Zheng's signcryption scheme [12,13]. The details of the signcryption scheme is explained in appendix A.

In the presented protocol, r and s is calculated in a similar way as that of Zheng's signcryption scheme. Comparing with unsigncryption process of Zheng's signcryption scheme, the presented protocol has distinctive features as follows.

1. The anonymity of the mobile user is provided:
   In the presented protocol, the base station compute the decryption key, $K_{MB} = hash(r_B||RT||(x_B \cdot T))$, without the knowledge of the mobile user's public key, P_M. However this is not applied to the Zheng's signcryption scheme as described in appendix A. It is noticeable that the disclosure of the $r_M \cdot G$ does not reduce the security of the signcryption scheme [12].

2. It has directly verifiable non-repudiation procedure:
   When the mobile user denies the fact that he signed a message, the base station forward a signature (r, s) and corresponding message m with the certificate of the mobile user, $Cert_M$, to a judge. Then the judge can settle this dispute just by verifying the followings.

$$k = s \cdot P_M + (s \cdot r) \cdot G \quad \text{and} \quad r = hash(m||k)$$

   However, Zheng assumes a trusted judge or the use of zero knowledge protocol to prevent the disclosure of x_B.

3. It prevents a key recovery attack by a judge:
   Petersen and Michels [14] showed that the judge can decrypt any further message after he once settled a dispute between the mobile user M and the base station B. In case of Zheng's signcryption scheme, the judge gets $E(\equiv u \cdot P_M + u \cdot (r \cdot G)$, where $u = s \cdot x_B \bmod q)$ after he settles a dispute. He can compute the temporary value $K_{DH} = s^{-1} \cdot E + (-r) \cdot P_B = x_M \cdot (x_B \cdot G)$. Then, for any signature (r^*, s^*) he can compute $E^* = s^* \cdot K_{DH} + s^* \cdot (r^* \cdot P_B)$ and $K^*_{MB} = h(E^*)$.
   However, the modified scheme prevents key recovery attack because the judge gets $E' = s \cdot P_A + s \cdot (r \cdot G)$ after he settles a dispute and can't compute K_{DH} with E'.

3.3   Considerations

An examination as to whether the requirements **S1** to **S6** are satisfied by the protocol follows:

- **S1:** Since M makes a signature on the random number r_B transmitted by B, the verifier B can authenticate M. While M can not authenticate B.
- **S2:** Explicit key authentication to B and implicit key authentication to M. Since only the one who knows r_M and x_M can compute K_{MB} and a signature (r, s), and moreover, the computed signature is encrypted with K_{MB}, the verifier B can assure that K_{MB} is actually computed by the signer M. Since the key K_{MB} can only be computed by a holder of the secret key x_B, the sender M can have an implicit key authentication.
- **S3:** Mutual agreement of session key between M and B. This is because the key is derived by randomly chosen numbers r_M and r_B by M and B, respectively.
- **S4:** Mutual assurance of key freshness. This is due to the fact that the session key is derived by randomly chosen number r_M and r_B by M and B, respectively. Accordingly, both entities ensure the freshness of the session key.
- **S5:** Anonymity of mobile user. Since any information on M is encrypted on the protocol, anonymity is satisfied.
- **S6:** non-repudiation of user. This property is based on using the signcryption scheme.

The above properties will be summarized in a table later when compared with the properties of the other protocols.

The following examines attack prevention:

- **Inclusion of RT in the signed part of message:** To prevent play-in-the-middle attack
- **Inclusion of ID_M in the signed part of message:** To prevent a parallel session attack
- **Inclusion of ID_B in the signed part of message:** To consider the payment protocol[5]
- **Inclusion of r_B in the signed part of message:** To preclude a time-memory trade-off attack[5]
- **Inclusion of random numbers r_M and r_B in the keying data:** To prevent a replay of an old key

The requirement problems **R1** to **R3**, which only occur in a mobile communication environment, are evaluated as follows:

- **R1:** Since the protocol has a 2-pass scheme, it has the minimum number of message exchanges
- **R2:** Since the protocol utilizes elliptic curve cryptographic algorithms, it can achieve as small a bandwidth as possible without any lack of security as compared with the RSA or ElGamal systems. The level of security assumed here is equivalent to the strength of ElGamal system having a finite field with a 1024 bit-length prime number, as shown Table 2. The length of the other parameters not specified in Table 2 are assumed to be the same as shown in Table 3.

Table 2. Parameter lengths of RSA, Rabin, ElGamal, and ECC systems.

	RSA $\mid n \mid = 1024$ $e = 2^{16} + 1$	Rabin $\mid n \mid = 1024$	ElGamal $\mid p \mid = 1024$ $\mid q \mid = 160$	ECC $E(F_q)$ $\mid q \mid = 160$
System parameter	-	-	2208	481
Public key	1041	1024	1024	161
Secret key	2048	1024	160	160
Signature(single)	1024	1024	1184	320
Encrypt.(single)	1024	1024	2048	321
Certificate (including 32 bit for identifier)	$2097 + Z^*$	$2080 + Z^*$	$2240 + Z^*$	$513 + Z^*$

Z^* : length of the common bits in each certificate type except the identifier, public key and signature of a corresponding entity.

Table 3. Length in bits of some parameters.

Parameters	ID	$RT(TS)$	Random number	Hash output
Length in bits	32	32	160	160

Based on Table 2 and Table 3, we can compute the bandwidth of the protocol as shown in Table 4. These are compared with those of the other protocols in the next section.

Table 4. Length in bits of exchanging messages of the protocol.

	$B \to M$	$M \to B$
Length in bits	$160 + 513 + Z + 32$	$161 + 32 + 320 + 513 + Z$

- **R3:** Off-line computation is desirable at the mobile user end due to limited performance. In the new protocol, M need to compute two multiplications on an EC system for computing K_{MB} and T. These two multiplications can be done in off-line state if the base station broadcast its public key information regularly and frequently. B then needs a single multiplication for K_{MB} and two multiplications for signature verification. All of these multiplications for the base station should be done in on-line state.

4   Comparison

In this section, the properties of the new protocol are compared with those of the PACS and Zheng's 1.5-move scheme. Flow diagrams of them are included in appendix B.

4.1   Security Properties

Based on the description of security properties given in section 2, the security features of all three protocols can be summarized as shown in Table 5.

Table 5. Comparison of security features.

Security features	The proposed		The PACS		The 1.5-move	
	$M \to B$	$B \to M$	$M \to B$	$B \to M$	$M \to B$	$B \to M$
Entity auth.	Y	N	Y	N	Y	N
Key auth.	YE	YI	YE	YI	YE	YI
Key agreement	Y	Y	N^1	N^1	N^1	N^1
Key freshness	Y	Y	N	N	N	N
Anonymity	Y	N	Y	N	Y	N
Non-repu.	Y	N	Y^2	N	N	N

Y: satisfied, N:not satisfied, YE:satisfied explicitly, YI:satisfied implicitly, 1: key transport scheme, 2: key revealed once the signature is publicly open

Since the PACS and the 1.5-move protocol use a value related the system time instead of a random number in generating a key, they do not satisfy the property of key freshness in strict sense, whereas the new proposal does. In addition, the new protocol utilizes a key agreement scheme, whereas the PACS and the 1.5-move protocol use a key transportation scheme. In general, a key agreement scheme can prevent either entity from generating a weakened or deliberate key.

4.2   Weaknesses

As shown in Table 5, some of partly satisfied properties, if put together, can cause some weaknesses as follows:

- **Key is revealed if the signature is open:** The PACS generates a signature, and uses itself as a session key. The key can play the role of the signature between two involving entities, yet not publicly as a general signature. This constraint creates a limitation in a variety of applications, including the initialization of the payment mechanism for value-added service.
- **Impersonation of the mobile user by an inside attacker is possible:** This attack was introduced by Xu and Wang [15] and was applied to Beller-Chang-Yacoby protocol [16]. Zheng's 1.5-move protocol has a severe weakness that an inside attacker who knows the secret key of the base station, x_B, can impersonate the mobile user without the knowledge of the mobile user's secret key, x_M. Authenticity of the mobile user is utterly depend on the fact that only the one who knows the mobile user's secret key can create the value, $P_B^{x_M + r_M} \bmod p$. However, inside attacker can also create this value, $(P_M \cdot g^{r_M})^{x_B} \bmod p$.

The presented protocol does not have a weakness up to our knowledge.

4.3   Bandwidth Efficiency

The lengths in bits of exchange messages were calculated for each protocol to compare their bandwidth efficiency. As small a bandwidth occupation as possible is required for a mobile communication environment. By adopting elliptic curve cryptographic systems(see Table 2), a high bandwidth efficiency and lower computational complexity can be achieved compared with the RSA or ElGamal cryptographic systems. It was assumed that each protocol here employed an elliptic curve cryptographic system with a generator of the order of 2^{160}.

Table 6 shows the bandwidth efficiency of each protocol, and the comparisons are based on the statistics used in Table 3. The proposed protocol had slightly longer exchange messages than the PACS. This is due to the usage of a random number and DH key exchange scheme variant, accordingly, the weaknesses of the PACS can be solved by adding these components. Since Z can include a version number, issuing party name, or some other information, there can be a slight difference in total length. The 1.5-move protocol has the shortest exchange message but it can not satisfy the non-repudiation of user. If the signature scheme were assumed for the 1.5-move protocol, total length of exchange message is modified into $1790 + 2Z$.

Table 6. Bandwidth efficiency[bits].

	The proposed	The PACS	The 1.5-move
$B \rightarrow M$	$705 + Z$	$545 + Z$	$545 + Z$
$M \rightarrow B$	$1026 + Z$	$1090 + Z$	$925 + Z$
Total	$1731 + 2Z$	$1635 + 2Z$	$1470 + 2Z$

In Table 6, it is assumed that the PACS uses ElGamal encryption of elliptic curve cryptographic system (see appendix C.3) instead of Rabin[17] encryption. If the PACS protocol uses Rabin encryption as described in [1], the length of the exchanging messages is much longer than that in Table 6.

4.4   Computational Load

The computational load was calculated in bit multiplications for each protocol. It is assumed that each protocol employed the same elliptic curve cryptosystem with a generator of the order of 2^{160}.

Most of the computation time was mainly spent on signature generation/verification, public key encryption/decryption, and multiplication for session key generation. Accordingly, the computational load related to hashing, symmetric key encryption, some additions and scalar multiplication were ignored if they were minor load. In addition, the load for certificate verification is also ignored as this is common to all protocols. The method of calculating the computational load of each associated transformation is described in appendix C.

Table 7 summarizes and compares the computational loads for each proto-col. The computational load for session key generation in Table 7 considers the multiplications included in $r_M \cdot G$, $r_M \cdot P_B$, and $x_B \cdot T$. The multiplications for generation/verification of tag (see appendix A) in the 1.5-move protocol are included in 'etc.'.

Table 7. Computational load [bit multiplication].

		The propsed	The PACS	The 1.5-move
	Signature generation	-	[36M]	-
	Public key encryption	-	[72M]	-
M	Session key generation	[72M]	-	[72M]
	etc.	-	-	[36M]
	Subtotal	[72M]	[108M]	[108M]
	Public key decryption	-	36M	-
	Session key generation	36M	-	36M
B	Signature verification	72M	72M	-
	etc.	-	-	36M
	Subtotal	108M	108M	72M

[·] : off-line computational load

The proposed protocol has the least computational load for mobile user. Even though reducing the computational load, which can be done in off-line state, can not decrease the latency time, it can do the power consumption of the mobile user device.

Considering only the computational load for the base station, the 1.5-move protocol has the least one. However, the 1.5-move protocol did not consider the non-repudiation of the user at all. If the signature scheme is applied to the 1.5-move protocol, the computational loads are modified into [144M] for the mobile user and 144M for the base station.

5   Conclusion

The new protocol for authentication and key establishment was presented, which was a 2-pass public key based scheme designed specifically for mobile communi-cations. ¿From the viewpoint of security, it was improved by solving the weak-nesses of the PACS and the 1.5-move scheme. It also achieves higher bandwidth efficiency and less computational load at mobile user device.

Acknowledgements

We are grateful to Colin Boyd and Dong-Gook Park of the Queensland University of Technology, Australia, for their helpful discussion.

References

1. JTC(Air)/94.12.15-119Rc, *Personal Communication Services, PACS Air Interface Specification*, PN3418.
2. Y. Zheng, "An authentication and security protocol for mobile computing," *proceedings of IFIP*, pp.249-257, 1996.
3. Bellocre, *Generic Criteria for Version 0.1 : Wireless Access Communications Systems (WACS)*, Technical Reference, TR - INS - 001313, Issue 1, Oct. 1993.
4. Reserch & Development Center for Radio Systems, *RCR STD-28, Personal Handyphone System, Version 1*, Dec. 1993.
5. G. Horn and B. Preneel, "Authentication and payment in future mobile systems," *Computer Security - ESORICS 98, LNCS 1485*, pp. 277-293, 1998.
6. G. Horn, K.M. Martin, and C.J. Mitchell, "Authentication protocols for mobile network environment value-added services," draft, available at http://isg.rhbnc.ac.uk/cjm/ChrisMitchell.htm
7. ANSI X9.63-199x, *Elliptic curve key agreement and key transport protocols*, Working draft, July 1998.
8. A.J. Manezes, P.C. van Oorshot, and S.A. Vanstone, *Handbook of applied cryptography*, CRC Press, 1996.
9. V. Miller, "Uses of elliptic curves in cryptography," In H. C. Williams, editor, *Advances in Cryptology – Crypto '85, LNCS 218*, pp. 417-426, 1985.
10. N. Koblitz, "Elliptic curve cryptosystems," *Mathematics of Computation*, vol. 48, no. 177, pp. 203-209, 1987.
11. A. J. Menezes, *Elliptic Curve Public Key Cryptosystems*, Boston, MA: Kluwer Academic Publishers, 1993.
12. Y. Zheng, "Digital signcryption or how to achieve cost (signature & encryption) << cost (signature) + cost (encryption)," *Advances in Cryptology–CRYPTO '97, LNCS 1294*, pp. 165-179, 1997.
13. Y. Zheng and H. Imai, "Efficient signcryption schemes on elliptic curves," Proceedings of the IFIP 14th International Information Security Conference (*IFIP/SEC'98*), 1998.
14. H. Petersen and M. Michels, "Cryptanalysis and improvement of signcryption schemes," *IEE Proc. Comput. Digit. Tech.*, vol. 145, no. 2. 1998.
15. S. Xu and X. Wang, "Cryptanalysis and two authentication and key distribution protocols in portable communications systems," IEEE International Symposium on Information Theory and its Applications (*ISIST'96*), vol. 1. pp.347-350, 1996.
16. M. J. Beller, L. F. Chang, and Y. Yacobi, "Privacy and authentication on a portable communications system," *IEEE GLOBECOM'91*, pp. 1922-1927, 1991.
17. M.O. Rabin, *Digitalized Signatures and Public-Key Functions as Intractable as Factorization*, MIT Lab. Comp. Sci., TR-212, Jan. 1979.

A   Signcryption Scheme

Zheng and Imai [13] specify signcryption schemes on elliptic curve. A signcryption scheme is a cryptographic method that fulfills both the functions of secure encryption and digital signature in a logical single step, but with a computational load smaller than that required by traditional signature-then-encryption. Figure 2 illustrates the signcryption scheme on elliptic curve. We assume the same notations as given in Table 1.

Signcryption of m by M the sender	Unsigncryption of (c, r, s) by B the recipient
$(k_1 \| k_2) = hash(r_M \cdot P_B)$	$u = s \cdot x_B \bmod q$
$c = E_{k_1}(m)$	$(k_1 \| k_2) = hash(u \cdot P_M + (u \cdot r) \cdot G)$
$r = KH_{k_2}(m, blind\_info) \Rightarrow c, r, s \Rightarrow$	$m = D_{k_1}(c)$
$s = \dfrac{r_M}{r + x_M} \bmod q$	$KH_{k_2}(m, blind\_info) = r$

Fig. 2. Implementations of signcryption on elliptic curve.

where KH means a keyed one-way hash function. The *blind_info* in the computation of r may contain the public keys or public key certificates of both A and B.

B   Flows of the PACS and the 1.5-Move Protocols

The PACS and the 1.5-move protocol are both described using the notation in Table 1.

B.1   PACS Protocol

The PACS protocol [1] is one of the North American PCS standard system. The flows of the protocol can be depicted as follows.

1. $B \to M : Cert_B \| RT$
2. $M \to B : Enc_{P_B}\{K \| ESN \| TID_M \| RT\} \| E_K, \{Cert_M\}$

where Enc_{P_B} denotes a public key encryption. TID_M and ESN denote a temporal identity of user and a serial number of user device, respectively. The session key is given by $K = Sig_M\{RT \| ID_B \| TID_M \| ESN\}$.

B.2   1.5-Move Protocol

Zheng's 1.5-move protocol [2] was designed for authentication and security for mobile computing. The flows of the protocol can be described as follows.

1. $B \to M : Cert_B \| TS$
2. $M \to B : c_1 (\equiv g^{r_M} \bmod p) \| c_2 (\equiv G(P_B^{r_M} \bmod p) \oplus (K \| TS \| Cert_M \| tag)),$

where $tag = hash(K \| TS \| Cert_M \| (P_B^{x_M + r_M} \bmod p))$ and $G(\cdot)$ is a cryptographically strong pseudo-random number generator.

C   Computational Load in Bit Multiplications

C.1   Single Exponentiation over a Finite Field

To compute single exponentiation of M^d mod n, where n, M ,and d are 160-bit numbers. On average the exponentiation requires 160 squarings and 80 multiplications. Accordingly, the number of bit multiplications to compute a single exponentiation is as follows.

$$number\ of\ bit\ multiplications = (1 + 0.5) \times 160 \times (160 \times 160)$$
$$= 6M$$

C.2   Elliptic Curve DSA Scheme

$k \cdot P$ is computed to execute a signature process, where P is a generator point of an EC system, and k is a randomly chosen integer. The lengths in bits of P and k are each 160. The multiplication requires 160 doublings and 80 point-additions. Single addition on the EC system requires three multiplications and one inverse over a finite field of a 160 bit number order. The inverse has a time load roughly equivalent to three multiplications over the finite field. Thus, single multiplication on EC system is equivalent to 6 multiplications over the finite field. In addition, we finally need two more multiplications and one inverse over the finite field to have a signature parameter. A single multiplication requires $160 \times 160 = 25K$ bit multiplications, yet this is ignored when compared to the main computational load. The total number in bit multiplications to generate a signature is as follows.

$$number\ of\ bit\ multiplications = (160 + 80) \times 6 \times (160 \times 160)$$
$$\cong 36M$$

To verify the signature, multiplication in the form of $k \cdot P$ is expected two times. Thus, the number of bit multiplications for a single signature verification is double that of $k \cdot P$, i.e. $72M$.

C.3   ElGamal Encryption in EC System [7]

To encrypt a message m using ElGamal encryption in an EC system, we compute $\Gamma = k \cdot G$ and $\delta = m \oplus \Psi$, where $\Psi = k \cdot (b \cdot G)$, k is a randomly chosen number, and G is a generator. The number of bit multiplications to encrypt a message is as follows.

$$number\ of\ bit\ multiplications = (160 + 80) \times 6 \times (160 \times 160) \times 2$$
$$= 72M$$

Since the decryption process needs single multiplication, it has $36M$ in bit multiplications.

Verifiable Secret Sharing and Time Capsules

Josef Pieprzyk[1] and Eiji Okamoto[2]

[1] Centre for Computer Security Research
School of IT and Computer Science
University of Wollongong
Wollongong, NSW 2522, Australia
josef@cs.uow.edu.au
[2] Center for Cryptography, Computer and Network Security
University of Wisconsin Milwaukee
Milwaukee, WI 53201, USA
okamoto@cs.uwm.edu

Abstract. The paper considers verifiable Shamir secret sharing and presents three schemes. The first scheme allows to validate secrets recovered. The second construction adds the cheater identification feature also called the share validation capability. The third scheme permits to share multiple secrets with secret validation. The constructions are based on hashing and for security evaluation, hashing is modelled as a random oracle with public description. We discuss an application of verifiable secret sharing for the design of cryptographic time capsules for time-release crypto.

1 Introduction

Tompa and Woll in [14] demonstrated how a dishonest participant can recover the secret in a Shamir threshold scheme, leaving other active participant with an invalid secret. They also suggested a prevention method in which both the share and the coordinate compose the secret share of a participant. Since that time, a main research effort has been concentrated into a broad area of the validation of secret. The results achieved so far are related to secret validation in both unconditionally and conditionally secure secret sharing.

In unconditionally secure secret sharing, there is no share verification to check whether or not participants have received their correct shares. Instead, shares are verified by the combiner who refuses to accept shares which have not passed verification process. If a share submitted by a participant fails verification, then the participant is identified as a cheater. Note that even an honest participant can be labelled a cheater if the share has been corrupted during transmission from the dealer (or if the assumption about honesty of the dealer does not hold). Rabin and Ben-Or [8] gave a solution which allows the combiner to verify shares provided by participants by checking whether they satisfy a system of linear equations. Carpentieri in [3] improved the above scheme by showing that the verification can be done with shorter shares.

JooSeok Song (Ed.): ICISC'99, LNCS 1787, pp. 169–183, 2000.

In conditionally secure secret sharing, two distinct secret verification problems have been addressed: noninteractive share verification and secret verification. The first problem relates to a misbehaving dealer or unreliable communication channel used for share transmission from the dealer to a participant. Feldman [4] gave a solution in which the dealer, for the polynomial $f(x) = a_0 + a_1 x + \ldots + a_{t-1} x^{t-1}$, broadcasts the exponents g^{a_i} for $i = 0, \ldots, t-1$ where g is a primitive element of a cyclic group from GF(q) (q is a large enough so the discrete logarithm problem becomes intractable). Pedersen [7] used a commitment scheme to obtain verifiable secret sharing. Note that share verification can be applied in two cases: (1) when a participant obtains their share from the dealer and (2) when a participant submits their share to the combiner.

There is also a class of publicly verifiable secret sharing (PVSS) schemes introduced by Stadler [12] (see also [10]). This class is not very interesting from our point of view as the verification of secret recovered can only be done by the participants (unless the secret becomes public !). Besides, the proposed PVSS schemes are very expensive to set up and to run.

The work is structured as follows. Limitations of verifiable secret sharing are discussed in Section 2. The motivation and the necessary background is introduced in Sections 3 and 4, respectively. Secret sharing with validation of secrets is studied in Section 5. Section 6 shows how secret sharing with secret validation can be upgrade to identify cheaters. Multisecret sharing with secret validation is presented in Section 7. Applications of the verifiable secret sharing for designing time capsules are considered in Section 8.

2   Limitations of Proposed Verifiable Secret Sharing Schemes

Unconditionally secure verification of shares suffers from the following drawbacks.

- Information needed for share verification is typically very long. In the Rabin-Ben-Or scheme, each participant holds $(3n - 2)$ additional shares. The Carpentieri scheme assigns $(2n + t - 1)$ additional shares, where n is the size of the group and t is the threshold.
- The burden of storing information needed for verification is shifted to participants.
- The secret recovery is performed collectively by all active participants (the combiner is distributed).
- Delegation of share verification and key recovery to a trusted combiner involves transfer of long verification information.

In conclusion, unconditionally secure verification is a valid option if the parameters t and n are relatively small. Note that if a threshold scheme is designed for a very large group ($t > 1000$), then this option is not very attractive. Conditionally secure verification based on the Pedersen scheme [7] offers some benefits:

- each participant holds a share of the length twice of the length of the original shares,
- verification information is public so the verification can be done without interaction with other participants.
- verification can be run in two different cases: by participants to check the validity of shares obtained from the dealer and, by the combiner to identify cheating,
- the scheme is perfect, i.e. any $(t-1)$ collaborating participants learn nothing about the secret.

Drawbacks of the Pedersen scheme include the following:

- to enable each participant to identify cheaters, the combiner must release original shares provided by participants (via secure channels) or alternatively, zero-knowledge proofs can be applied to check whether shares have been correct (see [10]). Both solutions are expensive.
- the length of a single verifying information (exponents) must be as long as the modulus q used for computations,
- the verifying information is long and must be either stored centrally (to save on storage) or by each participant (to avoid authenticated communication from the central trusted registry at the time of verification).

3   Motivation

Verifiable secret sharing addresses three different problems:

1. verification of secret recovered by the combiner,
2. verification of shares at the time of setting up the scheme [4, 7],
3. cheater identification – share verification at the secret reconstruction stage [3, 8].

Note that problems (2) and (3) are identical. The only difference is who is performing the verification. Note that verification of shares obtained from the dealer can be necessary in the two cases: (a) when the dealer misbehaves and (b) when the communication channels from dealer to participant are unreliable.

Our goal is to address the following two problems:

Problem 1: verification of secret recovered by the combiner.
Problem 2: cheater identification (share verification).

To make implementation efficient and fast, we use a collision-free hash function with public description. The use of hashing has the following advantages:

- selection of probabilistic parameters necessary in secret sharing can be done with the assistance of hash functions,
- security properties can be proved using the random oracle model,
- the length of shares do not need to be larger than the hashing block size (n.b. a typical hashing block size is 10 times shorter than the block size in the Pedersen scheme).
- computation of verification information using hashing is much faster then equivalent computations using exponentiation.

4 Building Blocks

Given a group of participants $\mathcal{P} = \{P_1, \ldots, P_n\}$. Assume that the group wish to collectively own a secret s in such a way that only a big enough subset of $\mathcal{P}$ can recover the secret. The definition of a big enough subset can be given by the enumeration of all smallest subsets which are still able to see the secret. This collection together with all possible supersets is called *the access structure* Γ. Threshold (t, n) scheme have a very simple access structure. It consists of all subgroups whose cardinality is equal or larger than t or

$$\Gamma = \{\mathcal{A} | \#\mathcal{A} \geq t\},$$

where $\#\mathcal{A}$ stands for cardinality of $\mathcal{A} \subset \mathcal{P}$.

Secret sharing is *perfect* if any subgroup $\mathcal{A}$ which does not belong to the access structure can learn nothing about the secret. More precisely, the entropy of secret is undiminished for any subset $\mathcal{A} \notin \Gamma$ or simply

$$\mathcal{H}(S) = \mathcal{H}(S|\mathcal{A})$$

Clearly, $\mathcal{H}(S|\mathcal{A}) = 0$ for any $\mathcal{A} \in \Gamma$. We define a weaker notion of perfectness which is useful in the conditionally secure setting.

Definition 1. *A secret sharing is computationally perfect if finding the secret involves the exhaustive search of the whole space from which the secret has been chosen (so if the space contains 2^k elements, an average search takes 2^{k-1} steps).*

Secret sharing is a pair of two algorithms: the dealer $\mathcal{D}$ and the combiner $\mathcal{C}$. The dealer sets up the scheme for a given or randomly chosen secret with requested security parameter defined by the size of shares. The combiner is activated at the reconstruction stage. Collaborating participants submit their shares to the combiner who recovers the secret if the currently active set of participants belongs to the access structure. Otherwise, the combiner fails.

The Shamir (t, n) secret sharing is a collection of the dealer and combiner [11]. Computations are done in $GF(q)$ and q specifies the size of the set from which shares and secrets are drawn. The dealer takes a secret $s \in GF(q)$ and the threshold parameter t. For them, the dealer chooses a polynomial

$$f(x) = s + a_1 x + \ldots + a_{t-1} x^{t-1}$$

for random elements $a_i \in GF(q)$ for $i = 1, \ldots, t-1$. The shares are $s_i = f(x_i)$ for $i = 1, \ldots, n$. The dealer sends s_i to P_i via a secure channel while x_i together with their assignment to participants are made public.

At the reconstruction stage, if t or more participants submit their shares to the combiner, then it applies the Lagrange interpolation to recover the polynomial $f(x)$ and the secret $s = f(0)$. If the number of participants is smaller than t, then the interpolation will produce a different polynomial $\tilde{f}(x) \neq f(x)$ and the combiner will recover invalid secret with an overwhelming probability.

For security evaluation, we need a mathematical model of hashing. We use the random oracle model [1]. The hash function in this model is a function

$$H : \Sigma^* \to \Sigma^k$$

which takes an input message of arbitrary length $m \in \Sigma^*$ and assigns a k-bit output. For a given message m, the k-bit output is selected randomly, independently and uniformly from the set Σ^k of all k-bit values. The hash function H is publicly accessible.

Lemma 1. *Given a hash function in the oracle model, a subset $S \subset \Sigma^*$ and a randomly chosen element $s \in S$ where $\#S = q$ and $q << 2^k$. Then the knowledge of $u = H(s)$ allows to identify s from S after searching through $q/2$ possible values of S, on the average.*

Instead of a formal proof, let us notice that this lemma represents a typical problem of searching through a sequence of unsorted entries with random numbers. The assumption that $q << 2^k$ "guarantees" that the value u is unique in the set S (with a very high probability). In this case, it is known that to identify a single and unique entry amongst a sequence of q unsorted entries takes $q/2$ searches, on the average.

Hashing in the random oracle model is also subject to the generic birthday attack which follows the well-known birthday paradox. For details see [6].

5   Secret Sharing with the Validation of Secrets Recovered

Given a (t, n) Shamir secret sharing based on a polynomial $f(x) = s + a_1 x + \ldots + a_{t-1} x^{t-1}$ in GF(q). There is also a publicly accessible hashing function $H : \Sigma^* \to \Sigma^k$ which for an arbitrary message $m \in \Sigma^*$ returns a k-bit hash value $H(m)$.

The Dealer

Given collection of participants $\mathcal{P} = \{P_1, \ldots, P_n\}$ and the threshold parameter t, the dealer

1. chooses the secret $s \in_R GF(q)$ at random and computes co-efficients of polynomial $f(x)$ according to the following

$$a_i = H(s, i, \mathcal{P}, t, T)$$

where s is the secret to be shared by the group $\mathcal{P}$, i is integer indicating the coefficient index, $\mathcal{P}$ specifies the membership of the group, t the threshold parameter and T is a timestamp. We assume that there is no pair of secret sharing with the same timestamp,

2. finds shares

$$s_i = f(x_i) \pmod q$$

3. distributed shares via confidential channel to P_i; $i = 1, \ldots, n$. The values x_i are public.

The Combiner

At the reconstruction stage, the combiner collects shares from participants. If the number of shares is equal to t, the combiner reconstructs the polynomial $\tilde{f}(x)$ using the Lagrange interpolation and computes $\tilde{s}$. To verify the recovered secret, the combiner re-computes coefficients $\tilde{a}_i$ from the public hash algorithm, i.e.

$$a_i \overset{?}{=} H(\tilde{s}, i, \mathcal{P}, t, T).$$

If the checks hold for all $i = 1, \ldots, t - 1$, the secret is accepted and sent (via confidential channels) to all active participants. They can repeat the verification process.

Theorem 1. *The secret sharing with validation of secret is computationally perfect in the random oracle model.*

Proof. In the assumed model of hashing, all random variables assigned to messages are independent as long as messages are different. Note that co-efficients a_i are, therefore, independent random variables as each message $m = (s, i, \mathcal{P}, t, T)$ differs from each other on the co-ordinate index i. Further, we suppose that there is $(t - 1)$ participants, say $P_1, \ldots, P_{t-1}$, who want to recover the secret and pool their shares together. They can ensemble the following system of linear equations:

$$\begin{bmatrix} s_1 - s \\ \vdots \\ s_{t-1} - s \end{bmatrix} = \begin{bmatrix} x_1 & \cdots & x_1^{t-1} \\ \vdots & & \vdots \\ x_{t-1} & \cdots & x_{t-1}^{t-1} \end{bmatrix} \begin{bmatrix} a_1 \\ \vdots \\ a_{t-1} \end{bmatrix} = X \times \begin{bmatrix} a_1 \\ \vdots \\ a_{t-1} \end{bmatrix}$$

and express coefficients a_i as

$$\begin{bmatrix} a_1 \\ \vdots \\ a_{t-1} \end{bmatrix} = X^{-1} \times \begin{bmatrix} s_1 - s \\ \vdots \\ s_{t-1} - s \end{bmatrix}$$

The above equation has as many solutions as there are different $s \in GF(q)$. As the secret is selected randomly and uniformly from $GF(q)$, it means that all values are equally probable. The scheme, however, allows them to checks if a currently selected secret is correct by trying to re-generate the coefficients of $f(x)$.

Now we can create a subset $S \subset \Sigma^*$ such that each element is of the form $(\alpha, i, \mathcal{P}, t, T)$, where $\alpha \in GF(q)$. Next we are going to identify $(t - 1)$ messages $\beta_1, \ldots, \beta_{t-1} \in S$ such that $H(\beta_i) = a_i$. Using Lemma 1 we can conclude that it will take $q/2$ on the average to find a unique sequence of matching β_i.

The scheme allows each active participant to confirm that the recovered secret is valid. If, however, the recovered secret is invalid then the scheme does not provide any facilities to identify cheaters. Note that a single cheater can recover the valid secret (and verify its validity) if he applies the method described by Tompa and Woll in [14]. Note further that honest participants are able to discover the fact that they have been cheated.

6   Secret Sharing with Cheater Identification

Verifiable secret sharing schemes allow for cheater detection or identification either at the same time when shares are pooled or later. This is certainly the case in unconditionally secure secret sharing when participants "simultaneously" reveal their shares permitting the active participants to verify their correctness. Consider the conditionally secure secret sharing. Feldman [4] makes $g^{f(x)}$ public so this allows to identify cheaters after they expose their shares as every body can check whether

$$g^{f(x_i)} \stackrel{?}{=} g^{\tilde{s}_i}$$

where $\tilde{s}_i$ is the share provided by P_i. Similar comments can be made about the Pedersen scheme [7].

In other words to identify cheaters, the combiner must know their shares. This has some very profound implications on the security of secret sharing. The combiner, after validation of shares, may not be able to compute the secret. This happens when t participants are active but at least one of them cheats. The combiner may either

– wait until the number of honest participants reaches t or
– forget the shares and abort the recovery of secret.

The first solution makes sense if there are many other participants who are ready to pool their shares on request from the combiner. An enemy who would like to access the secret controlled by secret sharing, certainly sees the combiner as an attractive target (instead of attacking t different participants, it is enough to break security of the combiner). The longer the combiner "lives" the higher risk of a successful attack. If the waiting time is too long, then active participants may decide to terminate the combiner. This obviously increases complexity of secret sharing as there must be an additional *abort* operation. This operation is used whenever the combiner detected cheating and the number of valid shares is not enough to reveal the secret.

So we have arrived at the second option which allows to abort the secret recovery by termination of the combiner. This could be done "safely" if the combiner is trusted in the sense that it does not disclose any valid shares to unauthorised persons and is memoryless so after its termination all secret elements are deleted.

Below we present a scheme which allows the combiner to:

– identify cheaters without revealing shares by active participants,
– abort key recovery before participants pool their shares to recover the secret. This occurs only when the number of honest participants is smaller than the threshold parameter t.
– repeat key recovery after abortion (this may be done by the same or a different combiner).

The scheme holds valid shares for the smallest possible time interval allowing the combiner to recover the secret only if there are t active and honest participants.

The Dealer

Given collection of participants $\mathcal{P} = \{P_1, \ldots, P_n\}$, the threshold parameter t, and the number of permitted abortions ℓ, the dealer

- selects co-efficients of polynomial $f(x)$ according to the following

$$a_i = H(s, i, \mathcal{P}, t, T)$$

creates the polynomial $f(x) = s + a_1 x + \ldots + a_{t-1} x^{t-1}$ and finds shares $s_i = f(x_i)$,

- assumes that $c_{i,0} = (x_i, s_i)$ is concatenation of x_i and s_i, and computes check values

$$c_{i,j} = H(c_{i,j-1}, P_i, j-1, \mathcal{P}, t, T)$$

for $i = 1, \ldots, n$ and $j = 1, \ldots, \ell$,

- computes a polynomial

$$G(x) = \prod_{i=1}^{n} (x - c_{i,\ell}),$$

- distributes shares and verification information via secure channel, i.e. P_i gets

$$s_i \text{ and } \{c_{i,j} | j = 0, \ldots, \ell - 1\}.$$

This information is known to P_i only; $i = 1, \ldots, n$,

- publishes the polynomial $G(x)$.

Share Validation

Assume that at the pooling time, there are v active participants, say $P_1, \ldots, P_v$. Each active participant P_i:

- announces $c_{i,\ell-1}$ by displaying the data on a public billboard (broadcasting) in the form of a signed message,
- if all check values are displayed on the billboard, participants compute hash values $\{c_{i,\ell} = H(c_{i,\ell-1}, P_i, \ell - 1, \mathcal{P}, t, T); i = 1, \ldots, v\}$ and check whether $G(c_{i,\ell}) = 0$ for all active participants. If there are t participants out of v active ones ($t \leq v$) whose check values have passed verification, then they call the combiner. Otherwise, the attempt is aborted.

Clearly, any failed attempt for secret recovery, causes that participants reveal part of their check values or in other words, they have disclosed pre-images of the hash values.

It is a security policy matter to decide what needs to be done when there is a group of dishonest participants. If the number of cheaters is small then the recovery of secret can go ahead if the number of honest participants is at least t. In general, one would expect that if a majority of participants has passed the share validation and it contains at least t members, then they will go for recovery of the secret.

The Combiner

The combiner is activated only if there is a group of t or more honest participants.

- The combiner asks the active and honest participants to submit their shares.
- Participants submit their shares. The combiner repeats computations of hash values and verifies whether the last check values are zeros of $G(x)$. This part can be substantially shortened if the combiner can access transcript of the public discussion carried out during share validation.
- Shares provided by cheaters are disregarded and if there is at least t shares, the combiner recovers the secret using the Lagrange polynomial interpolation.
- The combiner validates the recovered secret (see the secret sharing from the previous Section).
- Distributes it via secure channel to all active participants who may repeat the whole verification process.

Note that share validation can detect opponents who impersonate participants. A successful completion of the share validation does not guarantee that the active participants will not try to cheat at the recovery stage. It, however, reassures that all active participants hold their valid shares so the combiner will be able to recover the secret (perhaps after a "gentle" warning for misbehaving participants).

Consider the following attacks on the system.

- **Recovery of check information and shares.** This attack tries to reverse the hash function. To be successful the attacker must perform, on the average, $\approx 2^{k-1}$ steps if the hash function outputs k-bit digests. Clearly, after $\ell - 1$ abortions, the shares can be computed with the same computational overhead. Note that the public polynomial $G(x)$ gives out n digests expected to appear when the valid check values are produced. In the random oracle model, an attacker is able to find a value α such that $G(\alpha) = 0$ after $\approx \frac{2^k}{2n}$ steps. The polynomial $G(x)$ together with the hash function is a form of the well-known sibling intractable hash function. The security evaluation of such construction can be found in [16].
- **Cheating by the participants.** A participant P_i can try to find a colliding message $\tilde{s}_i$ for his valid share s_i such that

$$c_{i,1} = H(\tilde{s}_i) = H(s_i).$$

This will cost P_i on the average $\approx 2^{k/2}$ steps. Note that the cheating will not be detected by anybody except by the combiner (with the probability $1 - 2^{-k}$) at the validation of secret stage. If the combiner is implemented as a distributed system and the cheater can see shares of all active participants, then he may apply (successfully) the Tompa-Woll attack. If, however, the combiner is isolated from active participants and does not return secrets which have not passed the validation stage, then the secret is lost assuming the abortion is not an option.
- **Collusion of $t-1$ participants.** This attack has been discussed in previous Section. The scheme is computationally perfect - to find the secret it is necessary to exhaustively search 2^k possible values.

It can be argued that participants may have limited computing resources so multiple computation of hashing values can substantially slow down the secret recovery process. This is the case when participants are mobile agents who keep their shares on smart cards. It is possible to modify the secret sharing with cheater identification in such a way that check values are computed in "parallel" instead of in "sequence".

The dealer can be modified by calculating

$$c_{i,j} = H(x_i, s_i, j, ID_{P_i}, \mathcal{P}, t, T)$$

for $j = 1, \ldots, \ell$. Further, the dealer computes n distinct polynomials – each for every participant. So P_i is assigned $G_i(x)$ such that

$$G_i(x) = \prod_{j=1}^{\ell} (x - \alpha_{i,j})$$

where $\alpha_{i,j} = H(c_{i,j}, ID_{P_i}, \mathcal{P}, t, T)$. Polynomials $G_i(x)$ are public.

7   Multisecret Sharing with Validation

Given ℓ secrets say $(\kappa_1, \ldots, \kappa_\ell)$ which are supposed to be shared among the same group $\mathcal{P}$ of participants. Participants may have limited storage resources and they require to hold as few shares as possible. So the following problem arises. How to design a family of ℓ secret sharing schemes each of which allowing to recover a single secret assuming that any participant holds a single share only. The solution is based on the work [13] and is described for the case when $\ell = 2$ but can be easily extended for an arbitrary ℓ.

The Dealer

Given collection of participants $\mathcal{P} = \{P_1, \ldots, P_n\}$ and two secrets (κ_1, κ_2) which are supposed to be shared by $\mathcal{P}$ using two threshold schemes (t_1, n) and (t_2, n). Two cryptographically strong collision-free hash functions H_1 and H_2 with public description are given. The dealer

1. selects basic shares S_j for P_j at random from the large enough Galois field GF(q),
2. assigns a public x_j for each P_j,
3. for each basic share S_j, computes two shares

$$s_j^{(1)} = H_1(S_j, \mathcal{P}, t_1, T) \text{ and } s_j^{(2)} = H_2(S_j, \mathcal{P}, t_2, T)$$

4. applies the Lagrange approximation and finds the two polynomials

$$f_i(x) = \kappa_i + a_1^{(i)} x + \ldots + a_{n+1}^{(i)} x^{n+1}$$

which passes through the points $(0, \kappa_i)$, $(x_j, s_j^{(i)})$ and additionally $a_1^{(i)} = H_i(\kappa_i, \mathcal{P}, t_i, T)$; $i = 1, 2$,

5. sends via private channels basic shares to participants and broadcasts the coefficients $a_{t_i}^{(i)}, \ldots, a_{n+1}^{(i)}$.

The Combiner

At the reconstruction stage, t_i active participants pool their shares together. Knowing the co-efficients $a_{t_i}^{(i)}, \ldots, a_{n+1}^{(i)}$ and t_i points, the combiner can recover the polynomial $f_i(x)$ and the secret

$$\kappa_i = f_i(0)$$

and validate the secret by checking

$$a_1^{(i)} \stackrel{?}{=} H_i(\kappa_i, \mathcal{P}, t_i, T)$$

Note that this time validation of the secret recovered is done by the combiner. To allow participants to validate the secret, the combiner must provide the polynomial $f_i(x)$ to P_j so P_j can check her share $f_i(x_j) \stackrel{?}{=} s_j^{(i)}$ and the value $a_1^{(i)} \stackrel{?}{=} H_i(\kappa_i, \mathcal{P}, t_i, T)$.

8   Time Capsules

Rivest, Shamir and Wagner [9] consider time-lock puzzles which can be used to control the delay of execution of cryptographic operations (also see [5]). Applications of time-release crypto for key escrowing can be found in [2, 15]. There are two general approaches to the design of time capsules:

- algorithmic – the delay is measured by the time-complexity of a well-understood numerical problem,
- probabilistic – the delay is measured by the number of steps (typically in the exhaustive search) necessary to discover the secret.

The algorithmic approach suffers from an obvious drawback that the time-complexity of majority of "cryptographically" useful problems is not known. This typically tends to "shorten" the expected delay as our knowledge about algorithms and our computing technology progresses. On the other hand, probabilistic time capsules (timers) suffer from the inherited probabilistic nature which determines the delay in terms of the average rather than a precise value. Additionally, probabilistic timing can be easily run in parallel shortening it proportionally to the number of timer copies executed concurrently. Their strong point, however, is that time measure does not depend on the progress in Theory of Algorithms.

8.1   Time Capsules with a Single Timekeeper

Rivest et al. based their time capsules on repeated squaring [9] which is believed to be inherently sequential. Hashing is also inherently sequential and although is much faster than squaring, the delay can be controlled by the size of the message to be hashed. Assume that there is a collision-free hash function H whose implementation provides v hashings per second. Suppose further that the requested delay for a secret s recovery is τ seconds. The number of necessary hashing operations to introduce the delay is τv. The delay can be implemented in different ways which are equivalent in the sense that all must used hashing sequentially τv times. Consider the following two cases:

- $s = \underbrace{H(\ldots H(r)\ldots)}_{\tau v}$ – this is sequential hashing of the initial vector r,

- $s = H(\underbrace{r,\ldots,r}_{\tau v})$ – the initial vector r is used to create a sufficiently long message so its hashing consumes τv operations,

where s is the secret whose recovery must be delayed. The time capsule uses secret sharing with the validation of the secret and is a collection of the following algorithms:

- the dealer who sets up the capsule for the requested parameters. Those parameters include the number of agents who are authorised to switch on the capsule (to count down the delay), the threshold used to activate the capsule, and the requested delay,
- the timekeeper who collects shares from agents and having her own shares is able to recover the initial vector r and after the delay τ recover the secret s.

Consider a simple case when the time capsule is controlled by a single agent P_1 and a timekeeper P_2.

Dealer

Given a collection of participants $\mathcal{P} = \{P_1, P_2\}$, the threshold parameter $t = 2$, and the delay τ, the dealer

1. chooses the initial vector $r \in_R GF(q)$ at random and "winds up" the timer by finding the secret
$$s = H(\underbrace{r,\ldots,r}_{\tau v})$$

2. computes $a_1 = H(s, 1, \mathcal{P}, t, T)$, and finds the polynomial $f(x) = r + a_1 x$ together with two shares $s_1 = f(x_1)$ and $s_2 = f(x_2)$,

3. distributes the shares to participants so P_i knows her secret share s_i while x_i is public $(i = 1, 2)$.

Timekeeper

The time capsule is activated by the agent P_1 who gives her share to the timekeeper P_2. Knowing the two shares, P_2 can recover the polynomial and the initial vector $r = f(0)$ and start computing the secret which will be recovered when $a_1 \stackrel{?}{=} H(\tilde{s}, 1, \mathcal{P}, t, T)$ where $\tilde{s}$ is the current hash value after $i \leq \tau v$ hashing operations. This condition can be checked concurrently so it will not influence the delay τ.

Note that winding up the timer to the requested delay takes the same amount of time as the delay. This is clearly a problem especially when delays are to be longer (days, weeks, months). A simple solution to this problem is the use of ℓ concurrent winding threads. The first thread starts hashing from the initial vector $r_1 = r$ while others start from random vectors $r_i \in_R GF(q)$ $(i = 2, \ldots, \ell)$. After a precise number of τv iterations, outputs are collected from the threads – let them be $h_1, \ldots, h_\ell$. Now the parallel threads are "glued" together to create a single hashing stream which will delay the timekeeper by $\ell\tau$. To do this, the dealer computes

$$g_i = h_i \oplus r_{i+1}$$

for $i = 1, \ldots, \ell - 1$ and makes them public. Note that the secret $s = h_\ell$. To find out the secret s, the timekeeper must start from $r_1 = r$ and continue hashing the prescribed number of times τv. Then she collects the output h_1, recovers $r_2 = g_2 \oplus h_1$ and keeps hashing again τv times, getting h_2, recovering $r_3 = g_3 \oplus h_2$, etc. Finally, she gets $h_\ell = s$.

The major criterion for security evaluation seems to be the stability of the delay introduced by the capsule. There are two kinds of factors which influence the intended delay:

1. progress in technology – a faster (hardware, software) implementation of cryptographic primitives used in the time capsules,
2. advancement in cryptanalysis of cryptographic primitives used in the time capsules.

The first factor always shortens the delay. This may force the designers to create time capsules with grossly overestimated delay so when the timer is activated, the delay will be far too high than requested. This can be solved by designing a time capsule with multiple points of entry into the hashing sequence (controlled by agents).

The second factor relates to two aspects: collision freeness of our underlying hash function and the structure of the timer. It is easy to see that sequential execution of hashing can be circumvent if there is a collision in the sequence traversed by the timer.

8.2   Time Capsules with Multiple Timekeepers

It seems to be difficult (if not impossible) to design a stable time capsule with an arbitrary long delay. It is however possible to design a system in which multiple timekeepers are used and they can collectively recover the secret if a big enough collection has successfully completed their computations. The delay in recovery of the secret will be enforced by the slowest timekeeper in the group. Additionally, the secret will be recovered collectively so if the delay is shorter than assumed, all participants will be equally advantaged or disadvantaged.

Given a group of timekeepers $\mathcal{P} = \{P_1, \ldots, P_n\}$ and an agent A whose responsibility is to activate a time capsule. For this purpose, the agent keeps shares for each timekeeper so when the time comes, the agent broadcasts the shares and timekeepers can start counting down.

Dealer
The dealer sets up the whole system. The input parameters are the secret s and the requested delay.

1. First he designs n $(2,2)$ Shamir schemes which will be used by the pair (A, P_i) to collectively hold the initial vector r_i; $i = 1, \ldots, n$. Let the schemes be based on polynomials $f_i(x) = r_i + a_i x$ where

$$a_i = H(s, P_i, T)$$

   s is the secret and T is a timestamp. The dealer privately sends $f_i(x_{P_i})$ to P_i and $f_i(X_A)$ to A.
2. The dealer winds up the timer by computing

$$h_i = H_\tau(r_i)$$

   where H_τ is a collision free hash function with the delay τ.
3. Next he designs a (t, n) Shamir scheme which allows any t timekeepers to recover the secret s. To do this, he takes $n + 1$ points: (x_i, h_i); $i = 1, \ldots, n$, the point $(0, s)$, and finds a polynomial

$$g(x) = s + b_1 x + \ldots + b_n x^n.$$

4. Co-ordinates x_i and co-efficients $(b_t, \ldots, b_n)$ are made public.

Agent
The agent activates the timekeepers by broadcasting shares $f_i(X_A)$ to timekeepers.

Timekeepers
There are n independent timekeepers.

1. Each P_i takes the shares $f_i(X_A)$ and his own $f_i(x_{P_i})$, recovers the initial vector r_i and computes his h_i (after the delay τ).
2. Now a group of t active timekeepers collectively recover the secret s.

Acknowledgements. Authors thank anonymous reviewers and Hossein Ghodosi from James Cook University for their constructive comments which improved the final version of the paper.

References

1. M. Bellare and P. Rogaway. Random oracles are practical: a paradigm for design efficient protocols. In *Proceedings of the 1st ACM Conference on Computer and Communication Security, November 3-5, 1994, Fairfax, Virginia*, pages 62–73, 1993.
2. M. Burmester, Y. Desmedt, and J. Seberry. Equitable key escrow with limited time span. In K. Ohta and D. Pei, editors, *Advances in Cryptology - ASIACRYPT'98*, pages 380–391. Springer, 1998. Lecture Notes in Computer Science No. 1516.
3. M. Carpentieri. A perfect threshold secret sharing scheme to identify cheaters. *Designs, Codes and Cryptography*, 5(3):183–187, 1995.
4. P. Feldman. A practical scheme for non-interactive verifiable secret sharing. In *Proceedings of the 28th IEEE Symposium on Foundations of Computer Science*, pages 427–437. IEEE, 1987.
5. W. Mao. Send message into a definite future. In V. Varadharajan and Y. Mu, editors, *Information and Communication Security – Second International Conference, ICICS'99*, pages 244–251. Springer-Verlag, 1999. Lecture Notes in Computer Science No. 1726.
6. A. Menezes, P. van Oorschot, and S. Vanstone. *Handbook of Applied Cryptography*. CRC Press, Boca Raton, 1997.
7. T.P. Pedersen. Non-interactive and information-theoretic secure verifiable secret sharing. In J. Feigenbaum, editor, *Advances in Cryptology - CRYPTO'91*, pages 129–140. Springer, 1992. Lecture Notes in Computer Science No. 576.
8. T. Rabin and M. Ben-Or. Verifiable secret sharing and multiparty protocols with honest majority. In *Proceedings of 21st ACM Symposium on Theory of Computing*, pages 73–85, 1989.
9. R. Rivest, A. Shamir, and D. Wagner. Time-lock puzzles and time-release crypto. http://theory.lcs.mit.edu/rivest/RivestShamirWagner-timelock.ps, 1996.
10. B. Schoenmakers. A simple publicly verifiable secret sharing scheme and its application to electronic voting. In M. Wiener, editor, *Advances in Cryptology - CRYPTO'99*, pages 148–164. Springer, 1999. Lecture Notes in Computer Science No. 1666.
11. A. Shamir. How to share a secret. *Communications of the ACM*, 22:612–613, November 1979.
12. M. Stadler. Publicly verifiable secret sharing. In U. Maurer, editor, *Advances in Cryptology - EUROCRYPT'96*, pages 190–199. Springer, 1996. Lecture Notes in Computer Science No. 1070.
13. Y. Tamura and E. Okamoto. Concept and implementation of flexible secret sharing scheme. In *Proceedings of 1998 Computer Security Symposium(CSS'98)*, 1998.
14. M. Tompa and H. Woll. How to share a secret with cheaters. *Journal of Cryptology*, 1(2):133–138, 1988.
15. K. Viswanathan, C. Boyd, and E. Dawson. Publicly verifiable key escrow with limited time span. In *Proceedings of the Fourth Australasian Conference on Information Security and Privacy (ACISP99)*, volume 1587, pages 36–50. Springer-Verlag, 1999.
16. Y. Zheng, T. Hardjono, and J. Pieprzyk. The sibling intractable function family (SIFF): notion, construction and applications. *IEICE Trans. Fundamentals*, E76-A:4–13, January 1993.

A New Approach to Robust Threshold RSA Signature Schemes

Rei Safavi-Naini[1], Huaxiong Wang[2], and Kwok-Yan Lam[3]

[1] School of IT and CS, University of Wollongong, Australia
rei@uow.edu.au
[2] Department of Computer Science, National University of Singapore, Singapore
wanghx@comp.nus.edu.sg
[3] Department of Computer Science, National University of Singapore, Singapore
lamky@comp.nus.edu.sg

Abstract. In a threshold RSA signature scheme, dishonest participants can disrupt signature generation by submitting junk instead of their partial signatures. A threshold signature system is *robust* if it allows generation of correct signatures for a group of t honest participants, and in the presence of malicious participants. The purpose of this paper is two-fold. First we show that a robust (t,n) threshold RSA signature scheme, proposed by Rabin in Crypto'98, lacks an essential property of (t,n) threshold schemes and allows an adversary to forge signatures. Then we propose a new approach to the construction of t-robust (t,n) threshold RSA signature scheme which can be seen as the dual to Rabin's approach. We discuss the efficiency of our system and show that when t is small (compared to n) our scheme is much more efficient than other existing schemes.

1 Introduction

Threshold cryptography, and in particular threshold signature, was independently invented by Desmedt [13], Boyd [9], Croft and Harris [12]. The main goal of threshold cryptography is to replace a system entity - such as a transmitter - in a classical cryptosystem with a group of entities sharing the same power. A threshold cryptosystem must remain secure not only under the attacks on the original cryptosystem, but also new types of attacks that are introduced because of the distributed structure of the system.

In a (t,n) threshold signature scheme [18], signature generation requires collaboration of at least t members of a set of n participants. Although construction of threshold signature schemes generally uses a combination of secret sharing schemes and signature schemes, as noted in [14], a simplistic combination of the two primitives could result in a completely insecure systems that allows the members of an authorized group to recover the secret key of the signature scheme. In a secure threshold signature scheme the power of signature generation must be shared among n participants in such a way that t participants can collaborate to produce a valid signature for any given message whilst no subset of fewer than t

JooSeok Song (Ed.): ICISC'99, LNCS 1787, pp. 184–196, 2000.
© Springer-Verlag Berlin Heidelberg 2000

participants can forge a signature even if many signatures on different messages are known.

A major problem in the construction of threshold RSA signature schemes is that the secret exponent which must be shared among the participant, is an element of $\mathbf{Z}_{\phi(N)}$ which is an Abelian group, and not a field. This means that the majority of classical secret sharing schemes, such as Shamir's scheme, cannot be used. The first simple and elegant solution to distributed RSA signature is due to Boyd [9] and Frankel [21] and gives the share of the signature key d (the RSA exponent) to n signers $P_1, \ldots, P_n$ such that P_i holds d_i and $d = d_1 + \cdots + d_n$. To sign a message m, each signer P_i produces a partial signature m^{d_i} which is combined (multiplied)as $m^d = m^{d_1} \cdot \ldots \cdot m^{d_n}$ to create the signature on m. This is an (n, n) scheme and requires collaboration of every single member of the group for generating a signature. This can be seen as a drawback which drastically reduces availability of the system and in particular in cases where trust structure in the group permits signing even if t out of n group members collaborate. In order to implement a system with a threshold t, $t < n$, one can generalize the above (n, n) scheme using cumulative secret sharing scheme developed in [32], or adopting a protocol such as the one below. The dealer generates the shares of $\binom{n}{t}$ independent runs of a (t, t) additive secret sharing scheme for the same secret, the signing key d in this case, and gives appropriate shares to each signer. Now any t subset of the group has the complete set of shares for one run of the secret sharing scheme and can sign a message. The main drawback of schemes such as this is their inefficiency in the sense that each signer has to store shares which is in total $\binom{n}{t}$ times of the size of RSA signing key.

Desmedt and Frankel [18] initiated the study of efficient threshold RSA signature and gave a heuristic solution for it. The basic idea is generalizing Shamir's polynomial scheme over $\mathbf{Z}_{\phi(N)}$ by extending Lagrange polynomial interpolation over a finite field to a module over a ring, and then using it for the signature generation. The resulting scheme requires each participant to have a share whose size is n times the size of the secret, in this case RSA secret exponent. Another elegant approach which was implicitly proposed by Blackburn et al [6] achieves the threshold t by utilizing an appropriate perfect hash family to combine independent runs of a (t, m) RSA signature scheme where $n > m$. When t is small compared to n, this approach is much more efficient than Desmedt-Frankel scheme and in its optimal form can reduce the size of each singer's key to $O(\log n)$ times RSA signing key, which is much less that n times RSA signing key required by the Desmedt-Frankel.

In this paper we focus on another aspect of threshold signature systems: robustness against dishonesty of participants. In shared generation of signature, dishonest participants may submit some junk instead of their partial signatures. A (t, n) threshold RSA signature scheme is called *robust* if it can correctly compute the signature even in the presence of up to $t - 1$ arbitrary malicious signers. Robust threshold signature schemes have very important applications. Although distributing signature generation process effectively distributes responsibility of a trusted node among n local nodes, and hence removes system's bottleneck, there is always the danger of reduced availability because a faulty node can easily disrupt the system. Adding robustness ensures that distribution of trust is not at the expense of reduced availability. In a robust scheme partial signatures are verifiable and so combiner only uses pick correct compounds (i.e. partial signatures) into the computation; see also [25, 23].

Desmedt pointed out [15] that because an RSA signature can be publicly verified, it is possible to detect cheaters if more than t partial signatures are used and the number of cheaters is ℓ, where $0 \le \ell \le t - 1$. The basic idea is to use $\ell + t$ partial signatures, instead of t, and note that among all $\binom{\ell+t}{t}$ subsets of size t of the partial signatures, there is at least 1 subset with t correct partial signature that results in a signature that can be correctly verified. It is easy to see that the main condition for this construction to work is that $\ell + t \le n$ or $t \le \lfloor \frac{n}{2} \rfloor$.

Again the main drawback of the above solution is its inefficiency: for ℓ malicious users, in the worst case $\binom{\ell+t}{t}$ signature must be generated and verified and so the cost of such a system is prohibitive.

Gennaro et al [25], and independently Frankel et al [23], initiated the study of robust RSA signature schemes. Gennaro et al developed two methods to achieve robustness in such schemes. The first solution is an interactive protocol that uses zero-knowledge proof systems of [26] and the second one is a non-interactive scheme that is based on a novel technique of ICP (Information Checking Protocol) to verify the integrity of the partial signature.

Another approach to the construction of efficient robust threshold RSA signature scheme due to Frankel et al [23] is to extend the notion of *result-checking* due to Blum et al [7] to *witness-based cryptographic checking*. The work in [23] provides a more general theoretical framework, and also applies to RSA as a specialization.

Rabin [30] further studied the robustness in threshold RSA signature schemes which also provides the proactivity.

Rabin's (t, n) robust threshold RSA signature scheme [30] is simple and has low memory and computation cost, and can be viewed as a "ramp scheme" where the degradation in the threshold is allowed. Its simplicity is due to this relaxation which may allow every honest shareholder after $t - 1$ bad ones are exposed, to generate a signature on its own. However as we will show in section 3, the scheme is not (t, n) threshold and requires collaboration of n participants $((n, n)$ threshold) for generation of a signature. This followed by a new scheme which is both (t, n) threshold and robust against up to $t-1$ malicious adversaries. This construction is motivated by Blackburn et al's work [6], and builds a (t, n) threshold scheme from multiple runs of a (t, t) scheme, with cheater detection property, combined by using a perfect hash family. The scheme is particularly efficient for large groups with small threshold.

The paper is organized as follows. In Section 2 we describe the model. In Section 3 we review Rabin's robust threshold RSA signature scheme and show an attack that allows generation of valid signatures by a malicious participant. In Section 4 we present a (t, t) RSA signature with cheater detection and then propose our (t, n) robust threshold RSA signature scheme in Section 5. The paper is concluded in Section 6.

2   Preliminaries

The Communication Model. We will follow the model of [25] and [30]. We assume that the system consists of a set of n participants $\{P_1, \ldots, P_n\}$. They are connected by a complete network of private (preserving secrecy) point-to-point channels. In addition, they have access to a dedicated broadcast channel; by dedicated we mean that if P_i broadcasts a message, it will be recognized by other participants as coming from P_i.

The Adversary. We assume that an adversary can corrupt up to $t - 1$ of the n participants in the network. We consider the worse possible kind of adversary, i.e. a *malicious* adversary that learns all the information held by the corrupted participants and hears the broadcasted messages. He can cause corrupted players to behave in any possible malicious way.

The Dealer. In order to focus on the high-level description of the protocols, we further assume that there is a dealer, who sets up the keys. This includes generation of RSA key, and generation and distribution of shares to participants. The trusted dealer can be eliminated by a distributed key generation process of Boneh and Franklin [8].

Notation. For a positive integer k we denote $[k] = \{1, \ldots, k\}$. The public modulus is denoted by N. We assume $N = pq$, and p, q are safe primes. That is $p = 2p' + 1, q = 2q' + 1$, where $p < q$ and p, q, p', q' are prime numbers. We denote $\phi(N) = (p - 1)(q - 1)$ and $d \in [\phi(N)]$ the secret key of RSA.

3   Rabin Robust Threshold Scheme and Its Weakness

In Crypto 98, Rabin [30] suggested an approach to achieving robustness in (t, n) threshold RSA signature schemes through the usage of share back-ups. Rabin's scheme [30] is t-robust for $t < \lfloor \frac{n}{2} \rfloor$ which means that it can correctly compute signatures even in the presence of up to $t - 1$ malicious (corrupted) participants. However we will show that it is not really (t, n) threshold in the sense that it actually requires collaboration of all n participants for generating a signature. If one of the group members is unavailable, the only way of generating the signature is by reconstructing the absent participant's partial key, followed by a key proactivisation to give new keys to participants. We will show that allowing partial keys to be reconstructed leads to an attack that can effectively reveal the secret key of the RSA. The attack described in section 3.2.

3.1   Rabin's Scheme

Rabin [30] continued the work of Gennaro et al [25] and proposed a simplified approach to threshold robust RSA which reduces memory and computation cost of the system. In this approach an (n, n) RSA signature scheme in the additive form where each participant holds a secret input (his partial key) is used to construct the signature. Each participant's input is then shared among all other participants using a verifiable secret sharing scheme and forms the back-up information. In the event that one or more (up to $t - 1$) participants fail to cooperate, or are faulty, then their information can be reconstructed from the 'back-up' copy and incorporated into the computation.

A brief description of Rabin's scheme follows.

Key Generation. In the key generation phase, the dealer chooses the RSA secret key $d \in \mathbf{Z}_{\phi(N)}$ and performs the following steps:

1. Chooses and securely gives to P_i his secret signature generation key $d_i \in_R [-nN^2, \cdots, nN^2]$, for $1 \leq i \leq n$, and sets $d_{public} = d - \sum_{i=1}^{n} d_i$.
2. Computes witnesses w_i, given below, and broadcasts them

$$w_i = g^{d_i} \pmod{N}, \quad 1 \leq i \leq n,$$

3. Uses a (t, n) verifiable secret sharing similar to [20, 22] and described in [30], to share d_i among the n participants. We follow Rabin and refer to this protocol as (t, n) Feldman-$\mathcal{Z}_N$-VSSS.

Signature Generation. To sign a message m, participants carry out the following steps:

1. Participant P_i publishes his partial signature $\sigma_i = m^{d_i} \pmod{N}$.
2. The signature is calculated as $m^{d_{public}} \prod_{i=1}^{n} \sigma_i \pmod{N}$. If this signature can be correctly verified using the public key, the process finishes.
3. If an error is detected in the signature, each participant P_i must prove the correctness of his partial signature using Gennaro et al non-interactive partial signature verification protocol [25].
4. If P_i's proof fails, more than t participants reconstruct d_i using the share of Feldman-$\mathcal{Z}_N$-VSSS and obtain $\sigma_i = m^{d_i}$.
5. Compute the signature $SIG(m) = m^{d_{public}} \prod_{i=1}^{n} \sigma_i \pmod{N}$.

It is shown that the memory and computation cost of this scheme is less than the the protocols of [25] and [23] for robust threshold RSA signature scheme. However, although Rabin's scheme does allow construction of signatures if only t participants collaborate but in practice any time that less than n participants collaborate partial keys of all other participants are revealed and a proactivisation phase is required. In the following section it is shown that this could be used by an attacker to forge signatures.

3.2   A Weakness of Rabin's Scheme

We first review some basic notions underlying robust (t, n) threshold signature schemes. The aim of a (t, n) threshold signature scheme is to allow any subset of at least t participants to generate a valid signature for an arbitrary message m. Thus, an important feature of threshold schemes is that they increase availability of the system as only t active and honest participants are enough to produce a valid signature. Participants who do not take part in the signature generation (for m) are called *non-active* participants (for m). A participant is called *corrupted* if he does not follow the protocol. The scheme is $t - robust$ if it can correctly compute the valid signatures even in the presence of up to $t - 1$ corrupted participants. It should be emphasized that in robust threshold signature schemes, a non-active participant is not necessarily corrupted. This means that a (t, n) threshold signature scheme allows up to $n - t$ non-active participants, while in a t-robust scheme, the condition $t \leq \lfloor \frac{n}{2} \rfloor$ must be satisfied as the adversary who has corrupted $t - 1$ participants can perform *any* malicious action. That is in a (t, n) threshold signature scheme which is t-robust, if non-active participants are all corrupted, then $n - t + 1 \leq \lfloor \frac{n}{2} \rfloor$ and so $t > \lceil \frac{n}{2} \rceil$ which is a contradiction.

Next we give a simple attack to show that Rabin's scheme can be subverted to forge signatures. The adversary needs to only corrupt one participant: without loss of generality, assume that P_1 is the corrupted participant. P_1 will perform the following steps:

1. P_1 asks a subset $\{P_i, i \in A\}$ of participants such that $|A \cup \{1\}| = t$ to co-operate to sign a message m.
2. P_1 informs participants $P_i, i \in A$ that the participants $P_j, j \in \{1, \ldots, n\} \setminus A$ are non-active (as the scheme is assumed to be (t, n) threshold).
3. At the same time, P_1 starts a parallel session, this time asking $\{1, \ldots, n\} \setminus A$ to co-operate to sign another message m'. This time he will claim that participants in A are non-active. We note that $|\{1, \ldots, n\} \setminus A| \geq t$.
4. At the end of the two parallel runs, P_1 has all the partial keys and so can construct the secret key d.

It is worth noting that the two runs of the protocol are in parallel and result in the recovery of the secret key and so proactivisation will not be helpful as it will only change representation of the key and not its value.

This means that:

Rabin's robust threshold RSA signature [30] scheme is (n, n) threshold and t-robust, but not (t, n) threshold.

The weakness of Rabin's scheme is due to the fact that during signature generation, the keys of (non-active or, crashed) participants are revealed. In an RSA based threshold signature, it is important to keep participants' keys secret and only reveal partial signatures. For further discussion on this, see [17].

4 (t, t) RSA Signature Scheme with Cheater Detection

The idea of our approach is to use (t, t) schemes as building blocks in constructing (t, n) scheme, a similar approach of using (t, t) systems as building blocks can be found in [22] on the subject. To this end, in this section we describe a (t, t) RSA signature scheme that allows detection of cheaters which will then be used to build a robust (t, n) threshold scheme. The scheme uses Gennaro et al's partial signature verification protocol [25] to verify partial signatures in the *additive* (t, t) threshold RSA signature scheme. It consists of the following two phases.

1. **Key generation:** The dealer who has generated the RSA secret key $d \in [\phi(N)]$ carries out the following steps:
   - Chooses and secretly gives value $d_i \in_R [\phi(N)]$ to P_i for $1 \leq i \leq t$ such that $d = d_1 + \cdots + d_t \pmod{\phi(N)}$.
   - For each pair of participants P_i and P_j, the dealer generates and distributes secret information such that P_i and P_j can execute Gennaro et al partial signature verification protocol [25] to verify correctness of partial signatures At the the end of distribution phase each P_i holds the following values.
     (a) His share d_i;
     (b) Auxiliary authentication values $y_{i,1}, \ldots, y_{i,t}$, where $y_{i,j} \in \mathbf{Z}$ is used to prove the correctness of his partial signature to P_j;
     (c) Verification data $v_{1,i}, \ldots, v_{t,i}$, where $v_{j,i} = (b_{j,i}, c_{j,i})$ such that $b_{j,i} \in [N^{\delta_1}]$, $c_{j,i} \in [N^{1+\delta_1+\delta_2}]$, and $y_{j,i} = d_j b_{j,i} + c_{j,i}$. For each j, the 2-tuple value $v_{j,i}$ is used to verify correctness of P_j's partial signature.

   We will denote the collection of values held by P_i and containing the values in (a), (b) and (c), as $\hat{S}_i$.

2. **Signatures generation:** For each pair P_i and P_j, P_j can detect P_i's cheating in the following way.
   - P_i broadcasts his partial signature $\sigma_i = m^{d_i} \pmod{N}$ and the auxiliary value $Y_{i,j} = m^{y_{i,j}} \pmod{N}$, for $j = 1, \ldots, t$.
   - P_j accepts P_i as an honest participant if $\sigma_i^{b_{i,j}} m^{c_{i,j}} = Y_{i,j}$, and concludes that the partial signature of P_i is $m^{d_i} \pmod{N}$ or $-m^{d_i} \pmod{N}$; otherwise detects P_i as a cheater.
   - After performing the verification phase and given that a participant accepts t partial signatures, he can generate a signature for m, $SIG(m) = \sigma_1 \ldots \sigma_t$ or $-\sigma_1 \ldots \sigma_t$. Using the public key, it can be easily determined which one is the correct signature.

Theorem 1. *([25]) In the above scheme, the followings hold:*

Completeness: *If t participants follow the protocol, then they will always generate a correct signature $SIG(m)$ for a message m;*

Soundness: *A cheating P_i can convince P_j to accept $\sigma_i \neq \pm m^{d_i} \pmod{N}$ with probability at most $\frac{1}{p'} + \frac{1}{N^{\delta_1}} + \frac{1}{N^{\delta_2}}$;*

Zero-Knowledge: *Up to $t - 1$ cheating participants do not learn any more information about another participant's key other than his partial signature.*

The proof is straightforward from Theorem 5 in [25].

5   A (t, n) Robust Threshold RSA Signature Scheme

Now we present an efficient solution to threshold RSA signature that also provides robustness. In order to achieve threshold, we use the above (t, t) threshold RSA signature scheme as the building block and 'lift' it to (t, n) threshold by applying a perfect hash family. While Rabin used a (n, n) threshold RSA signature scheme as the underlying scheme and then used a (t, n) secret sharing scheme to 'break' the underlying scheme to achieve (t, n) threshold signature, our approach can be considered as the opposite direction of Rabin. Both two approaches utilize the simplicity of the additive threshold scheme, and rely on Gennaro et al partial signature verification protocol to achieve the robustness.

Before we describe the details of our scheme, we briefly review some basic notions and results on perfect hash families. An *(n, m, t)-perfect hash family* is a set of functions $\mathcal{F}$ such that

$$f : \{1, \ldots, n\} \longrightarrow \{1, \ldots, m\}$$

for each $f \in \mathcal{F}$, and for any $X \subseteq \{1, \ldots, n\}$ such that $|X| = t$, there exists at least one $f \in \mathcal{F}$ such that $f|_X$ is one-to-one.

We use the notation $PHF(W; n, m, t)$ for an (n, m, t)-perfect hash family with $|\mathcal{F}| = W$ and will write $\mathcal{F} = \{f_1, \ldots, f_W\}$. When $m = t$, $PHF(W; n, t, t)$ is called a *minimal perfect hashing family*. Perfect hashing families originally arose as part of compiler design - see [28] for a summary of the early results, and [11] for a survey of recent results. Perfect hashing families have found numerous applications in circuit complexity of threshold functions and to the design of deterministic analogue to probabilistic algorithms; see [1]. Numerous constructions for perfect hash families using finite geometries, designs theory and error-correcting codes are known.

The connection between perfect hash families and construction of efficient threshold secret sharing schemes was implicitly discovered by Blackburn et al in [6]. The utilization of perfect hash family for efficient construction of threshold signature schemes was later noted in the survey papers [15] and [4]. While [15] and [4] are only concerned with providing threshold property, we are also concerned with robustness. In order to be able to exploit the simple *additive* (t, t) threshold scheme, we will use a particular class of perfect hash families– *minimal perfect hash family*.

Our (t, n) robust threshold RSA signature schemes consists of two phases: *Key generation* and *Signature generation*. In the key generation, a dealer D, who knows the secret key d of RSA generates and distributes the secret key information (*shares*) to all participants involved in the system. The secret key of each participant consists of three parts: (1) the value for generating partial signature; (2) the values for proving the correctness of his partial signature to other participants; and (3) the value for verifying partial signatures from other participants. During signature generation, participants i) generate partial signatures; ii) prove correctness of their own signature and verify validity of other participants' partial signatures; and iii) combine the correct partial signatures to generate a full signature.

Another way of interpreting our approach can be described as follows. We execute multiple rounds of a (t, t) threshold RSA signature scheme with cheater detection, and then apply a perfect hash family to assign secret key to the n participants in such a way that every t out of n participants have the complete set of shares for at least a (t,t) scheme and so can sign any message. However because of the existence of corrupted participants we need to ensure robustness. Our basic approach is to employ the cheater detection capability of the underlying (t,t) scheme to 'filter-out' cheating participants by successively forming groups of size t, executing the (t,t) scheme with cheater detection to detect cheaters and substitute them with new participants. This procedure will finish when a set of t honest participants, which is known to exist, is found. At this stage a correct signature can be generated. In essence the scheme uses the basic approach of trying to form a group of t unfaulty players but instead of forming all verifying all signatures that can be formed by a set of $2t-1$ partial signatures as described before (Section 1), it removes cheaters in the group and substitutes them with new participants. In the worst case, it requires $t-1$ times in performing the underlying (t,t) schemes and has be significantly reduced from the trivial solution that requires $\binom{2t-1}{t}$ times of the underlying (t,t) scheme in the worst case. Moreover, when t is small (relative to n), it is known that $PHF(W;n,t,t)$ with $W = O(\log n)$ exist, thus the scheme achieves significantly higher storage and computation efficiency through the use of perfect hash families.

We now present our robust (t,n) threshold RSA signature scheme.

Key Generation. Assume that $\mathcal{F} = \{f_1, \ldots, f_W\}$ is a $W(n,t,t)$ minimal perfect hash family and $0 \le \delta_1, \delta_2 \le 1$ are two security parameters. The hash family and security parameters are both publicly known. Let $d \in [\phi(N)]$ be the RSA secret key which is chosen by the dealer and kept secret. The dealer generates partial keys of W independent runs of the (t,t) RSA signature scheme with cheater detection (given in section 5) for the *same RSA secret key* d. That is

$$\hat{S}^1 = (\hat{s}_1^1, \ldots, \hat{s}_t^1), \ldots, (\hat{s}_1^W, \ldots, \hat{s}_t^W)$$

where $\hat{s}_i^k$ denotes the key of ith participant in the kth run of the underlying (t,t) RSA signature, for $1 \le i \le t$ and $1 \le k \le W$. This can be expressed as a $t \times t$ array consisting of a single row, the ith row, and a single column, the ith, given below,

$$\hat{s}_i^k = \begin{bmatrix} & & (b_{1,i}^k, c_{1,i}^k) & & \\ & & \vdots & & \\ & & (b_{i-1,i}^k, c_{i-1,i}^k) & & \\ y_{i,1}^k \cdots y_{i,i-1}^k & & d_i^k & & y_{i,i+1}^k \cdots y_{i,t}^k \\ & & (b_{i+1,i}^k, c_{i+1,i}^k) & & \\ & & \vdots & & \\ & & (b_{t,i}^k, c_{t,i}^k) & & \end{bmatrix},$$

such that the following conditions are satisfied
- $b_{i,j}^k \in [N^{\delta_1}]$, $c_{i,j}^k \in [N^{1+\delta_1+\delta_2}]$ and $y_{i,j}^k = d_i^k b_{i,j}^k + c_{i,j}^k$, for all $1 \le i, j \le t$ and all $1 \le k \le W$;
- $d_1^k + d_2^k + \cdots + d_t^k = d$ for all $1 \le k \le W$.

The dealer then distributes the secret keys of the n participants $P_1, \ldots, P_n$ in such a way that P_ℓ, $1 \leq \ell \leq n$, holds

$$\hat{S}_\ell = [\hat{s}^1_{f_1(\ell)}, \hat{s}^2_{f_2(\ell)}, \cdots, \hat{s}^W_{f_W(\ell)}].$$

Signature Generation. Generation and verification of partial signatures is by using the perfect hash family and reducing to the underlying (t, t) scheme. Assume that participants $\{P_i; i \in A\}$, where $A \subseteq \{1, \ldots, n\}$ and $|A| \geq t$, want to generate the signature for a message m. Without loss of generality, we assume that $|A| = t$ and $A = \{1, \ldots, t\}$ [1]. From the property of the minimal perfect hash family it follows that there exists a function $f_k \in \mathcal{F}$ such that f_k restricted to A is one-to-one. To generate a signature for a message m, any pair of participants P_i and P_j in the group carry out the following procedure.

- P_i broadcasts $m^{d^k_{f_k(i)}}$ and the auxiliary values $m^{y^k_{f_k(i),1}}, \ldots, m^{y^k_{f_k(i),t}}$.
- P_j verifies if $(m^{d^k_{f_k(i)}})^{b^k_{f_k(i),j}} m^{c^k_{f_k(i),j}} = m^{y^k_{f_k(i),j}}$. If yes, P_j accepts $m^{d^k_{f_k(i)}}$ as a correct partial signature for m from P_i.
- 'Filter-out' the cheating participants and substitutes them with new participants to form a new group of size t, and perform the process according to the above two steps until t correct partial signatures are formed. The RSA signature is obtained by multiplying the correct partial signatures.

Theorem 2. *Under the assumption that factoring is intractable the above scheme is a secure t-robust and (t, n) threshold RSA signature scheme for any $t \leq \lfloor \frac{n}{2} \rfloor$.*

Proof. (sketch:) Completeness and soundness is straightforward from the properties of perfect hash families and the completeness and soundness of the underlying (t, t) threshold signature schemes.

We are left to prove the security of the scheme. The proof of security is by using a simulation argument for the view of the adversary and showing that an adversary who has access to all the key information of the corrupted participants and the signature on m could generate by itself all the other public information produced by the protocol. Observe that due to the independence of the W runs of the underlying (t, t) threshold signature schemes and the properties of the perfect hash family, an adversary who corrupted up to $t - 1$ participants will have no advantage with respect to an individual run of (t, t) scheme. Thus it is sufficient to show that the adversary can not break any of the (t, t) schemes.

In the following we show the security of the underlying (t, t) scheme. Without loss of generality, assume that an instance of the underlying (t, t) scheme consists of t participants $P_1, \ldots, P_t$ and an adversary $\mathcal{A}$ who has corrupted the first $t - 1$ participants $P_1, \ldots, P_{t-1}$ and has learned their secrets. We give a simulator $SIMU$ for our scheme. The input to the simulator $SIMU$ is the message m and

---

[1] The assumption effectively means that if $|A| > t$ we can choose a $B \subseteq A$ such that $|B| = t$ and apply the protocol to B. Also, the t participants are not necessarily the first t participants

its signature m^d (mod N). However, the secret information held by P_t is never exposed and is not simulated.

The $SIMU$ works as follows.

1. **Key Generation**
   - Chooses $\tilde{d}_1, \ldots, \tilde{d}_{t-1} \in_R [\phi(N)]$,
   - Executes simulation of key generation of Gennaro et al partial signature verification protocol. At the end of the distribution phase of the simulation, each P_i holds the following values, where $i = 1, \ldots, t-1$.
     (a) $\tilde{d}_i \in_R \mathbf{Z}_{\phi(N)}$;
     (b) Auxiliary authentication values $\tilde{y}_{i,1}, \ldots, \tilde{y}_{i,t} \in \mathbf{Z}$;
     (c) Verification data $\tilde{v}_{1,i}, \ldots, \tilde{v}_{t,i}$ where $\tilde{v}_{j,i} = (\tilde{b}_{j,i}, \tilde{c}_{j,i})$ such that $\tilde{b}_{j,i} \in_R [N^{\delta_1}], \tilde{c}_{j,i} \in_R [N^{1+\delta_1+\delta_2}]$ and $\tilde{y}_{j,i} = \tilde{d}_j \tilde{b}_{j,i} + \tilde{c}_{j,i}$. Note that $\tilde{v}_{j,i}$ is used to verify the correctness of partial signature of P_j.

2. **Signature generation**
   - Computes partial signature $\tilde{\sigma}_i = m^{\tilde{d}_i}$ (mod N), $i = 1, \ldots, t-1$ and the auxiliary value $\tilde{Y}_{i,j} = m^{\tilde{y}_{i,j}}$ (mod N), $i = 1, \ldots, t-1, j = 1, \ldots, t$.
   - Sets $\tilde{\sigma}_t = m^d / \prod_{i=1}^{t-1} \tilde{\sigma}_i$ (mod N) and the auxiliary values $\tilde{Y}_{t,1}, \ldots, \tilde{Y}_{t,t-1}$ by

$$\tilde{Y}_{t,k} = (\tilde{\sigma}_t)^{\tilde{b}_{t,k}} m^{\tilde{c}_{t,k}}, \quad k = 1, \ldots, t-1.$$

   - Executes simulation of partial signature verification protocol.

It is straightforward to verify that the view of the adversary $\mathcal{A}$ on execution of the protocol, and its view on execution of $SIMU$ are statistically indistinguishable, and so the result follows.

Efficiency

Now consider the efficiency of our scheme. The size of the key used for signature generation is $\log N$ which is the same as the regular RSA key, and is much better than that of the Rabin's scheme [30] which is $2 \log nN$ and that of Gennaro et al scheme [25] which is $n \log N$.

The memory requirement for each participant in our scheme is $2W(1 + \delta_1 + \delta_2)t \log N$, where W is the size of the minimal perfect hash family, and for Gennaro et al's and Rabin's scheme are $2(1 + \delta_1 + \delta_2)n^2 \log N$ and $2(2 + \delta_1 + \delta_2)n \log nN)$, respectively. Thus the memory requirements of our scheme depends on the size of the perfect hash family. It is known that when t is small (relative to n), minimal perfect hash families with $W = O(\log n)$ exists. This means that for large group sizes with small threshold values, storage requirements of our scheme is much better than those of Gennaro et al's and Rabin's schemes.

6   Conclusions

In this paper we addressed a weakness of Rabin's robust (t, n) threshold RSA signature scheme and proposed a new approach to the construction of efficient and robust threshold RSA signature schemes. Our approach provides an efficient construction for robust (t, n) threshold RSA signature only when the threshold t is small compared to n. It remains open for the efficient solution to the generic systems, in particular, to the optimal resilient case, that is $n = 2t + 1$.

The existing RSA threshold signature schemes require a trusted dealer, who knows the secret exponent and $\phi(N)$ of the RSA system. Eliminating trusted dealer from distributed RSA is studied by Boneh and Franklin [8], Blackburn, Blake-Wilson, Burmester and Galbraith [5], Frankel, MacKenzie and Yung [24], and Miyazaki, Sakurai and Yung [29].

An interesting extension of our scheme is to distribute the computation and provide efficient construction for systems without the trusted dealer.

Acknowledgments

We would like to thank one of the program committee members for the very constructive suggestions which improve the presentation of the paper.

References

1. N. Alon and M. Naor, Derandomization, witnesses for Boolean matrix multiplication and construction of perfect hash functions, 1996.
2. M. Atici, S.S. Magliveras, D. R. Stinson and W.D. Wei, Some Recursive Constructions for Perfect Hash Families. *Journal of Combinatorial l Designs* 4(1996), 353-363.
3. M. Ben-Or, S. Goldwasser, and A. Wigderson, Completeness Theorems for Non cryptographic Fault-Tolerant Distributed Computations. In *Proc. 20th Annual Symp. on the Theory of Computing*, pages 1-10, ACM, 1988.
4. S. R. Blackburn, Combinatorics and Threshold Cryptology, in *Combinatorial Designs and their Applications*(Pitman Research Notes in Mathematics), to appear.
5. S.R. Blackburn, S. Blake-Wilson, M. Burmester and S. Galbraith, Shared generation of shared RSA keys, *Tech. Report CORR98-19, University of Waterloo*.
6. S. R. Blackburn, M. Burmester, Y. Desmedt and P. R. Wild, Efficient multiplicative sharing schemes, in *Advance in Cryptology–Eurocrypt '96*, LNCS, **1070**(1996), 107-118.
7. M. Blum and S. Kannan, Program correctness checking and the design of programs that check their work, In *Proc. of the 21st ACM Symposium on Theory of Computing*, 1989.
8. D. Boneh and M. Franklin, Efficient generation of shared RSA keys, *Adv. in Cryptology - Crypto'97*, Lecture Notes in Comput. Sci., **1294**, 425-439.
9. C. Boyd, Digital multisignatures, *Cryptography and coding* (Beker and Piper eds.), Clarendon Press, 1989, 241-246.
10. E. Brickell and D. Stinson, "The Detection of Cheaters in Threshold Schemes," in *Advances in Cryptology - Proceedings of CRYPTO '88* (S. Goldwasser, ed.), vol. 403 of *Lecture Notes in Computer Science*, pp. 564–577, Springer-Verlag, 1990.
11. Z. J. Czech, G. Havas and B.S. Majewski, Perfect Hasing, *Theoretical Computer Science* **182**(1997), 1-143.
12. R.A. Croft and S.P. Harris, Public-key cryptography and re-usable shared secrets,*Cryptography and coding* (Beker and Piper eds.), Clarendon Press, 1989, 241-246.

13. Y. Desmedt, Society and group oriented cryptology: a new concept, *Advances in Cryptography–Crypto'87*, Lecture Notes in Comput. Sci. **293**, 1988, 120-127.

14. Y. Desmedt, Threshold cryptography, *European Trans. on Telecommunications*, 5(4), 1994, 449-457.

15. Y. Desmedt, Some recent research aspects of threshold cryptography, 1997 *Information Security Workshop, Japan* (JSW '97), LNCS 1396, 99-114.

16. Y. Desmedt, G. Di Crescenzo and M. Burmester, Multiplicative non-abelian sharing schemes and their application to threshold cryptography, *Advances in Cryptology–Asiacrypt '94* Lecture Notes in Comput. Sci. **435**, 1994, 21-32.

17. A. De Santis, Y. Desmedt, Y. Frankel and M. Yung, How to share a function securely, In *Proc. 26th Annual Symp. on the Theory of Computing*, pages 522-533, ACM, 1994.

18. Y. Desmedt and Y. Frankel, Homomorphic zero-knowledge threshold schemes over any finite group, *SIAM J. Disc. Math.* **7** 4(1994), 667-679.

19. Y. Desmedt, B. King, W. Kishimoto and K. Kurosawa, A comment on the efficiency of secret sharing scheme over any finite abelian group, *Information Security and Privacy*, ACISP'98 (Third Australasian Conference on Information Security and Privacy), Lecture Notes in Comput. Sci. **1438**, 1998, 391-402.

20. P. Feldman, A practical scheme for non-interactive verifiable secret sharing. In *Proc. 28th Annual FOCS*, 437-437, IEEE, 1987.

21. Y. Frankel, A practical protocol for large group oriented networks, In *Eurocrypt '89*, pages 56-61, 1989. Springer-Verlag. LNCS 434.

22. Y. Frankel, P. Gemmell, P. MacKenzie and M. Yung, Optimal resilience proactive public-key cryptosystems, in *Proc. 38th FOCS*, 384-393, IEEE, 1997.

23. Y. Frankel, P. Gemmell and M. Yung, Witness-based Cryptographic Program Checking and Robust Function Sharing, In *Proc. 28th STOC*, 499-508, ACM, 1996.

24. Y. Frankel, P. MacKenzie and M. Yung, Robust efficient distributed RSA-key generation, in *Proc. 30th STOC*, 663-672, ACM, 1998.

25. R. Gennaro, S. Jarecki, H. Krawczyk and T. Rabin, Robust and efficient sharing of RSA functions, In *Crypto'96*, pages 157-172, 1996. Springer-Verlag, LNCS 1109.

26. O. Goldreich, S. Micali and A. Wigderson, Proofs that yield Nothing but the validity of the assertion, and a methodology of cryptographic protocol design, In *Proceeding 27th Annual Symposium on the Foundations of Computer Science*, pages 174-187. ACM, 1986.

27. E. Karnin, J. Greene, and M. Hellman, "On Secret Sharing Systems," *IEEE Transactions on Information Theory*, vol. IT-29, pp. 35–41, Jan. 1983.

28. K. Mehlhorn, *Data Structures and Algorithms*, Vol. 1, Springer-Verlag, 1984.

29. S. Miyazaki, K. Sakurai and M. Yung, On threshold RSA-signing with no dealer, In *PreProc. of ICISC99*, 187-197, 1999 (see also this proceedings).

30. T. Rabin, A simplified Approach to Threshold and Proactive RSA, In *Crypto'98*, pages 89-104, 1998. Springer-Verlag, LNCS 1109.

31. A. Shamir, "How to Share a Secret," *Communications of the ACM*, vol. 22, pp. 612–613, Nov. 1979.

32. G. Simmons, W.-A. Jackson and K. Martin, The Geometry of Shared Secret Schemes, *Bulletin of the Institute of Combinatorics and its Applications (ICA)*, Vol.1, pp. 71-88, 1991.

33. M. Tompa and H. Woll, "How To Share a Secret with Cheaters," *Journal of Cryptology*, vol. 1, no. 2, pp. 133–138, 1988.

On Threshold RSA-Signing with no Dealer

Shingo Miyazaki[1], Kouichi Sakurai[2], and Moti Yung[3]

[1] Toshiba Corporation, System Integration Technology Center
3-22, Katamachi, Fuchu-shi, Tokyo 183-8512, Japan
shingo@sitc.toshiba.co.jp
[2] Kyushu Univ. Department of Computer Science
6-10-1, Hakozaki, Higashi-ku, Fukuoka, 812-8581 Japan
sakurai@csce.kyushu-u.ac.jp
[3] CertCo, NY, USA
moti@certco.com, moti@cs.columbia.edu

Abstract. We consider methods for threshold RSA decryption among distributed agencies without any dealer or trusted party. We present two methods: One is based on the previous two techniques by [FMY98] and [FGMY97]. It demonstrates the feasibility of combining the distributed key generation and the RSA secure function application. The other method [MS99] is newly developed technique based on [FMY98] and further inspired by Simmons' protocol-failure of RSA (we believe that it is very interesting that a "protocol failure attack" be turned into a constructive method!). The latter requires less "distributed computation" as the key is being set up and it can be more smoothly incorporated into the existing distributed key generation techniques.

1 Introduction

The area of distributed cryptography has been very active in the last few years. In particular threshold cryptosystems and proactive cryptosystems have been developed to allow for distribution of the power to perform signatures or decryption in an organization. It is a very interesting key management technique where the outside world does not get exposed to the internals of the organization (see surveys in [Des92,FY98]).

The distributed RSA systems were developed but (unlike the case of Discrete Log based distributed systems which was known for a while but was corrected recently) the key generation was assumed to be done by a centralized dealer [DF91,DDFY94,FGY96,GJKR96,FGMY97,R98]. The issue of initiating and further operating using a distributed parties was open for a while.

Boneh and Franklin [BF97] changed this situation and showed how a set of three or more participants can generate an RSA function distributedly. Their solution assumed honest parties. It was generalized to withstand faults by Frankel, MacKenzie and Yung [FMY98] extends Boneh-Franklin's scheme of (n, n) to robust (t, n) threshold scheme. In the scheme of Frankel-MacKenzie-Yung. The scheme was motivated as an initiation of a distributed RSA service where the

JooSeok Song (Ed.): ICISC'99, LNCS 1787, pp. 197–207, 2000.
© Springer-Verlag Berlin Heidelberg 2000

key is never held at a single location. How to connect the scheme of generation to further signing was not explicitly expressed in their work. We take the step to validate that indeed we can connect the signing methods with the generation methods. Indeed, with the strong "share representation" modification techniques from [FGMY97] changing key representation and adapting representations between schemes looks doable. In fact, one of our method is essentially realizing such a scheme, demonstrating how to start a distributed RSA service from "key birth" to its actual usage.

Our Contribution:

We believe that closing the various gaps which were left over and assuring that a complete distributed RSA service can be performed is quite an interesting issue, which we cover in this work. We, in fact, give two techniques to treat the problem of bridging the "distributed RSA application" to a starting step of "distributed key generation" (i.e., with no dealer). We give RSA function application (signing and decrypting) which is initiated directly from the distributed generated RSA key.

We present two methods. The first is based on the previous two techniques by [FMY98] and [FGMY97]; this checks carefully the feasibility and feels the details on how to start the RSA service from the distributed generation. The other method is newly developed technique which, interestingly, is inspired by Simmons' protocol-failure attack on RSA.

The latter method requires less "distributed computation": especially in adapting the key generation to function application period. Thus, our newly proposed scheme seems overall more efficient compared to the combination of the previous methods. The former method, however, requires less computation at the combiner– so may be useful for small devices at the combining function.

Note that such comparison of these two schemes indicates a new measure of the performance of a distributed cryptographic protocol that consists of multiple stages, as we point at the need to look at the combined performance. Tradeoffs have to be assessed based on the computational context and we follow this.

Related Works:

Shoup [Sho99] independently presented threshold RSA-signatures by using the GCD-computation as our second method. Shoup's model requires a trusted dealer, whereas ours assumes no trusted dealer. In Shoup's schemes, the size of an individual signature is bounded by a constant times the size of the RSA modulus. Our second method achieves the similar property, while the size of an individual share increase as the number of the distributed agencies and as the number of the threshold.

2   The Starting Point: Threshold (Proactive) RSA with a (Trusted) Dealer

Frankel, Gemmell, MacKenzie and Yung [FGMY97] presented a protocol for RSA function sharing in the dealer-model. We start be reviewing their scheme. (See [FGMY97] for the detail.)

2.1   Key Distribution

The dealer distribute a secret key to the n shareholders with threshold t.

The dealer first generates a (variant of) RSA public key (e, N) and its secret key d. Let $L = (n-1)!$ and $H = \gcd(e, L^2)$. (Indeed [FGMY97] and [FMY98] describe $L = n!$. However, a smaller $(n-1)!$ is OK.). Then, the dealer computes $(P, \tilde{s})$ such that $eP + \frac{L^2}{H^2}\tilde{s} = 1$ by using the extended Euclidian algorithm. Next, the dealer computes k such that $d \equiv P + L^2 k$ (mod $\phi(N)$). Note that the relation $k \equiv d\tilde{s}H^{-1}$ (mod $\phi(N)$) holds.

Now the dealer chooses a random polynomial

$$f(x) = f_0 + f_1 x + f_2 x^2 + \ldots + f_{t-1} x^{t-1},$$

where $f_j \in_R \{0, L, \ldots, 2L^3 n^{2+\epsilon} t\} (1 \le j \le t-1)$ and $f(0) = L^2 k$.

Finally, the dealer computes $s_j = f(j)$, then secretly sends (P, s_j) to each shareholder P_j.

2.2   Distributed Decryption

We consider how to decrypt by a set of t shareholders among the n shareholders.

Suppose a client has a ciphertext

$$C = M^e \pmod{N}$$

encrypted with the dealer's public key (e, N). The client wants to decrypt this encrypted message C by asking t shareholders, Λ.

Step.1: The client sends C to each shareholder P_j ($\in \Lambda$).

Step.2: Each shareholder P_j first computes $\sigma_j = s_j \lambda_{j,\Lambda}$ with his (distributed) secret s_j, where

$$\lambda_{j,\Lambda} = \prod_{l \in \Lambda \setminus \{j\}} \frac{l}{l - j}$$

, then computes $Z_j = C^{\sigma_j} \pmod{N}$. Each shareholder P_j secretly sends Z_j to the client via secure channel.

Some shareholder should send also $C^P \pmod{N}$ to the client (Note that this might be redundant option, because anybody can compute $(P, \tilde{s})$.)

Step.3: From received pieces Z_j and $M^P \pmod{N}$, the client decrypts the message M as follows:

$$C^P \prod Z_j = C^{P + \sum_j s_j \lambda_{j,\Lambda}}$$
$$= M^{ed}$$
$$= M \pmod{N}.$$

We remark that the scheme described above can be applicable to RSA-signature schemes, where the client asks to get agencies signature M^d on his selected message M (thanks to the same mechanism of RSA-signing as RSA-decoding).

3   What Is the Technical Obstruction ?

We would like to start distributed decryption and signature based on the distributed generation (of course, without reconstructing the key at a centralized dealer!).

Now we discuss the model without a dealer. The scheme in [FGMY97] as is cannot be applicable to our case, because [FGMY97] assumes a trusted dealer.

Frankel, MacKenzie and Yung [FMY98] proposed methods for distributed key generating and (t, n)-sharing based on the Boneh-Franklin's technique [BF97]: first key generating of (n, n)-sharing, then compute a polynomial of sum w.r.t. the secret key d. Each each shareholder P_j plays as a dealer to the shared secret d_j and do the transform of "Sum-to-Poly," After (t, n)-sharing of the secret d, any t members of the shareholders can recover the secret key d.

Now again suppose that a client has a ciphertext

$$C = M^e \pmod{N}$$

encrypted with the dealer's public key (e, N). The client wants to decrypt this encrypted message C by asking t shareholders, Λ. In the direct method by [FGMY97] to this case, each shareholder P_j needs to compute the coefficients of Lagrange's interpolation

$$\lambda_{j,\Lambda} = \prod_{l \in \Lambda \setminus \{j\}} \frac{l}{l - j} \pmod{\phi(N)}$$

Note, however, that no single shareholder knows the complete factorization of N. Thus, shareholders cannot compute the inverse of $l - j\phi(N)$. Namely, $\lambda_{j,\Lambda}$ is not obtained. Then, simply as described in [FMY98] (without some additional measures), shareholder cannot execute the distributed computation:

$$\prod_{j \in \Lambda} C^{s_j \cdot \lambda_{j,\Lambda}} = C^d = M \pmod{N}$$

The paper is dedicated to feeling this gap and to showing how to actually do it.

4   Scheme A: Combining [FMY98] with [FGMY97]

In our first designed scheme, we combine the previous two schemes [FGMY97], [FMY98]. This checks carefully the fact that indeed the distributed key generation can serve as a starting point for a distributed RSA service where the key is never known to any party. This feasibility demonstrates the strength of the "sharing representation" modification techniques. The method takes the following steps. First, each agency uses the [FMY98]-method for (n, n)-sharing in the model of no-dealer. Second, each agency plays as the dealer for his shared secret by using the [FGMY97]-method.

4.1 Key Generation and Secret Key Sharing

Step 1: Do distributing key generation of (n, n)-sharing in the model of no-dealer by using [FMY98]-method. Now, each agency $\mathcal{P}_j$ keeps his shared secret d_j satisfying

$$d = d_1 + d_2 + \cdots d_n$$

Step 2: Each agencies $\mathcal{P}_j$ plays as the dealer for his shared secret key d_j by using the technique of [FGMY97]: $\mathcal{P}_j$ computes (k_j, v_j) such that $d_j = L^2 k_j + v_j$, then choose a polynomial

$$f_j(x) = L^2 k_j + f_{j,1} x + \cdots + f_{j,t-1} x^{t-1}$$

$\mathcal{P}_j$ sends $s_{j,i} = f_j(i)$ to $\mathcal{P}_i$ $(1 \leq i \leq n)$. (In fact, commitment scheme and Pedersen's sharing which are both unconditionally hiding, are used). For each shareholder but the last, v_j is sent to the last player with a public commitment. The last player adds all the shares it got and this is his new share. He computes P using the extended Euclid and his share is represented as $d_n = L^2 k_n + P$, he distributes $L^2 k_n$ and proves that it equals the summed shares plus its own original minus P.

Step 3: Each agency $\mathcal{P}_j$ verifies the correctness of $s_{i,j}$ received from n agencies. If this is OK, computes $S_j = \sum_{i=1}^{n} s_{i,j}$ and keep it. Note that S_j is the value $S_j = F(j)$ of the polynomial $F(x)$

$$F(x) = L^2 K + F_1 x + \cdots F_{t-1} x,$$

where $K = \sum_{j=1}^{n} k_j$ and $F_i = \sum_{j=1}^{n} f_{j,i}$ $(1 \leq i \leq t - 1)$

4.2 Distributed Decryption/Signature

Any t shareholders of n can decrypt/sign. Let Λ be one group of t shareholders Set $V = \sum_{j=1}^{n} v_j$.

Step 1: $\mathcal{U}$ sends the ciphertext C to each shareholder $\mathcal{P}_j$.

Step 2: Each shareholder $\mathcal{P}_j$ $(j \in \Lambda)$ execute the following:

$$\lambda_{j,\Lambda} = \prod_{l \in \Lambda \setminus \{j\}} \frac{l}{l - j}$$
$$\sigma_j = s_j \cdot \lambda_{j,\lambda}$$
$$\sigma_j = sum - to - sum(\sigma_i, i \in \Lambda)$$
$$Z_j = C^{\sigma_j} \pmod{N}$$

(In the above, the sum-to-sum assures that the individual fresh shares have the property that their indeed sum of the σ-s stays the same, but each sub-sum is completely random).

Then, send the result Z_j to $\mathcal{U}$.

Step 3: $\mathcal{U}$ recovers M from the public information V and t results:

$$C^P \prod_{j \in \Lambda} Z_j = C^{L^2 K + V}$$

$$= C^{\sum_{j=1}^n (L^2 k_j + v_j)}$$

$$= M^{e \sum_{j=1}^n d_j}$$

$$= M \pmod{N}$$

4.3   Notes

- Actually not n but only $t + 1$ players need to generate the key. If one misbehaves it is eliminated and the process restart. There are at most t misbehaving parties and at the end $t + 1$ where one of them is honest distributes to the n parties the shares of d.
- Decryption/signing is as in [FGMY97].
- Note that, in signing version, the ciphertext C is replaced by the message M.
- In Scheme-A, at the key generation and the secret sharing, on **Step.3** each shareholder should check the correctness of all $s_{i,j}$ that received from $\mathcal{P}_i$ (for all i). This is for Robustness. Also share randomization needs to be supported by robustness tools (see [FGMY97]).

4.4   Security

The initial distribution is secure, due to the distributed key generation of [FMY98] (given an adversary we can simulate the view). Then the distribution of the d_i is robust and results in a t-out-of-n sharing of a $L^2 k$ and the same P is the public part. This is simulatable using the sum-to-poly arguments in [FGMY97] and the public unconditionally concealing commitments. The adversary controlling at most t agents has a view which is simulatable. The signing operation is simulatable as well as in [FGMY97].

The operations are robust, from [FGMY97,FMY98], we can add checks and elimination of misbehaving parties.

5   Scheme-B: Based on GCD-Decoding

In our second presented scheme, the client decodes the ciphertext by using the extend Euclidian algorithm. This trick is quite known as in Simmons RSA protocol-failure [Sim83]. For the signing part, we start without robustness assuming honest but curious behavior) and then we add robustness.

5.1   The Common Modulus Protocol Failure in RSA

We review the common modulus protocol-failure in RSA schemes remarked by Simmons [Sim83]. The common modulus protocol-failure is that if the same message is ever encrypted with two different exponents under the same modulus, and those two exponents are relatively prime, then the plaintext can be recovered without either of the decryption exponents by using the extended Euclidean algorithm.

Let m be the plaintext message. The two different encryption keys are e_1 and e_2. The common modulus is N. The corresponding two ciphertext are $c_1 = m^{e_1} \pmod{N}$, $c_2 = m^{e_2} \pmod{N}$. We consider the cryptanalyst who knows N, e_1, e_2, c_1 and c_2. The cryptanalyst is able to recovers in the following way. Since e_1 and e_2 are relatively prime, the extended Euclidean algorithm can find r and s satisfying the relation $re_1 + se_2 = 1$. Note that either r or s has to be negative, so we assume that r is negative. Then, the cryptanalyst can use again the extended Euclidean algorithm for computing c_1^{-1}. Thus, the plaintext message m is recovered by the computation that $(c_1)^{-r}(c_2)^s = m \pmod{N}$.

Thus, each party must choose its own RSA modulus. This protocol failure is a negative aspect of RSA scheme, however, our new technique shows that the trick has a positive cryptographic application.

5.2   Distributed Key Generation and Secret Key Sharing

In the model of non-dealer, by using Frankel-MacKenzie-Yung's secret key sharing scheme [FMY98], n shareholders $\mathcal{P}_j$ jointly generates RSA public-key (e, N) and share the secret key d with (t, n)-secret sharing which is simulatable (over a subset of the integers) where $a_1, \ldots, a_{t-1} \in Z$, and:

$$f(x) = d + a_1 x + a_2 x^2 + \cdots + a_{t-1} x^{t-1}$$

At this point they are ready to perform distributed operations. This part is robust and secure based on [FMY98].

5.3   Distributed Decryption Protocol

The client $\mathcal{U}$ asks t-shareholders, Λ to decrypt the ciphertext C. Recall that $L = (n-1)!$ and $\gcd(e, L^2) = 1$.

Step 1: $\mathcal{U}$ sends C to each shareholders $\mathcal{P}_j (\in \Lambda)$.
Step 2: Each shareholders $\mathcal{P}_j$ computes:

$$\lambda_{j,\Lambda} = \prod_{l \in \Lambda \setminus \{j\}} \frac{l}{l - j}$$
$$\sigma_j = s_j \cdot L^2 \cdot \lambda_{j,\lambda}$$
$$\sigma_j = sum - to - sum(\sigma_i, i \in \Lambda)$$
$$Z_j = C^{\sigma_j} \pmod{N}$$

then send partial-result Z_j to the combiner $\mathcal{U}$.

Step 3: $\mathcal{U}$ computes based on the actual t shareholders: $M^{L^2} \pmod{N}$:

$$\prod_{j \in \Lambda} Z_j = (M^e)^{L^2 \cdot \sum_{j \in \Lambda}(s_j \cdot \lambda_{j,\Lambda})} = M^{L^2} \pmod{N}$$

Step 4: $\mathcal{U}$ decrypts M from a pair of the different ciphertexts $(C_1, C_2) = (M^{L^2}, M^e)$:

[4a]: $a_1 = (L^2)^{-1} \pmod{e}$
[4b]: $a_2 = (a_1 L^2 - 1)/e$
[4c]: $M = C_1^{a_1}(C_2^{a_2})^{-1} \pmod{N}$

Note that this procedure is sound because

$$C_1^{a_1}(C_2^{a_2})^{-1} = M^{L^2 a_1 - e a_2} = M \pmod{N}$$

5.4   Security

The initial distribution of d using Pedersen's (in large enough subset of the integers) based sum-to-poly is secure and simulatable as was shown in [GMY98] (so that t-1 shares can be picked at random. For arguing the security of signing: Given the outcome M (which the simulator has), then $M^{L^2} = C_1$ can be computed by the simulator and given $t - 1$ partial results Z_j (which are $t - 1$ wise independent) we can use the equation in step 3 to compute the missing Z_t (say), simply by dividing mod N the result M^{L^2} by the partial results.

5.5   A Version for Signing

Scheme-B can be modified for threshold signing. Instead of the ciphertext C, we give a message M to t agencies. Then, $\mathcal{U}$ get the following two data:

$$\begin{cases} M = (M^d)^e \pmod{N} \\ \prod_{j \in \Lambda} C^{\sigma_j} = (M^d)^{L^2} \pmod{N} \end{cases}$$

$\mathcal{U}$ computes (α, β), by using the extended Euclidean algorithm, satisfying

$$e\alpha + L^2\beta = 1$$

$\mathcal{U}$ can get the signature with (α, β):

$$M^\alpha \left(\prod_{j \in \Lambda} C^{\sigma_j}\right)^\beta = (M^d)^{e\alpha + L^2\beta} \Rightarrow M^d$$

5.6   Robust-Version

Of course, key distribution and re-distribution are robust as in [FMY98], [FGMY97]. The client adds the following protocol to Step 2 for checking whether each shareholder P_j computes the partial result Z_j with its share s_j correctly. Let g be a generator of Z_N^*. Now, each party P_j publishes $h_j = g^{s_j} \pmod{N}$ and the client has the partial result Z_j from P_j. The client can verify the validity of each partial result by applying Pedersen's technique [Ped91a].

Step 1: $\mathcal{U}$ chooses $r_{j1}, r_{j2} \in Z$ randomly and then computes $\delta_j = L^2 \lambda_{j,\Lambda}$ and $X_j = g^{r_{j1}} C^{\delta_j r_{j2}} \pmod{N}$. $\mathcal{U}$ sends X_j to each party P_j.

Step 2: P_j generates a random number $\alpha_j \in Z$ and computes $Y_{j1} = X_j^{\alpha_j} \pmod{N}$ and $Y_{j2} = Y_{j1}^{s_j} \pmod{N}$. P_j transfers (Y_{j1}, Y_{j2}) to $\mathcal{U}$.

Step 3: $\mathcal{U}$ sends (r_{j1}, r_{j2}) to P_j.

Step 4: P_j check that $X_j = g^{r_{j1}} C^{\delta_j r_{j2}} \pmod{N}$ and transfers α_j to $\mathcal{U}$ if and only if the validity of X_j is accepted.

Step 5: $\mathcal{U}$ can verify the validity of each partial result Z_j via the following formula:

$$(h_j^{r_{j1}} Z_j^{r_{j2}})^{\alpha_j} = Y_{j2} \pmod{N}.$$

6   Comparison: Scheme-A vs. Scheme-B

We discuss the comparison between Scheme-A and Scheme-B.

6.1   Extend Euclidian Algorithm

In Scheme-B, the ciphertext is decoded by using the extend Euclidian algorithm at the last stage of the client as the same mechanism as the protocol-failure in RSA [Sim83]. We should note that the extend Euclidian algorithm plays an important role also in Scheme-A (and in [FGMY97]), in the key distribution stage. So, an apparent difference between these schemes is the stage when the extend Euclidian algorithm is executed. (see Table on the comparison). Scheme-A requires maintenance of long term keys as multiples of L^2 and Scheme-B gets rid of this requirement.

	[FGMY97]	Scheme A	Scheme B
Model	Dealer	Dealer/Non-Dealer	Dealer/Non-Dealer
Euclidean formula	$eP + \frac{L^2}{H^2}\bar{s} = 1$	same as FGMY97 (public P)	$ea_1 - L^2 a_2 = 1$
When is EA done ?	At key distribution	At key distribution	Every Dec./Sig.

EA: the (extended) Euclidean Algorithm

6.2   Computation by Multi-parties vs by Single-party

With respect to key generation/distribution stages, Scheme-B is somewhat simpler than Scheme-A. Scheme-A requires much distributed computation in its set-up (due to divisibility conditions). Given that a system is initiated distributedly where d is shared in the right domain, Scheme-B is simpler. Only in cases (e.g. palm-top devices [BD99]) when we may want to perform combining while minimizing multiplication and exponentiation, Scheme-A may be useful, because GCD combining is somewhat more costly than simply combining the partial results in scheme-A where only one exponentiation and $t - 1$ multiplications are performed.

References

[BD99]   D. Boneh and N. Daswani. "Experimenting with electronic commerce on the PalmPilot," Proc. of Financial Cryptography '99, LNCS, Vol. 1648, Springer-Verlag, pp. 1–16, 1999.

[BF97]   D. Boneh and M. Franklin, *"Efficient generation of shared RSA keys,"* Advances in Cryptology – CRYPTO '97, LNCS 1294, pp. 425-439, 1997.

[DDFY94] A. De Santis, Y. Desmedt, Y. Frankel, and M. Yung. *"How to share a function securely (extended summary)."* In *Proceedings of the Twenty-Sixth Annual ACM Symposium on the Theory of Computing*, pages 522–533, Montréal, Québec, Canada, 23–25 May 1994.

[Des92]   Y. Desmedt. *"Threshold cryptosystems" (a survey).* In Auscrypt'92, LNCS 718, pages 3–14.

[DF91]   Y. Desmedt and Y. Frankel. *Shared generation of authenticators and signatures (extended abstract).* In CRYPTO'91 CRYPTO91, pages 457–469.

[DF89]   Y. Desmedt and Y. Frankel, *"Threshold cryptosystems,"* Advances in Cryptology – CRYPTO '89, LNCS 435, pp. 307-315, 1989.

[FGMY97] Y. Frankel, P. Gemmell, P. D. MacKenzie and M. Yung, *"Optimal-resilience proactive public-key cryptosystems,"* 38th Annual Symposium on Foundations of Computer Science, pp. 384-393, 1997.

[FGY96]   Y. Frankel, P. Gemmell, and M. Yung. *Witness-based cryptographic program checking and robust function sharing.* In STOC'96, pages 499–508.

[FMY98]   Y. Frankel, P. D. MacKenzie and M. Yung, *"Robust efficient distributed RSA-key generation,"* Proceedings of the thirtieth annual ACM symposium on theory of computing, pp. 663-672, 1998.

[FMY99]   Y. Frankel, P. D. MacKenzie and M. Yung, *"Pseudorandom Intermixing": A Tool for Shared Cryptography,"* To appear in Proc. of PKC'2000, International Workshop on Practice and Theory in Public Key Cryptography, January 18-20, 2000, Melbourne, Australia.

[FY98]   Y. Frankel and M. Yung, *"Distributed public key cryptosystems,"* (Invited Paper) Public Key Cryptography, LNCS 1431, pp. 1-13, 1998.

[GJKR96] R. Gennaro, S. Jarecki, H. Krawczyk, and T. Rabin. *"Robust and efficient sharing of RSA functions."* In *Advances in Cryptology—CRYPTO '96*, volume 1109 of *Lecture Notes in Computer Science*, pages 157–172. Springer-Verlag, 18–22 Aug. 1996.

[MS99]   S. Miyazaki and K. Sakurai, *"Notes on thershold schemes in distributed RSA cryptosystems"* (in Japanese), Proc. the 1999 Symposium of Cryptography and Information Security (SCIS'99), T1-3.1, pp.451-456. (June 1999).

[Oka97]   T. Okamoto, *"Threshold key-recovery systems for RSA,"* Security Protocols, LNCS 1361, pp. 191-200, 1997.

[Ped91a] T. P. Pedersen, *"Distributed provers with applications to undeniable signatures,"* Advances in Cryptology – Eurocrypt '91, LNCS 547, pp. 221-238, 1991.

[Ped91b] T. P. Pedersen, "A threshold cryptosystem without a trusted party," Advances in Cryptology – Eurocrypt '91, LNCS 547, pp. 522-526, 1991.

[R98]   T. Rabin. A simplified approach to threshold and proactive rsa. In Krawczyk CRYPTO98, pages 89–104.

[Sha79]   A. Shamir, *"How to share a secret,"* Communication ACM, 22, pp. 612-613, 1979.

[Sho99]   V. Shoup, "Practical threshold signatures," June 24, 1999, A revision of IBM Research Report RZ 3121 (APril 19, 1999) available from Theory of Cryptography Library 99-11, http://philby.ucsd.edu/cryptolib/1999.html.

[Sim83]   G. J. Simmons, "A 'weak' privacy protocol using the RSA cryptoalgorithm," Cryptologia, vol. 7, pp. 180-182, 1983.

A New Approach to Efficient Verifiable Secret Sharing for Threshold KCDSA Signature

Ho-Sun Yoon and Heung-Youl Youm

Department of Electrical and Electronic Engineering,
College of Engineering, Soonchunhyang University, Korea
hsyoon@elec.sch.ac.kr, hyyoum@asan.sch.ac.kr

Abstract. This paper is to propose a threshold KCDSA (Korean Certification-based Digital Signature Algorithm) signature scheme. For this goal we present some new secret sharing schemes and a robust multiplication protocol based on the non-interactive ZK proof. The secret sharing scheme using a hash function can verify the validity of the received data easily, therefore we can reduce the complexity of the protocol. Also we present a robust multiplication protocol of two shared secrets based on non-interactive ZK proof scheme. Players check the validity of the broadcasted shares by using a non-interactive ZK proof, and accept only the correct shares, and then interpolate the polynomial by using the accepted shares. Finally we propose a threshold KCDSA signature, which is composed of key-sharing protocol and signature generation protocol. We prove that the proposed KCDSA signature is a (t, n)-robust threshold protocol, which tolerates up to t eavesdropping and halting faults if the number of players is $n \geq 2t + 1$.

1 Introduction

The highly advanced cryptography and the network security techniques have been developed during the past twenty years. In particular a number of cryptographic techniques have been introduced to multiparty computation problem for the secrecy and integrity of information. The combined techniques become extremely powerful tools for the cryptographic application, but some of these results have a lack in practical feasibility.

Shamir firstly introduces the concept of secret sharing scheme in [2]. In Shamir's scheme a misbehaving dealer can send the inconsistent shares to any participants, from which they will not be able to reconstruct a correct secret collaboratively. To prevent such a malicious behavior of the dealer, Feldman and Pedersen's verifiable secret sharing schemes in [3,4,5] are proposed respectively.

For the long lifetime of the shared secret, a proactive secret sharing scheme is presented in [6]. This technique refreshes the shared secrets to improve the security of the shared secrets during a time period. Namely, what is actually required to protect the secrecy of the information is to be able to periodically update the shares without changing the original content of a secret. Proactive secret sharing scheme is consisted of three protocols, i.e., the private key renewal protocol, the share recovery protocol, and the share renewal protocol. Also the on-line secret sharing scheme dynamically

JooSeok Song (Ed.): ICISC'99, LNCS 1787, pp. 208-220, 2000.

can add more participants on line, without having to redistribute new shares secretly to the current participants [7,8,9].

These Secret sharing schemes are developed for the various applications. The threshold signature system for DSS, the Digital Signature Standard, was studied in [1,10] etc. Also, in particular approach to public key cryptosystem, the results of some studies were published, namely [11,12,13] for the case of RSA signatures, and [14] for ElGamal type of signatures.

In this paper, we present new verifiable secret sharing scheme using a hash function and a robust multiplication scheme for two secrets. The presented secret sharing scheme allows for verifying the validity of the received values easily, but it can not check whether the received values are derived from some polynomial or not. However, in reconstruction phase we can check whether the shares are correct or not. If players find some problems, then a group of players perform a sharing phase and a reconstruction phase to detect the misbehaving players by using Pedersen's VSS or Feldman's VSS. Also we present a robust protocol to compute a multiplication protocol of two shared secrets. This protocol is using the non-interactive ZK proof.

We describe the existing secret sharing schemes in section 2, and some basic protocols, such as a verifiable secret sharing using a hash function, and a robust multiplication protocol based on the non-interactive ZK proof are presented in section 3. The basic operation of KCDSA is presented in section 4, and in section 5 we show how the presented schemes are combined jointly and securely to generate the KCDSA private/public key generation, and we present a robust threshold KCDSA signature scheme. Finally section 6 summarizes the results.

2   Existing Secret Sharing Schemes

In this section we review a few known secret techniques.

2.1   Shamir's Secret Sharing

This scheme has a trusted dealer having authority in distributing his share to each player. If the dealer wants to distribute a secret α, then he chooses randomly a polynomial $f(x)$ of degree t, such that $f(0)= \alpha$. Dealer sends the computed shares $P_i = f(i)$, to players, for $i=\{1, \ldots , n\}$. The coalition of t (or more) players can interpolate a polynomial to evaluate secret α. In general, this secret sharing scheme is denoted by $(\alpha_1, \cdots, \alpha_n) \xleftarrow{(t,n)} \alpha$.

2.2   Verifiable Secret Sharing

In this scheme, each player wants to verify the validity of the received information from other players. A lot of VSS schemes have been proposed so far. Theses schemes adopt the methods that commit to all coefficients of polynomials, and then each player verifies the received information from dealer (or others players) using the committed information. The VSS scheme is based on computational secrecy, or on information theoretic secrecy.

2.3   Joint Random Secret Sharing

In a joint random secret sharing scheme all players act as dealer, i.e. all players collectively choose their shares corresponding to a (t, n) threshold secret sharing of a random value. All players obtain their shares by adding the received partial information, and then the coalition of t (or more) players can interpolate to evaluate secret. Each player may verify the validity of the received information by using VSS.

2.4   Multiplication of Two Secrets

Given two secrets α and β that are both shared among the players, compute the product $\alpha\beta$, while maintaining both of original values secret. In (t, n) threshold secret sharing scheme, given that α and β are each shared by a polynomial of degree t+1, each player can locally multiply his shares of α and β, and the result will be a share of $\alpha\beta$ on polynomial of degree 2t. Namely, coalition of 2t+1 (or more) players can interpolate the polynomial to evaluate secret $\alpha\beta$, further this scheme needs joint zero secret sharing for re-randomizing procedure.

To overcome the above shortcoming, a new efficient protocol that reduces degree of polynomials and don't have to use the re-randomizing polynomials in a single step have been proposed[1]. However, this scheme needs three joint random secret sharing and an inverse computation of a (2t+1) by (2t+1) matrix. 2t+1 players must cooperate to generate the secret.

2.5   Exponential Interpolation Scheme

Given every player has a share α_i of the secret information α, the exponential interpolation scheme is to interpolate the g^α instead of α. Each player distributes his share using the Shamir's secret sharing scheme. The coalition of t+1 (or more) players can interpolate to evaluate secret g^α by using Lagrange interpolation coefficients.

2.6   Proactive Secret Sharing Scheme

In a (t, n)-threshold scheme, an adversary needs to compromise more than t players in order to learn the secret, and corrupt at least n-t shares in order to destroy the information. If the lifetime of secret is long, then an adversary may attack over a long period of time. Therefore, to protect long-lived secrets we need to periodically refresh the secrets. Proactive secret sharing scheme satisfying the requirements was presented in [6]. Proactive secret sharing scheme consists of three protocols as follows:

1. The private key renewal protocol
2. The share recovery protocol (including lost share detection)
3. The share renewal protocol

One can keep the KCDSA signature key securely for a long time by using the proactive secret sharing scheme while its shares can be refreshed periodically. An adversary trying to break the threshold signature scheme needs then to corrupt t servers in one single period of time, as opposed to having the whole lifetime of the key to do so.

3   Basic Building Blocks

In this section we present some basic VSS protocols. These protocols can be applied to the threshold KCDSA signature or other threshold applications.

3.1   A VSS Using Hash Function

In general, VSS based on Feldman or Pedersen's secret sharing scheme will allow the players to check the validity of the received shares from the others players, but VSS have the shortcoming that amount of computation is too much. Feldman's scheme guarantees that the security of the secret is only computationally secure; on the other hand Pedersen's scheme guarantees a theoretic secrecy for the shared secret.

For our protocol we need some assumptions for a hash function as follows:

1. $H(\bullet)$ indicates a commitment function. We can denote a commitment function by a hash function.
2. It is infeasible to find two strings x and y such that $H(x) = H(y)$. This property is known as a collision-resistant property of the hash function.
3. $H(\bullet)$ should be easy to compute.
4. $H(\bullet)$ should be one way. That is, given $y=H(x)$ it is impossible to find x computationally, but y can compute easily in $y=H(x)$.

We assume that $H(\bullet)$ is one way hash function satisfying above conditions. The security of $H(\bullet)$ can however be only conjectured on the basis of the collision resistance of the hash function. Namely, we assume that $H(\bullet)$, i.e. one way hash function, is secure.

In this subsection we present a VSS using hash function to reduce computational complexity. This protocol can verify the correctness of received shares by using the hash function, but it can't prove that the committed values are obtained from polynomial f(x) of degree at most t, such that f(0)=s. In Figure 1, H means a general arbitrary hash function.

1. Sharing Phase
- Choose two random polynomials of degree t.

$$f(x) = a_t x^t + \cdots + a_1 x + a_0$$

where, $a_t = H(a_{t-1}, \cdots, a_1, a_0)$

$$g(x) = b_t x^t + \cdots + b_1 x + b_0$$

where, $b_t = H(b_{t-1}, \cdots, b_1, b_0)$

- Compute the following values.

$$\alpha_i = f(i), \quad e_i = g(i)$$

- Compute $E_j = H(\alpha_j, e_j), j = 1, \cdots, n$.

Compute $A_j = g^{a_i} h^{a_i} \bmod p, \; for\, i = 0, \cdots, t \; j = 1, \cdots, n$.

- Broadcast E_j, A_j

2. Reconstruction Phase
- Collect more t+1 shares.
- Interpolate $\hat{f}(x)$ and $\hat{g}(x)$
- Compute $\hat{\alpha}_i = \hat{f}(i)$ and $\hat{e}_i = \hat{g}(i)$
- Check the following equations.

$$a_t \stackrel{?}{=} H(a_{t-1}, \cdots, a_1, a_0)$$
$$E_j \stackrel{?}{=} H(\hat{\alpha}_j, \hat{e}_j) \qquad j = 1, \cdots, n$$

If this test passes, then the recovered secret is regarded as $\alpha = \hat{f}(0)$, else, check the following equation to identify who is a false player using the published information.

$$g^{\alpha_i} h^{e_i} \stackrel{?}{=} \prod_{j=0}^{t} A_j^{i^j} \qquad i = 1, \cdots, n$$

Fig. 1. Our VSS using Hash Function

In the following we will refer to this protocol as VSS-H. The detailed protocol described in Figure 1.

3.2   Joint Unconditionally Secure RSS Based on Pedersen's VSS

In this section we present a modified joint unconditionally secure random secret sharing scheme based on Pedersen's VSS. In this scheme we commit to the values of polynomial instead of the coefficients of polynomial. We use a hash function to commit to the values of polynomial. VSS using a hash function can verify the correctness of received shares easily, but it can't prove that the committed values are obtained from the specified polynomial of degree at most t. Players commit to coefficients of polynomial, because the participants which take part in a reconstruction phase want to verify the broadcasted shares by t+1 (or more) players.

1. At sharing Phase, each player
   - Choose two random polynomials as follows.

   $$f_i(x) = a_{i,t} x^t + \cdots + a_{i,1} x + a_{i,0}$$

   $$g_i(x) = b_{i,t} x^t + \cdots + b_{i,1} x + b_{i,0}$$

   - Compute the values, $\alpha_{i,j}$ and $\beta_{i,j}$ as follows.

   $$\alpha_{i,j} = f_i(j), \ \beta_{i,j} = g_i(j)$$

   - Compute the following two values and broadcast them.

   $$E_{i,j} = H(\alpha_{i,j}, \beta_{i,j}), \ A_{i,j} = g^{a_{i,j}} h^{b_{i,j}} \quad (1 \leq i \leq n, 1 \leq j \leq n)$$

   - Send $\alpha_{i,j}$ and $\beta_{i,j}$ to the other players and at the same time receive $\alpha_{j,i}$
   and $\beta_{j,i}$ from the other players. $(1 \leq i \leq n, 1 \leq j \leq n)$

   - Check the correctness of the received values by using the verification
   equation, i.e. $E_{j,i} \overset{?}{=} H(\alpha_{j,i}, \beta_{j,i})$. If this test passes, then proceed the
   next step, else check the following equation. If the test doesn't pass, the
   player opens his share and stops the protocol.

   $$g^{\alpha_{j,i}} h^{\beta_{j,i}} \overset{?}{=} \prod_{k=0}^{t} (A_{j,k})^{i^k}$$

   - Compute his own shares

   $$s_i = \sum_{j=1}^{n} \alpha_{j,i} \overset{def}{\leftrightarrow} s(x), \ e_i = \sum_{j=1}^{n} \beta_{j,i} \overset{def}{\leftrightarrow} e(x)$$

2. At reconstruction phase,
   - Each player broadcasts the values s_i, e_i.
   - Each player check the received s_i, e_i by using the following equation.

   $$g^{s_i} h^{e_i} \overset{?}{=} \prod_{j=0}^{t} (A_{1,j})^{i^j} \cdots \prod_{j=0}^{t} (A_{n,j})^{i^j}$$

   - Collect the (t + 1) shares which were passed the above test.
   - Interpolate $\hat{s}(i)$, $\hat{e}(i)$, $\hat{s}(x)$, $\hat{e}(x)$ and compute $\hat{f}(i), \hat{g}(i)$.
   - Check the following equation again.

   $$g^{\hat{s}(i)} h^{\hat{e}(i)} \overset{?}{=} \prod_{j=0}^{t} (A_{1,j})^{i^j} \cdots \prod_{j=0}^{t} (A_{n,j})^{i^j}$$

   - Pass $\hat{s}(0)$ as a secret information

Fig. 2. Joint-Uncond-Secure-RSS based on Pedersen's VSS

The presented scheme can check the validity of the received data using hash function
in sharing phase, and then verify the validity of the broadcasted shares using
Pedersen's VSS in reconstruction phase. Namely, if no faults occur then players don't
have to compute the exponentiation operation in sharing phase. In the following we

will refer to this protocol by Pedersen-VSSH. The detailed protocol described in Figure 2.

3.3   ZK Proofs for Multiplication of Committed Shares

The robust multiplication protocol was presented based on an interactive ZKIP[1]. In this section we describe their idea again, and extend their scheme to generate a non-interactive robust multiplication scheme. The protocol is to prove that a player proves to verifier that he knows the discrete logarithm value and the committed value is constructed properly as shown in the Figure 3. This scheme can be used in a non-interactive robust multiplication protocol in case of the number of player involving in that protocol is 2t+1.

3.3.1   Interactive ZK Proof

The prover wants to prove to the verifier that he knows how to open such commitments and the opening of C that he knows is really the product of the values he committed to in A and B. Firstly, the prover publishes A, B, and C whose values are $g^\alpha h^\rho$, $g^\beta h^\sigma$, and $g^{\alpha\beta} h^\tau$ respectively, and then he proves to the verifier that he knows the polynomial representation of $A = g^\alpha h^\rho$. To verify that the prover knows the polynomial representation of $A = g^\alpha h^\rho$ which is known to verifier, two players performs the following protocol:

1. The prover chooses d and s at random. He sends to the verifier the message $M = g^d h^s$, for $d, s \in_R Z_q$
2. The verifier chooses a random number e, for $e \in_R Z_q$, and sends it to the prover.
3. The prover computes the values, y and w, as follows:
   $$y = d + e\beta, \quad w = s + e\sigma$$
4. The verifier checks the equation as follows:
   $$g^z h^{w_1} \overset{?}{=} M_1 A^e$$

Also, the prover can prove to the verifier that he knows the polynomial representation of $B = g^\beta h^\sigma$ in a similar way as the above procedure. The prover proves that he knows the polynomial representation of $C = g^{\alpha\beta} h^\tau$ and that the exponent is the multiplication of two known secret as Figure 3. Figure 3 describes the detailed protocol.

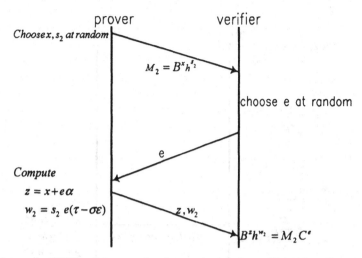

Fig. 3. Interactive ZK Proof to prove that the exponent is the multiply of two known values

3.3.2   Non-interactive ZK Proof

We present a non-interactive ZK proof as shown in Figure 4. Figure 4 is based on computational secrecy, and Figure 5 is based on information theoretic secrecy. In the following we will refer to protocol in Figure 5 as Non-interactive-ZK. The detailed protocol describes in Figure 4 and Figure 5.

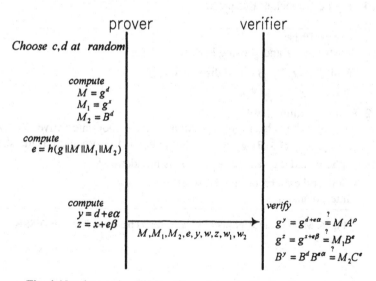

Fig. 4. Non-interactive ZK Proof based on computational secrecy

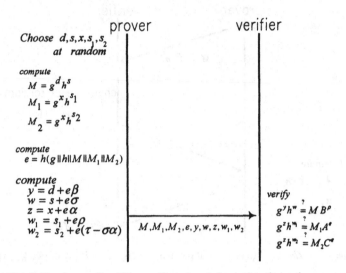

Fig. 5. Non-interactive ZK proof based on information theoretic secrecy

3.4   Robust Multiplication Protocol Based on Pedersen's VSS

In this section we show how to carry out the robust multiplication protocol based on Pedersen-VSSH and Pedersen-ZK. In the following we will refer to this protocol by Mult-ZK. Figure 6 indicates this protocol.

---

1. Sharing Phase
   - Distribute α and β using Pedersen-VSSH
   - Multiply α_i by β_i , and distribute $\alpha_i \beta_i$
   - Broadcast $C = g^{\alpha_i \beta_i} h^{\tau_i}$.

2. Reconstruction Phase
   - Perform the robust multiplication protocol, non-interactive-ZK, where $A = g^{\alpha_i} h^{\rho_i}$ and $B = g^{\beta_i} h^{\sigma_i}$ are the published information at the sharing phase, and the value $C = g^{\alpha_i \beta_i} h^{\tau_i}$ is broadcasted.
   - Collect the correct ($2t + 1$) shares.
   - Interpolate $\alpha\beta$.

Fig. 6. Multiplication protocol of two secrets based on Pedersen-VSSH

4 KCDSA Signature

KCDSA (Korean Certification-based Digital Signature Algorithm) is Korea standard digital signature algorithm [16]. In this section, A KCDSA is composed of public information p, q, g, a public key y and a secret key x, where:

1. p : a large prime that the length of it is 512+256i, for i={0, ... , 6}.
2. q : a prime factor of p-1 that the length of it is 128+32j, for j={0, ... , 4}.
3. g : a base element of order q (mod p).
4. x : the signer's private signature key such that $x \in_r Z_q$.
5. y : the signer's public verification key, such that $y = g^{x^{-1}} \bmod p$
6. H : collision resistant hash function, such that the length of output is q.
7. Z : hash value of the signer's certification data.

The detailed protocol is shown in Figure 7. The signature pair of the hashed message m is a (c,s) , where Z is the certificate of a verifier. The signature generation procedure of KCDSA is described in left part of Figure 7, and the verification of signature is described the right part of Figure 7.

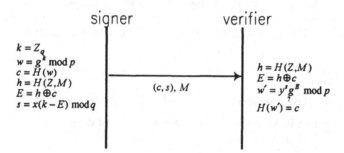

Fig. 7. KCDSA Signature Generation and Verification

5 Robust Threshold KCDSA Protocol

In this section we present a robust threshold KCDSA protocol for generating a distributed KCDSA signature. The robust threshold KCDSA protocol consists of two phases, one is a key distribution phase, and the other is a signature distribution phase.

5.1 Distributed Key-Pair Generation Phase

In this phase each player shares his private key x and public key y ($= g^{x^{-1}} \bmod p$). The detailed protocol describes in Figure 8.

1. Distribute private key x

   The players generate his share corresponding to a secret x, which is uniformly distributed in Z_q, with a polynomial of degree t, by using Pedersen-VSSH

   $$(x_1, \cdots, x_n) \xleftrightarrow{(t,n)} x \bmod q$$

2. Compute and broadcast the public key, $y = g^{x^{-1}} \bmod p$

   (a) The players generate a random value a, uniformly distributed in Z_q^* with a polynomial of degree t, using Pedersen-VSSH.

   (b) The players generate random polynomials of degree $2t$ with constant term 0, Denote the shares by $\{b_i\}$ $i=\{1, \cdots, n\}$

   (c) Player P_i broadcasts $V_i = x_i a_i + b_i \bmod q$ using Mult-ZK.

   (d) Player P_i computes each player.

   – Interpolate $V = ax \bmod q$

   – Compute $V^{-1} = (ax)^{-1} \bmod q$

   – Compute $(g^a)^{V^{-1}} = (g^a)^{(ax)^{-1}} = g^{x^{-1}} \bmod p$

Fig. 8. A Distributed Key-pair Generation Phase of KCDSA

1. Generate k

   The players generate a secret value k, uniformly distributed in Z_q, by running Pedersen-VSSH with a polynomial of degree t

   $$(k_1, \cdots, k_n) \xleftrightarrow{(t,n)} k \bmod q$$

2. Perform Joint Zero Secret Sharing

   The players generate random polynomials of degree $2t$ with constant term 0, Denote the shares created in these protocols as $\{c_i\}$ $i=\{1, \cdots, n\}$

3. Compute the values

   - Interpolate $w = g^k \bmod p$
   - $r = H(w)$, $h = H(Z, M)$, $E = h \oplus r$

4. Generate $s = x(k - E) \bmod q$

   - Player P_i computes $s_i = x_i(k_i - E) + c_i \bmod q$ using Mult-ZK
   - Interpolate the values, $s = x(k - E) \bmod q$

5. Output the pair (r, s) as the signature for M

Fig. 9. A Robust Distributed Signature Phase of KCDSA

5.2   Distributed Signature Phase of KCDSA

In this phase each player establishes the signature for a message M. The detailed protocol describes in Figure 9.

Lemma 1. KCDSA-Thresh is a (t, n)-robust threshold KCDSA signature protocol, that is, it tolerates up to t eavesdropping and halting faults if the number of players is $n \geq 2t + 1$.

Proof. Reviewing the proposed protocol can easily prove this lemma.

6   Conclusion

Threshold signature scheme allows a group of players to produce a signature rather than by one player. In this signature scheme, the secret key is shared by a group of player. And the public key is published by a group of players. Each payer should generates his partial signature to produce a complete signature given message m. This signature scheme is applied to a certification authority in a public-key infrastructure, where each signature system in a certification authority has a shared secret key corresponding to his public-key.

In this paper we present some basic protocol, i.e. a verifiable secret sharing scheme using hash function and a multiplication protocol of two secrets based on a non-interactive ZK proof. Also we apply these protocols to threshold KCDSA signature scheme. KCDSA-Thresh is a (t, n)-robust threshold KCDSA signature protocol, that is, it tolerates up to t eavesdropping and halting faults if the number of players is $n \geq 2t + 1$. The proposed VSS and robust multiplication protocol can be applied to any threshold cryptosystem.

Reference

1. Rosario Gennaro, Michael Rabin, Tal Rabin, "Simplified VSS and Fast-Track Multiparty Computations with Applications to Threshold cryptography", http://theory.lcs.mit.edu/~rosario/ research.html, preprint, 1998.
2. A. Shamir, "How to Share a Secret", Communication of the ACM, 22:612-613, 1979.
3. P. Feldman, "A Practical Scheme for Non-Interactive Verifiable Secret Sharing", In Proc. 28th Annual Symposium Foundations of Computer Science, pp 427-437, IEEE, 1987.
4. T. Pedersen, "Non-interactive and Information-theoretic Secure Verifiable Secret Sharing", Crypto'91, pp 129-140, 1991.
5. T. Pedersen, "A Threshold Cryptosystem Without a Trusted Party", Eurocrypt '91, pp 522-526, 1991.
6. A. Herzberg, S. Jarecki, H. Krawczyk, M. Yung, "Proactive Secret Sharing Or : How to Cope With Perpetual Leakage", Crypto '95, 1995.
7. C. Cachin, "On-line secret sharing", Proceeding of the 5th IMA conference on Cryptography and Coding, pp 190-198, 1995.
8. R.G.E. Pinch, "On-line Multiple Secret Sharing", Electronic Letters, pp 1087-1088, 1996.

9. Chan Yeob Yeun, Chris J. Mitchell, Mike Burmester, "An Online Secret Sharing which Identifies All Cheater", Proceeding of NORDSEC '98, 1998.
10. S. Langford, "Threshold DSS Signatures Without a Trusted Party", Crypto '95, pp 397-409, 1995.
11. D. Boneh, M. Franklin, "Efficient Generation of Shared RSA keys", Crypto '97, pp 425-439, 1997.
12. M. Malkin, T. Wu, D. Boneh, "Experimenting with Shared Generation of RSA keys", http://theory.stanford.edu/~dabo/ pubs.html, preprint
13. T. Rabin, "A simplified approach to threshold and proactive RSA", Crypto '98, 1998
14. C. Park, K. Kurosawa, "New ElGamal Type Threshold Digital Signature Scheme", IEICE Trans. Fundamentals, E79-A(1):86-93, 1996.
15. R. Gennaro, S. Jarecki, H. Krawczyk, T. Rabin, "Robust Threshold DSS signature", Eurocrypt '96, pp 354-371, 1996.
16. C.H. Lim and P.J. Lee, "A study on the proposed Korean Digital Signature Algorithm," Advances in Cryptology - Asiacrypt'98, Lecture Notes in Computer Science (LNCS), Vol.1514, Springer-Verlag, pp.175-186, 1998

A Hardware-Oriented Algorithm for Computing in Jacobians and Its Implementation for Hyperelliptic Curve Cryptosystems

Tetsuya Tamura[1], Kouichi Sakurai[2], and Tsutomu Matsumoto[3]

[1] IBM Research, Tokyo Research Laboratory, IBM Japan Ltd.,
1623-14, Shimotsuruma, Yamato-shi, Kanagawa 242-0002, Japan
ttamura@jp.ibm.com
[2] Kyushu University,
6-10-1 Hakozaki, Higashi-ku, Fukuoka 812-8581, Japan
sakurai@csce.kyushu-u.ac.jp
[3] Yokohama National University
79-5 Tokiwadai, Yokohama-shi, Kanagawa, 240-8501, Japan
Tsutomu@mlab.dnj.ynu.ac.jp

Abstract. In this paper, we present algorithms, suitable for hardware implementation, for computation in the Jacobian of a hyperelliptic curve defined over $GF(2^n)$. We take curves of genus 3 and 6, designed by using 0.27-um CMOS gate array technology, and estimate the number of multiplication operations and the size and speed of hardware based on the proposed algorithm. It is shown that hardware for genus 6 curves computes an addition (resp. doubling) operation in 100 (resp. 29) clock cycles and can work at clock frequencies of up to 83 MHz We also compare a hyperelliptic curve cryptosystem with RSA and elliptic curve cryptosystems from the viewpoint of hardware implementation.

1. Introduction

Koblitz [Ko88, Ko89] investigated the Jacobians of hyperelliptic curves defined over finite fields, and proposed hyperelliptic curve cryptosystems. Frey [FR94] showed that the discrete-logarithm problem of Koblitz's hyperelliptic cryptosystems can be solved in sub-exponential time, and Sakai, Sakurai, and Ishizuka [SSI98] and Smart [Sm99] studied the Jacobians of hyperelliptic curves and found Jacobians that are secure against all known attacks.

Explicit formulas for addition in the Jacobians of hyperelliptic curves were introduced by Cantor [Ca87] and Koblitz [Ko88, Ko89]. Although the formulas for addition in Jacobians are more complicated than those for the addition in points on an elliptic curve, if the order of a Jacobian of a hyperelliptic curve has the same size as the order of points on an elliptic curve, the ground field of the Jacobian is smaller than that of the elliptic curve. This is an advantageous feature for hardware implementation. In addition, the multiplication operation for polynomials used in the formulas can be effectively performed by parallel-processing hardware.

JooSeok Song (Ed.): ICISC'99, LNCS 1787, pp. 221-235, 2000.

The objective of this paper is to investigate how effectively hyperelliptic curve cryptosystems can be implemented by hardware means.

First, we explain the algorithm[TM99] based on the algorithm introduced by [Ca87] and [Ko88, Ko89], and discuss the number of multiplication operations from the viewpoint of hardware. For explanatory purposes, the Jacobians associated with curves C: $y^2 + y = x^7$ /GF(2) of genus 3 and C: $y^2 + y = x^{13} + x^{11} + x^7 + x^3 + 1$ /GF(2) of genus 6 were chosen from the Jacobians proposed in [SSI98] and [Sm99].

Recently, Gaudry gave a new algorithm(Gaudry's variant) for discrete logarithm problem on hyperelliptic curves [Ga99]. Duursma, Gaudry, and Morain [DGM99] also presented a method for speeding up discrete log computations on curves having automorphisms of large order. This method uses a parallel collision search and obtains a speed of $\sqrt{m}$ if there exists an automorphism of order m. Its author's heuristic analysis says that the attack would be effective for curves with genus > 4. Therefore, hyperelliptic cryptosystems defined over Jacobians with the curve of genus 3 has the same level of security as 160-bit-key elliptic curve cryptosystems(ECC), but hyperelliptic cryptosystems defined over Jacobians with the curve of genus 6 is weaker than 160-bit-key ECC. However, our result are independent from the specific structure of such a curve with large automorphisms and the security levels of the cryptosystems, and any technique of our implementation is valid for the curves without automorphisms of large order.

Next, we describe the result of the logic design and synthesis, which uses 0.27-um CMOS gate array technology to estimate the size and speed of the hardware.

Finally, we analyze the dependency of hardware efficiency on the genus of the curve. The efficiency is defined as the result of dividing the speed by (size)*(power consumption) and compare our results with RSA and to elliptic curve cryptosystems.

2. Preliminaries

Let K be a field, and let $\overline{K}$ denote its algebraic closure. We define a hyperelliptic curve C of genus g over K to be an equation of the form $y^2 + h(x)y = f(x)$, where $h(x)$ is a polynomial of degree at most g and $f(x)$ is a monic polynomial of degree $2g + 1$. We are concerned in this paper with finite fields of characteristic 2. The point $P(x,y)$ generates the free group of divisors. A divisor D is a finite formal sum of $\overline{K}$-points $D = \sum m_i P_i$, $m_i \in Z$. We define the degree of D as $\deg(D) = \sum m_i$. The divisors form an additive group, in which divisors of degree 0 form a subgroup D0. The rational function r has a finite number of zeros and poles on C. We associate r with its divisor $(r) = \sum m_i P_i$, where P_i are poles or zeros with multiplicities m_i. A divisor of a nonzero function, such as (r), is called a principal. The principal divisors form a subgroup of D0. The Jacobian variety is defined as the quotient group of $J_C(K) = D0/P$

Let F_q be a finite field with q elements. The discrete logarithm problem of $J_C(F_{q^n})$ is the problem, given two divisors D_1 and D_2 defined over F_{q^n}, of determining an integer m such that $D_2 = mD_1$, if such an m exists.

3. Proposed Algorithm

3.1 Computing in Jacobians

An element of Jacobian varieties can be represented uniquely by a reduced divisor. Any reduced divisor is regarded as a pair of polynomials satisfying deg b < deg a and deg a ≤ g. We give a brief description of an algorithm introduced by Cantor [Ca87] and [Ko88, Ko89] for addition: $D_3 = D_1 + D_2$, where $D_3 = \mathrm{div}(a_3, b_3)$, $D_2 = \mathrm{div}(a_2, b_2)$ and $D_1 = \mathrm{div}(a_1, b_1)$.

First, we compute the greatest common divisor (GCD) of polynomials a_1 and a_2. Note that the case in which $\gcd(a_1, a_2) = 1$ is extremely likely if the ground field K is large and a_1 and a_2 are the coordinates of two randomly chosen elements of the Jacobian. The case in which a_1 and a_2 are not prime does not strongly affect the performance. Therefore, this paper investigates only the case in which $\gcd(a_1, a_2) = 1$ and the doubling $D_1 = D_2$ case. We assume that other cases are processed by software with the assistance of hardware.

We use the extended Euclidean algorithm and compute $d = \gcd(a_1, a_2)$ and two polynomials s_1 and s_2 satisfying the equation $s_1 a_1 + s_2 a_2 = d$. For convenience, s_1 and s_2 are divided by d to meet the condition that $s_1 a_1 + s_2 a_2 = 1$.

The extended Euclidean algorithm is also used when we compute the error location and evaluation polynomials from the syndrome in decoding of Reed Solomon code. This is a powerful error correction code that is widely used by storage devices and communications, and is frequently implemented by hardware means. In decoding of Reed Solomon code, the GCD, d, is an error evaluation polynomial, s_1 and s_2 is an error location polynomial, and only one of s_1 and s_2 is needed. The difference between decoding and addition is that decoding of Reed Solomon code requires only one of s_1 and s_2, whereas, addition in a Jacobian requires both s_1 and s_2.

There are several different implementations for decoders using the Euclidean algorithm [IDI95], [JS99]. Here, we take the simplest one, to maximize the parallelism. The hardware for addition in Jacobians of genus-g hyperelliptic curves consists of four register sets: Ureg, Xreg, Yreg, and Zreg. Ureg and Xreg have (g+1) registers for storing the coefficients of a polynomial of degree g, while Yreg and Zreg have g registers for storing the coefficients of a polynomial of degree (g-1). A Galois field multiplier is placed in each register of Ureg and Yreg and one is also placed in the inversion operator in the circuit. The circuit contains a total of (4g+2) coefficients registers, (2g+1) Galois field multipliers, and one inversion operator (see Fig. 1).

No explanation of the hardware computation of the GCD by hardware is included here, because the process is not related to the objective of this paper, We assume that Ureg and Xreg store the GCD, d, and s_1 that satisfies $s_1 a_1 + s_2 a_2 = 1$ after the computation.

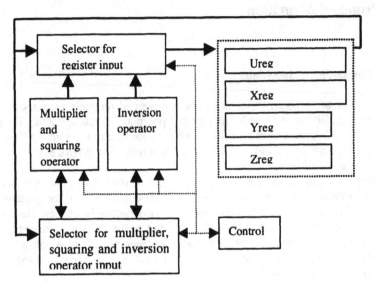

Fig. 1. Hardware Configuration

When a_1 and a_2 are prime, the algorithm for the addition is as follows [Ca87],[SSI98]:

```
Algorithm 1 (Addition)

Input     a₁, a₂, b₁, b₂, s₁, s₂
Output    a', b'
          a₃ = a₁ a₂                              Step A1
          b₃ =(s₁*a₁*b₂ + s₂*a₂*b₁) mod a3
          a₄ =(f + b₃ + b₃²) /a₃                  Step A2
          a₄ = a₄/(leading coefficient of a₄)
          b₄ =(b₃ + 1) mod a₄
          a' = a₄; b' = b₄;
          while(deg a₄> g) {                      Step A3
             a₅ =(f + b₄ + b₄²)/a₄
             a₅ = a₅ /(leading coefficient of a₅)
             b₅ =(b₄ + 1) mod a₅
             a' = a₅; b' = b₅;
             a₄ = a₅; b4 = b₅;
          }
          Return[a', b']
          End
```

- Step A1   Since the input polynomials a_1 and a_2 have degree g, and s_1 and s_2 have degree (g-1), a_3 and b_3 have degrees 2g and (2g - 1), respectively. Furthermore, ($s_1a_1b_2 + s_2a_2b_1$) has degree (3g - 2), and therefore step A1 takes a total of $(13g^2 - 12g + 2)$ field multiplications [SSI98].

- Step A2 a_3 and b_3 have degrees $2g$ and $(2g - 1)$, respectively. Since the degree of f is $(2g+1)$, a_4 and b_4 have degrees $(2g - 2)$ and $(2g - 3)$. Step A2 takes a total of $(16g^2 - 14g + 3)$ field multiplications.

- Step A3   In the first iteration, a_4 and b_4 have degrees $(2g - 2)$ and $(2g - 3)$, respectively. In the case of a genus 3 curve, the number of iterations is 1, and in the case of a genus 6 curve, the number of iterations is 2.

In Steps A1 and A2, computation of the multiplication and division operations for polynomials of degree $2g$ takes a considerable time. To reduce the amount of computation, we introduce the polynomial $q(x) = s_1 (b_1 + b_2) \bmod a_2$.

Lemma 1. *In step A2, using $q(x)$, we can express a_4 as follows:*

$$a_4 = Q(q^2 a_1, a_2) + Q(f, a_3)$$

Here, $Q(u, v)$ is a function that gives the quotient of dividing u by v.

Proof. First, we show that $b_3 = q\, a_1 + b_1$. Note that $s_1 a_1 + s_1 a_2 = 1$ and that $\deg a_1 a_2 > \deg b_1$. We can compute b_3 as follows:

$$b_3 = (s_1 a_1 (b_1 + b_2)) \bmod (a_1 a_2) + b_1 = \{s_1(b_1 + b_2) \bmod a_2\}\, a_1 + b_1 \qquad (1)$$
$$= q\, a_1 + b_1$$

Next, since the division for computing a_4 has no remainders, and $Q(b_3, a_3)=0$ from $\deg b_3 < \deg a_3$,

$$a_4 = Q(f + b_3 + b_3^2, a_3) = Q(f, a_3) + Q(b_3^2, a_3) \qquad (2)$$

Substituting (1) into the second term of (2) and noting that $Q(b_3, a_3)=0$ from $\deg b_1^2 < \deg a_3$, we obtain

$$Q(b_3^2, a_3) = Q(q^2 a_1 + b_1^2, a_1 a_2) = Q(q^2 a_1^2, a_1 a_2) = Q(q^2 a_1, a_2) \qquad (3)$$

Combining equations (2) and (3), we get $a_4 = Q(q^2 a_1, a_2) + Q(f, a_3)$.

We take the curve C: $y^2 + y = x^7 /GF(2)$ of genus 3 to show an application of Lemma 1. By applying $f(x) = x^7$, we obtain $Q(f, a_3) = x + c_2 + e_2$. The new Algorithm 2 is as follows:

```
Algorithm 2 (Addition)
Input     a₁, a₂, b₁, b₂, s₁, s₂
Output    a, b
          q  = s₁*(b₁+b₂) mod a₂                    Step A1'
          a₄ = Q(q²*a₁, a₂)+ x + c₂ + e₂            Step A2'
          a₄ = a₄ /(leading Coefficient of a₄)
          b₄ = (q*a₁ + b₁ + 1) mod a₄
          a' = a₄; b' = b₄;
          while(deg a₄> g) {                        Step A3'
              a₅ = Q(x⁷ + b₄²,a₄)
```

```
               a₅   = a₅/(leading coefficient of a₅)
               b₅   =(b₄ + 1) mod a₅
               a'   = a₅; b'  = b₅;
               a₄   = a₅; b4  = b₅;
               }
               Return[a', b']
               End
```

Here, $a_1(x) = x^3 + c_2 x^2 + c_1 x + c_0$ and $a_2(x) = x^3 + e_2 x^2 + e_1 x + e_0$. In the computation of the polynomial a_5, The polynomials a_3 and b_3 do not appear in Algorithm 2. The RTL behavior of the hardware for Algorithm 2 is described in the appendix.

When $D_1 = D_2$, i,e., for doubling, we can compute as follows.

```
Algorithm 3 (Doubling)
Input     a₁, b₁
Output    a', b'
               a₃   = a₁²                                    Step D1
               b₃   =(b₁² + f) mod a₃
               a₄   =(f + b₃ + b₃²)/a₃                        Step D2
               a₄   = a₄/(leading coefficient of a₄)
               b₄   =(b₃ + 1) mod a₄
               a'   = a₄; b'  = b₄;
               while(deg a₄> g) {                            Step D3
                 a₅   =(f + b₄ + b₄²)/a₄
                 a₅   = a₅/(Leading Coefficient of a₅)
                 b₅   =(b₄ + 1) mod a₅
                 a'   = a₅; b'  = b₅;
                 a₄   = a₅; b₄  = b₅;
               }
               Return[a', b']
               End
```

- Step D1   The polynomials a_1 and b_1 have degree g and $(g - 1)$ repectively. Therefore, the computation of a_3 and b_3 takes g^2 and $(g - 1)^2$ field multiplications, respectively. Step D1 takes a total of $(6g^2 + 1)$ field multiplications.

- Step D2   Step D2 is the same as Step A2 and takes $(16g^2 - 14g + 3)$ field multiplications.

- Step D3   Step D3 is the same as Step A3.

In Steps D1 and D2, the computation of multiplication and division operations for polynomials of degree 2g takes a considerable time. To reduce the amount of computation, we introduce the polynomial $q(x) = Q(b_3, a_1)$.

Lemma 2. *In step D2, we can express a_4 as follows:*

$$a_4 = q^2 + Q(f, a_3)$$

Here, $Q(u, v)$ is a function that gives the quotient of dividing u by v.

Proof. $Q(b_3, a_3)$ is equal to 0, because deg $b_3 <$ deg a_3; therefore,

$$a_4 = Q(b_3^2, a_3) + Q(f, a_3) \tag{4}$$

Let the result of dividing b_3 by a_1 be equal to $q + (r/a_1)$, where q is a quotient polynomial and r is a remainder polynomial. Since the characteristic is 2, we get the following equation:

$$b_3^2 / a_1^2 = q^2 + (r_1^2 / a_1^2) \tag{5}$$

Combining equations (4) and (5), we get $a_4 = q^2 + Q(f, a_3)$.

We also take the curve C: $y^2 + y = x^7$ /GF(2) of genus 3 to show the application of Lemma 2 to Algorithm 3.

```
Algorithm 4 (Doubling)
Input     a₁, b₁
Output    a', b'
          a₃ = a₁²                                   Step D1'
          b₃ = b₁² + x(a₁ - x³)²
          q = Q₂(b₃, a₁)                             Step D2'
          a₄ = q² + Q(f,a₃)
          a₄ = a₄/(leading coefficient of a₄)
          b₄ = (b₃ + 1) mod a₄
          a' = a₄; b' = b₄;
          while(deg a₄ > g) {                        Step D3'
              a₅ = Q(x⁷ + b₄², a₄)
              a₅ = a₅/(Leading Coefficient of a₅)
              b₅ = (b₄ + 1) mod a₅
              a' = a₅; b' = b₅;
              a₄ = a₅; b4 = b₅;
          }
          Return[a', b']
End
```

Here, $f \bmod a_3 = x (a_1 - x^3)^2$ and $b_1^2 \bmod a_3 = b_1^2$ are used in computing b_3. From the viewpoint of hardware implementation, registers for storing a_3 and b_3 are not really needed, since the hardware size of the squaring operator is less than that of the register. If the hardware has squaring operations, we can neglect Step D1', that is, the computation of polynomials a_3 and b_3. In addition, since the characteristic of the ground field is 2, the computation of q^2 requires only squaring operations, and the computation of a_4 is simplified.

3.2 Computational Complexity of the New Algorithm

To investigate the case of hardware implementation, we use two notations, m and M, for multiplication. As explained in the previous section, we assume that the hardware has (2g + 1) Galois field multipliers and can execute (2g + 1) multiplications simultaneously. We denote the execution on hardware by M, the multiplication itself by m, and the inversion by I. We estimate the computational complexity by using m, M, and I, and neglect the squaring. The estimation was made by using two Jacobian varieties: $J_C(GF(2^{59}))$, given by the genus 3 curve C: $y^2 + y = x^7$ /GF(2) and $J_C(GF(2^{29}))$, given by the genus 6 curve C: $y^2 + y = x^{13} + x^{11} + x^7 + x^3 + 1$ /GF(2). The results are given in Tables 1 - 4.

Table 1. Number of field operations (g = 3 addition)

Computation		Computation time	Computation time (H/W)	Processing time (H/W)
GCD		3I+23m	3I+9M	3t(I)+9t(M)
Step A1'		15m	4M	4t(M)
Step A2'	a_4	I+20m	I+6M	t(I)+ t(M)
	b_4	17m	5M	5t(M)
Step A3'	a_5	3m	2M	2t(M)
	b_5	3m	M	t(M)
Total		4I+81m	4I+27M	4t(I)+22t(M)

Table 2. Number of field operations (g = 3 doubling)

Computation		Computation time	Computation time (H/W)	Processing time (H/W)
Step D2'	q	3m	2M	0
	a_4	I+ 2m	I+ M	t(I)+ t(M)
	b_4	8m	2M	2t(M)
Step D3'	a_5	3m	2M	2t(M)
	b_5	3m	M	t(M)
Total		I+19m	I+8M	t(I)+6t(M)

Table 3. Number of field operations (g = 6 addition)

Computation		Computation time	Computation time (H/W)	Processing time (H/W)
GCD		6I+ 86m	6I+ 21M	6t(I)+21t(M)
Step A1'		66m	6M	6t(M)
Step A2'	a_4	I+ 85m	I+ 11M	11t(M)
	b_4	I+ 56m	6M	6t(M)
Step A3'	a_5	I+ 44m	I+ 9M	9t(M)
	b_5	16m	2M	2t(M)
Step A3'	a_6	I+ 27m	I+ 7M	t(I)+ t(M)
	b_6	12m	2M	2t(M)
Total		9I+392m	9I+ 64M	7t(I)+58t(M)

Table 4. Number of field operations (g = 6 doubling)

Computation		Computation time	Computation time (H/W)	Processing time (H/W)
Step D2'	q	15m	5M	0
	a_4	I+ 5m	I+ M	t(I)+ t(M)
	b_4	20m	2M	2t(M)
Step D3'	a_5	I+44m	I+ 9M	9t(M)
	b_5	16m	2M	2t(M)
Step D3'	a_6	I+27m	I+ 7M	t(I)+ t(M)
	b_6	12m	2M	2t(M)
Total		3I+139m	3I+28M	2t(I)+17t(M)

Here, t(M) and t(I) represent the processing times for multiplication and inversion. In table 1, it is assumed that t(I) > 5t(M) and that the computation of a_4 and its inversion can be executed simultaneously. In Table 2, it is assumed that t(I) > 2t(M) and that the computation of q and its inversion can be executed simultaneously. Similarly, it is assumed that t(I) < 11t(M) in table 3 and that t(I) > 5 t(M) in Table 4.

An efficient implementation of arithmetic in $GF(2^n)$ is discussed in [ITT86]. An inversion in $GF(2^{59})$ takes 8 multiplications and an inversion in $GF(2^{29})$ takes 6 multiplications. All the assumptions mentioned above are true when the method described in [ITT86] is used. Applying t(I) = 8t(M) or t(I) = 6t(M) to Tables 1-4, we can get the following results:

Table 5. Summary of the number of field operations and time

	Computation time	Processing time
g =3 addition	113m (4I+ 81m)	54t(M)
g =3 doubling	27m (I+ 19m)	14t(M)
g =6 addition	446m (9I+392m)	100t(M)
g =6 doubling	157m (3I+139m)	29t(M)

The data in table 5 show that the processing time is proportional to the genus, even though the computation time is proportional to the square of the genus or a higher order.

In [SSI98], the number of field operations is also estimated in the case of software implementation. An addition takes 401 multiplications and a doubling takes 265 multiplications in the Jacobian $J_C(GF(2^{59}))$, given by the genus 3 curve C: $y^2 + y = x^7$ /GF(2). In comparison, we found that the proposed algorithm is 3.5 (resp. 10) times better with respect to the total computation time and 7 (resp. 19) times faster with respect to the total processing time in the case of genus 3 addition (resp. doubling).

Assuming that k has 160 bits and that the number of doublings is equal to 160 and the number of additions is equal to 80, Table 5 gives the following results for the time taken to compute kD in Jacobians:

Table 6. Average processing time to compute kD

Operating frequency		Number of clocks for multiplication	
		Case A t(M) = 8clocks	Case B t(M) = 1clock
g =3	20 MHz	2.624 ms	0.328 ms
g =3	40 MHz	1.312 ms	0.164 ms
g =3	80 MHz	0.656 ms	0.082 ms
g =6	20 MHz	5,056 ms	0.632 ms
g =6	40 MHz	2.528 ms	0.316 ms
g =6	80 MHz	1.264 ms	0.158 ms

Here, to estimate the time in the case of hardware implementation, two parameters are introduced: (1) the operating frequency of the hardware, and (2) the number of clocks for a multiplication operation.

4. Hardware Implementation

4. 1 Hardware Efficiency

Here, we estimate the maximum operating frequency and the size of the hardware. Table 6 shows two cases for comparison: (1) g = 3 Case A multiplication and (2) g = 6 Case B multiplication. As is well known, the advantage of arithmetic in GF(2^n) are as follows:

1. The multiplication and addition can be executed by a small amount of hardware.
2. The 2^n power operation (n=1,2, ...) is very simple. When normal bases are used, only the bit shift operation is required; even if polynomial bases are used, the hardware size is still small.

Thus, no custom design methodology is needed for the Galois field arithmetic unit; it is sufficient to design the unit by using gate array methodology. We designed the hardware in VHDL and performed simulation using IBM's Booledozer [BLD] and Model Technology's ModelSim [MTS]. Circuits were synthesized by using IBM CMOS 5SE gate array technology with an effective channel length Leff = 0.27-um [CDB]. There are no optimal bases[MOV89] in GF(2^{59}) and GF(2^{29}), and we take $p(x) = x^{59} + x^6 + x^5 + x^4 + x^3 + x + 1$ or $p(x) = x^{29} + x^2 + 1$, respectively, as the primitive polynomials of the Galois field [LN87].

The results of design and synthesis are given in the table below. Here, we force the constraint that the delay on the critical path between registers is at most 12 ns; that is, the maximum operating frequency is 83 MHz.

Table 7. Size of hardware

Block	Size (cells)			
	g = 3 Case A		g = 6 Case B	
Multiplier	34265	[7]	66196	[13]
Squaring Operator	1344	[3]	495	[11]
Inversion Operator	27414	[1]	8580	[1]
Register	18408	[59bit x26]	17400	[29bit x50]
Control	9749		7395	
Selector for register	37140		53939	
Selector for operators	17402		16851	
Total	145722		170856	
(After optimization)	140647		165743	

The numbers in square brackets represent Galois field arithmetic units. Note that the sizes of 1-bit latch, 2-way XOR and 2-way NAND are 12 cells, 3 cells, and 2 cells, respectively.

The register block has 4g registers for storing the input polynomials, and consists of (8g + 2) coefficient registers. Note that the total numbers of register bits are nearly equal in both cases.

The cell size of one multiplier is 4895 in g = 3/Case A and 5092 in g = 6 /case B. The numbers are almost the same, but the implementation is different. In case A, the multiplier has registers and computes every one-byte input by using a division circuit(digit serial). On the other hand, the multiplier in Case B has no registers(bit parallel). The cell size of the multiplier in $GF(2^{59})$ is about 20K cells for t(M) = 1 in the case of genus 3. Since there are a total of 8 multipliers in the case of multipliers in the case of genus 3 curves, the total cell size will be about 250K cells if the multiplier of case B (t(M) = 1) is used. The data given in tables 6 and 7 show that the total size for g = 6/Case B is about 25K cells larger than that for g = 3/Case A, but that the computation time is four times shorter.

[En99] gives a formula for estimating the number of multiplication and inversion operations for addition and doubling. In the case where g ≤ 10, the formula uses Gauss reduction, and the degree of the leading term is 3. In the case where g ≥ 12, it uses Legendre reduction, and the degree of the leading term is 2. But since the coefficient of degree 3 is small, the term of degree 2 is dominant. This can be easily seen in Tables 1-4. The number of multiplication operations can therefore be approximated by $o(g^2)$.

We analyze the dependency of the hardware efficiency on the genus of the curves. Here, for simplicity, we consider only the Galois field multiplier and ignore other parts of the circuits. We define the efficiency as the result of dividing the speed by (hardware size)*(power consumption). Let n be the extension degree of the ground field over which a curve is defined. n is proportional to $1/g$. Since the power consumption of one multiplier is proportional to n^2 and the number of multiplication is proportional to g^2, the total power consumption does not depend on g. Since the delay of the multiplier of case B is proportional to $\log_2 n$, the computation time is proportional to $g^2 \log_2 n$ without any parallel operation.

As a result, the efficiency is proportional to $1/\log_2 [C/g]$, where C is a constant and C/g is n. This function increases slowly for small $g(> 0)$, and therefore, the efficiency does not vary considerably with g. The data given in Table 7 agrees well with this result. The size of the multiplier in the case of $g = 3$/Case B is estimated to be about 160 K cells and the size of the multiplier in $g = 6$/Case B is 70K cells. The computation time in $g = 6$/Case B is twice that in $g = 3$/Case B. Since it can be considered that the total power consumption in both cases is the same, we can get almost the same values for the efficiency. But it is impossible to design a multiplier for which the product of the size and speed is constant at any speed. For example, the size of the multiplier in case of $g=3$/Case A is 1/4, but the size is 1/8 of that in $g = 3$/Case B. The efficiency in $g = 6$/Case B is 2.3 times higher than that in $g = 3$/Case A if we consider only the multiplier.

4.2 Comparison of the Efficiency of Hyperelliptic Cryptosystems and Others

Here, we compare the efficiency of hyperelliptic cryptosystems with that of other cryptosystems. The results of hardware implementation of RSA have been reported in [HTAA90], [IMI92], and [SKNOM97]. The results of 512-bit-key RSA are summarized in the table below.

Table 8. Performance of RSA hardware

	Size	Computation time @20MHz
[HTAAA90]	1050Kbit RAM + 305K gates	2.0ms
[IMI92]	198K gates	2.5ms
[SKNOM97]	-	14.0ms

We want to compare by using 1024-bit-key RSA, which has the same level of security as a 160-bit-key ECC(elliptic curve cryptosystem). But it is not described in [HTAA90] and [IMI92]. Moreover, [SKNOM97] can not be used directly, because its arithmetic unit was designed by using a custom design methodology. Therefore, we assume that the computation time is proportional to the square of the key length, and multiply the time in Table 8 by 4 for 1024-bit-key RSA. The data given in Table 6, 7, and 8 show that the speed of hyperelliptic curve cryptosystems is 3 (resp. 12.6) times faster than that of RSA and that the size is smaller in $g = 3$/case A (resp. $g=6$/case B).

Next, we consider a computation with elliptic curve cryptosystems. Since the computation of ECCs is more complicated that that of RSA, a co-processor approach that gives the basic Galois field arithmetic is often used for the implementation [AVN93]. In [TOH98], it is reported that an average processing time of 32 ms is needed for hardware with 22K gates to compute kD at an operating frequency of 20 MHz. In comparison with [TOH98], hyperelliptic curve cryptosystems defined over Jacobians of genus 3 or 6 are about 50 or 100 times faster when the multiplier of case B is used at a frequency of 20MHz, though the sizes are 8 or 12 times larger.

The numbers of multiplication and inversion operations have been well studied in elliptic curve cryptosystems. IEEE P1363 [P1363] proposes Jacobians coordinates. In binary case, the numbers of multiplication needed to compute projective elliptic addition and doubling are 15 and 5, respectively. Assuming that the hardware of a 160-bit-key elliptic curve cryptosystem has registers and only one multiplier, which takes one clock cycle for computation using the normal bases, the size of the hardware will exceed to 270K cells, because the size of the multiplier is larger than 256K cells and the size of four input and three working registers is at least 14K cells.

In comparison with g = 3/Case B, the elliptic curve cryptosystem is larger but about 3.3 times faster, because the number of multiplications for addition in the case of hyperelliptic cryptosystem is 54t(M), while that in the case of an ECC is 15t(M). In [SSI98], the performance of hyperelliptic curve cryptosystems is compared with that of RSA and elliptic curve cryptosystems from the viewpoint of software implementation and it is shown that the efficiency of hyperelliptic curve cryptosystems is better than that of RSA but worse than that of elliptic curve cryptosystems. This agrees well with our analysis of hardware implementation.

5. Concluding Remarks

The discussion and results of hardware implementation presented in the previous section show that the hardware efficiency is almost the same for curves of small genus, and that it is possible to design the hardware by using the proposed algorithm without assuming unreasonable conditions as regards as the current semiconductor technology. In other words, we can choose a curve of any genus to realize the same level of security by hardware means with the same effort, as long as the genus of the curve is less than 5. This offers an inducement for the use of hyperelliptic curve cryptosystems.

Acknowledgement

The first author thanks Dr. Shigenori Shimizu for his encouragement and support. The Second author thanks supported by Scientific Research Grant, Ministry of Education, Japan, No. 11558033.

References

[AMV93]   G.B. Agnew, R.C. Mullin, and S.A. Vanstone, „An inplementation of Elliptic curve Cryptosystems Over F_2^{155}", IEEE Jounal on selected areas in communications, 11, No.5, 1993.

[BLD]   IBM Corp., „Booledozer User's Manual", 1999

[Ca87]   D.G. Cantor, „Computing in the Jacobians of Hyperelliptic curve", Math. Comp., 48, No.177, pp.95-101, 1987.

[CDB]   IBM Corp., „CMOS5SE Logic Product Data Book", 1995

[DGM99]   I. Duursma, P. Gaudry, F. Morain, „Speeding up the discrete log computation on curves with automorphisms", to appear at ASIACRYPT'99.

[En99]   A. Enge, „The extended Euclidean algorithm on polynomials and the efficiency of hyperelliptic cryptosystems", Preprint 1999.

[FR94]   G. Frey and H.G. Rück, „A Remarkable Concerning m-Divisibility and the Discrete Logarithm in the Divisor Class Group of Curves", Math. Comp, 62, No.206,pp.865-874,1994.

[Ga99]   P. Gaudry, „A Variant of the Adleman-DeMarris-Huang algorithm and its application to small genera," Conference on The Mathematics of Public Key Cryptography, Toronoto, 1999.

[HTAA90]   T. Hasebe, T.Torii, M. Azuma, R. Akiyama, „Implementation of high speed modular exponentiation calculator", 1990 Spring National Convention Record IEICE, A-284.

[IDI95]   K. Iwamura, Y.Dohi, and H.Imai, „A design of Reed Solomon decoder with systolic array structure", IEEE Trans. Comput., 44, No.1, pp.118-122, 1995.

[ITT89]   T. Itoh, O.Teechai, and S. Tsujii, „A fast algorithm for computing multiplicative inverse in $GF(2^t)$ using normal bases"(in Japanse), J.Society for Electronic Communications(Japan),44, pp.31-36,1989.

[JS99]   Jee Ho Ryu, and Seung Jun Lee, „Implementation of Euclidean calculation circuit with two way addressing method for Reed Solomon decoder" , J. Inst. Electron. Eng. Korea C(South Korea), bol.36-C, No.6, pp.37-43, 1999.

[Ko88]   N. Koblitz „A Family of Jacobians Suitable for Discrete Log Cryptosystems", Advances in Cryptology-CRYPTO'88, LNCS, 403, pp.151-182,Springer, 1988.

[Ko89]   N. Koblitz „Hyperelliptic Cryptosystems", J. Cryptology, 1, pp.139-150, 1989.

[LN87]   R.Lidl and H. Niederreiter, „Finite Fields", Encyclopedia of Mathematics and its application

[MTS]   Model Technology, „ModelSim/EE/PLUS User'z Manual", 1997.

[MOV89]   R. C. Mullin, I. M. Onyszhk, and S.A. Vanstone, „Optimal Normal Bases in $GF(p^n)$", Discrete Applied mathematics, 83, No.1, pp.149-161, 1989.

[P1363]   IEEE/P1363/D11, Standard for Public-Key Cryptography, draft standard, July,29,1999.

[SKNOMS97] A. Satoh, Y.Kobayashi, H. Niihima, N. Ohba, S. Munetoh, and S. Sone,"A High-Speed Small RSA Encryption LSI with Low Power Dissipation", Information Security, Springer-Verlag, LNCS 1396, pp.174-187, 1997.

[SKHN75]   Sugiyamam Y., Kasahara, M., Hirasawa, S. and Namekawa, T. „A Method for Solving Key Equation for Deocoding Goppa Codes," Inform, And Control,27,pp.87-99,1975.

[Sm99]   N. P. Smart, „On the Performance of Hyperelliptic Cryptosystems", Advances in Cryptology-EUROCRYPT'99, Springe-Verlag, LNCS 1592, pp.165-175. 1999.

[SSI98]   Y. Sakai, K. Sakurai, and H. Ishizuka, „Secure Hyperelliptic Cryptosysetems and Their Performance", PKC'98, Springer-Verlag, LNCS 1431, pp.164-181, 1998.

[TM99]   Tetsuya Tamura and Tsutomu Matsumoto, "A Hardware–Oriented Algorithm for Computing Jacobian on Hyperelliptic Curve", IEICE Trans. A vol.J82-A, no.8, pp.1307-1311, Aug. 1999.

[TOH98]   N. Torii, S.Okada, and T. Hasebe, „A chip Implementation of elliptic curve cryptosystems", Proc. 1998, Engineering Science Conference, IEICE A-7-1.

Appendix: RTL Behavior in Computing Addition

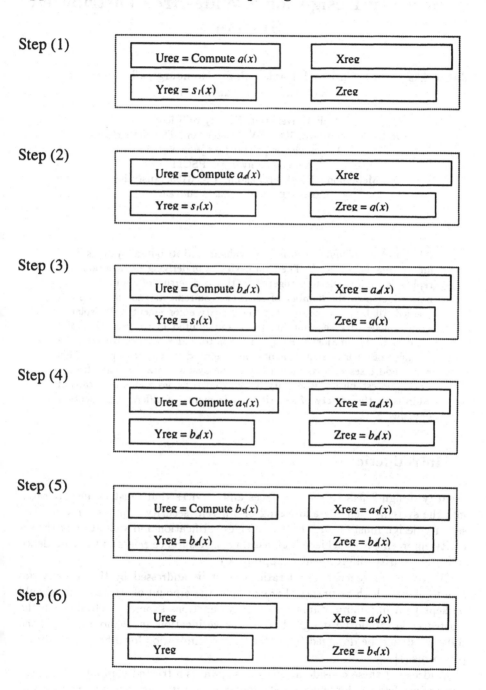

A Security Design for a Wide-Area Distributed System

Jussipekka Leiwo[1], Christoph Hänle[1], Philip Homburg[1], Chandana Gamage[2], and Andrew S. Tanenbaum[1]

[1] Vrije Universiteit, Faculty of Sciences
De Boelelaan 1081a, 1081 HV Amsterdam, The Netherlands
{leiwo,chris,philip,ast}@cs.vu.nl
[2] Monash University, PSCIT
McMahon's Road, Frankston, VIC 3199, Australia
chandag@pscit.monash.edu.au

Abstract. Designing security of wide-area distributed systems is a highly complicated task. The complexity of underlying distribution and replication infrastructures together with the diversity of application scenarios increases the number of security requirements that must be addressed. High assurance requires the security enforcement to be isolated from non-security relevant functions and limited in the size of implementation. The major challenge in the is to find a balance between the diversity of security requirements and the need for high assurance. This paper addresses this conflict using Globe system as a reference framework, and establishes a security design that provides a flexible means of addressing the variety of security requirements of different application domains.

1 Introduction

Security design refers to the interfaces and services that must be incorporated into the system to enable addressing of different security requirements [5]. The security design must be such that it enables verification and validation of the security enforcement to achieve high assurance. Assurance refers to the confidence that the security enforcement is appropriate.

A number of generic considerations must be addressed by the security design to achieve high assurance. For example, the amount of trusted code should be kept to a minimum, duplicate security functions should be eliminated, the trusted code should be designed to enable verification and validation, and the software should be designed to enable code optimization for different hardware platforms [1].

Addressing these considerations leads towards a trusted computing base capable of verifiably enforcing a small number of security requirements. Diversity of security requirements is often a prohibitive factor for high assurance [8]. However, security requirements of distributed object systems are of a very high diversity.

JooSeok Song (Ed.): ICISC'99, LNCS 1787, pp. 236–256, 2000.
© Springer-Verlag Berlin Heidelberg 2000

Not only secure communication requirements, e.g. encryption and authentication, but also requirements of secure operating systems, e.g. secure method or function execution and access control, and those of secure communication systems, e.g. traffic filtering and client behavior monitoring for intrusion and misuse detection, must be addressed.

Designing security on development platforms for distributed shared objects (DSO), such as Globe [15], that provide transparent distribution and replication of objects over multiple physical locations further complicates the security design. The major advantage of Globe type of systems over, say, CORBA is the scalability. CORBA, DCOM and other existing distributed object technologies assume a static replication model, whereas Globe enables per-object replication strategies allowing an increased flexibility in designing global distributed object systems [3].

Security design must also be established in a manner that allows object-specific security policies being established and maintained without limiting the range of applications [16].

A number of security architectures, such as Kerberos, Sesame, and DSSE, have been established for distributed and networked systems [9, 12]. Such architectures are mostly concerned with the development of secure applications on networked and distributed environments. They do not address the security of the inherently complicated distribution infrastructure itself. Scalability to global systems is also questionable.

This paper examines the problems associated with security designs of global DSO architectures in general, and Globe in particular. We begin by identifying the security requirements of distributed object systems and proceed by examining the challenges that the security designer faces when attempting to address these requirements. This is followed by a comparison of two possible security designs. The implementation aspects of the chosen design shall then be discussed. Finally, conclusions shall be drawn and directions highlighted for future work.

2   Security Requirements in Distributed Object Systems

Security requirements of distributed systems must be addressed through communication security, operating system security, and network security requirements. There are also object life-time requirements dealing with creation, binding and disposal of objects. Security management must be addressed, as well as educational and operational security requirements, and other pervasive security requirements.

In the following, examples of security requirements are given at each category. The list is comprehensive, yet it is questionable whether it can ever be complete. Also, the identified requirements are partially overlapping and each one may contribute to more than one security objective.

For example, message semantics based filtering of protocol messages is a means of achieving access control. If the traffic filtering is enforced at the application level, there is a close relationship to the access control based on method execution request. This is to be addressed at the operating system level. However, in most cases the traffic filtering is applied at lower levels, for example at the network layer, when it must be considered as a means of network security instead of a means of operating systems security. Both ways ways still contribute towards access control

2.1 Communication Security Requirements

Communication security is mostly concerned with cryptographic techniques to achieve confidentiality, integrity, authenticity, and non-repudiation of communicated messages. The communication security requirements of distributed object systems are not fundamentally different from general communication security requirements, for example those of the ISO OSI standard [9].

Depending on the type of communication, confidentiality can be addressed through **Connection-oriented confidentiality requirement** or **Connectionless confidentiality requirement**. Not all the protocol fields may require equal security, and it may become appropriate to address the **Selective field confidentiality requirement**. In some environment, even the fact that communication between certain hosts occurs may be sensitive, and the **Traffic flow confidentiality requirement** must be addressed.

Integrity measures may be implemented with or without recovery. Those supporting recovery allow the reconstruction of the message from the integrity check, whereas those without recovery can only be used for verifying the correctness of pairs of messages and integrity checks. As integrity can be addressed by entire protocol messages or selective fields, **Message integrity requirement with recovery**, **Message integrity requirement without recovery**, **Selective field integrity requirement with recovery**, and **Selective field integrity requirement without recovery** must be addressed.

Authentication can be applied either to a peer object or to the origin of data. Additional measures can be provided for client and user authentication. However, they should not be addressed at the technical infrastructure of object distribution. **Peer object authentication requirement** must be addressed when data from a communicating software module, such as a protocol stack implementation, must be authenticated to the peer object in communications. Closely related is the **Peer object integrity requirement** where assurance must be provided to a peer object in a communicating system of the peer object implementation not being altered by, for example, a trojan horse. **Data origin authentication requirement** addresses the authentication of the communicating hosts as sources of protocol messages.

Non-repudiation of origin requirement addresses concerns of a sender of a message repudiation participation in communication. **Non-repudiation of receipt requirement** provides means to verify that a particular recipient has in fact received a message. **Non-repudiation of delivery requirement** is more complicated. Usually, it can not be addressed at the applications level.

2.2   Operating Systems Security Requirements

Traditionally, operating system security has been concerned with access control to protect files from unauthorized reading or modification, or to prevent unauthorized execution of system administration software. In distributed object systems, object can refer to objects of any granularity. Therefore, **Object integrity requirement** and **Object confidentiality requirement** can refer to any operating system object, or to the distributed shared object as a whole. **Object level access control requirement** refers to the measures to determine which accesses are allowed within the DSO.

From the DSO point of view, object confidentiality, integrity, and access control requirements refer to the state of the entire distributed shared object remains confidential or unaltered. Access control requirements deal with which clients are allowed to bind to the DSO. At such a coarse granularity, security issues must be addressed through object life-time security measures, addressing them at the operating system level is hard.

With the emerge of new networking technologies and new programming languages, the scope of operating systems security extends into more active control of, for example, program execution. A typical example of extended operating system security functionality is the Java Virtual Machine (JVM). Such requirements can be addressed through **Secure method execution requirements**.

In addition to addressing the communication security requirements to prevent method invocations from remote hosts being tampered with, the **Method integrity requirement** must be addressed to assure remote clients with the correctness of methods stored and executed in remote hosts. This is different from the peer object integrity requirement in a sense that it addresses the actual methods provided by the DSO, not the integrity of the methods of replication and distribution infrastructure.

2.3   Network Security Requirements

Network security is addressed to provide assurance of the correct operating of the distributed system that employs the above security technologies. The firewall capacities are provided by the **Message semantics filtering requirement** that addresses the selective transmission of protocol messages to block unwanted protocol messages from being processed by the distributed object.

At the distributed object level this means selective processing of control messages and remote method invocations. At the underlying communication infrastructure, this means selective forwarding of suspicious datagrams.

Client behavior monitoring requirement and **Method execution sequence monitoring** are means to detect intrusions and misuse and trigger appropriate alarms.

2.4   Object Life-Time Security Requirements

There are a number of security requirements that must be addressed through the object life-time instead of during the operation of the object. Most importantly, secure binding and secure disposal of the components of a DSO. When a new client wishes to connect to a DSO, it must initiate the binding procedure.

The first step in binding is that the candidate client contacts a name server to request the unique object identity that matches the symbolic name of a DSO where the connection is to be established. The name service must be protected by addressing the **secure name service** requirements.

Once in the possession of the unique object identity, the new client proceeds with the binding by contacting the location server. The location server returns the contact address of the DSO as a pair of network address and port to connect to. This phase must be protected by addressing the **secure location service** requirements.

In the following step, the new client contacts the implementation repository where the program code is loaded to construct the local representative of the DSO in a local address space. This step must be protected by the **secure implementation repository** requirement. In fact, there are a number of open issues in the security of downloading executable content that must also be addressed in case of non-local implementation repositories.

With the implementation of the local representative, the client can proceed with the binding to the DSO using the newly created local representative and the contact address received from the location service. **Secure connection establishment** requirement must address the issues related to the establishment of the connection between the local representative and the DSO. This is also the phase where end-user security requirements, such as **user authentication** must be addressed.

At the end of the life time, the local representative disconnects from the DSO. This includes disposal of the local code as well as the state of the DSO and the security state of the communication. This must be addressed through the **secure object disposal** requirement.

2.5 Pervasive Security Requirements

Pervasive security requirements can be studied from a number of points of view. One point of view is the security requirements related to the management of information security. These requirements include, for example, security planning and security maintenance. Also, a number of security education and awareness requirements must be addressed. However, as these bear no significance to the establishment of a security design for DSO systems, they shall not be further studied herein.

Other pervasive security requirements are those that are needed for properly implementing specific security requirements. For example, security labels, trusted implementation of security functions and so further must be addressed once implementing the security design in a particular application scenario.

Finally, there are a number of implementation requirements that have security relevance, even though they are not addressed through actual security measures. A typical example is the total ordering of method invocations to pre-

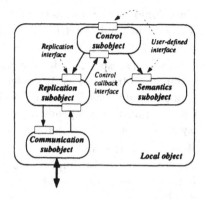

Fig. 1. Implementation of a Globe local object

vent inconsistent states of a DSO by incorrect method execution sequences. A comprehensive treatment is provided, for example, by Birman [4].

These are often considered to be issues addressed by the reliability engineering and dependable computing point of view. They are more concerned with continuity and correctness of services in the presence of random and independent failures, not in the presence of active attacks, i.e. selective failures. Therefore, they shall not be made a part of the security design.

3   Challenges of Security Design

To examine the challenges of security design in wide area distributed systems, the Globe object architecture will be used in this paper as a concrete example. After the introduction of the Globe object architecture, three security design challenges shall be addressed: selection of the security design model, coping with the diversity of security requirements, and the placement of security measures within the object architecture.

3.1   Globe Object Architecture

A central construct in the Globe architecture[1] is a distributed object. A distributed object is built from a number of local objects that reside in a single address space and communicate with local objects residing in other address spaces.

Together, the local objects form the implementation of a particular DSO. Local objects consist of the actual interface and the distribution mechanism, as illustrated in Fig. 1. The distribution mechanism enables transparent distribution and replication of objects, hiding details from application developers.

---

[1] More details available at http://www.cs.vu.nl/globe/

The semantics subobject contains the methods for the functionality of the DSO. This is the only subobject the application developer must develop himself. It is as objects in middleware architectures such as DCOM and CORBA.

The communication subobject is responsible for the communication between local objects residing at different address spaces. It implements a standard interface but can have several implementations depending on the particular communication needs and provides a platform-independent abstraction of underlying networks and operating systems providing the communication services.

The replication subobject replicates and caches the local objects and constructs the DSO from local objects. It also implements coherence protocols to decide when methods of the local semantics subobject can be invoked without violating the consistency policy.

The control subobject invokes the semantics subobject's methods. It also marshalls and unmarshalls invocation requests passed between itself and the replication subobject.

3.2   Security Design Model

Several security design models have been established over years. Most of them, however, focus on multilevel secure systems and databases (e.g. [1, 6, 14]) instead on the security of conventional applications on general purpose operating systems. High dependence on risk analysis further limits the applicability of many models (e.g. [2, 13]).

Limitations of risk analysis become evident in the design of the security of system development and distribution platforms. Since underlying implementation technologies and operational environment are not known at the time of security design, neither threats nor losses can be estimated. However, a systematic approach, such as [10], is required for guiding the security design.

The major advantages of [10] is that it reduces risk analysis into a decision making tool and is heavily based on the Common Criteria for security evaluation [7]. It divides security development into three stages. First one deals with the specification of all relevant security functions capable of satisfying a certain security objective. The second stage aims at selecting a subset of all possible security requirements to be implemented in a particular system. Selected measures are implemented and evaluated at the third stage.

This paper deals with the first stage of the model. A possible set of security requirements of a distributed object system are identified and a security design is established to aid in the implementation of security measures to address those requirements. When designing applications using Globe, the particular operating system and communication mechanisms can be selected, and a subset of possible security countermeasures implemented to match the specific application level security policy.

Table 1. Possible security requirements of Globe as divided to the underlying communication infrastructure (UCI), communication subobject (CoS), replication subobject (RS), control subobject (CS), application (AL), and operating system (OS)

Requirement	UCI	CoS	RS	CS	AL	OS
Connection-oriented confidentiality requirement	X	X	X	X	X	X
Connectionless confidentiality requirement	X	X	X	X	X	X
Selective field confidentiality requirement	X	X	X	X	X	X
Traffic flow confidentiality requirement	X	X	X	X	X	X
Message integrity requirement with recovery	X	X	X	X	X	X
Message integrity requirement without recovery	X	X	X	X	X	X
Selective field integrity requirement with recovery	X	X	X	X	X	X
Selective field integrity requirement without recovery	X	X	X	X	X	X
Peer-object integrity requirement	X	X	X	X	X	X
Message semantics filtering requirement	X	X	X	X	X	X
Peer entity authentication requirement	X	X	X	X	X	X
Data origin authentication requirement	X	X	X	X	X	X
Non-repudiation of origin requirement	X	X	X	X	X	X
Non-repudiation of receipt requirement	X	X	X	X	X	X
Non-repudiation of delivery requirement	X					
Object level access control requirement				X	X	X
Client behavior monitoring requirement				X	X	X
Method execution sequence monitoring requirement				X	X	X
Method integrity requirement				X		X
Object integrity requirement				X		X
Object confidentiality requirement				X		X
Secure method execution requirement					X	X

3.3   Diversity of Security Requirements

The ideal case of security in distributed object systems is a dedicated security subobject that implements security measures similarly to a traditional reference monitor. However, it is not obvious how this can be achieved in practice taking into account the high diversity of security requirements and the possibility of security requirement being addressed at different subobjects. The possible components where different security requirements of distributed object systems can be addressed within the Globe object architecture are illustrated in Table 1.

The semantics subobject does not have security relevance to the Globe architecture since it does not participate in replication and distribution. However, a high number of security requirements may be addressed at the application level through the semantics subobject.

The security architecture is also independent of the underlying communication architecture. A TCP/IP network could implement packet confidentiality and authenticity in form of IP SEC standard, or a transport layer security by SSL or SSH. In more advanced scenarios, network traffic could be authenticated and access control provided by Kerberos, Sesame, or DSSA/SPX. Since the objective of Globe is flexibility and platform independence, no assumptions of the available security services can be made. All communication security measures may need to be implemented at the communication subobject.

The communication subobject security is concerned with secure communication channels between local objects. Requirements are those of secure communication, i.e. confidentiality, integrity, authenticity, and non repudiation. Access control can be enforced through traffic filtering based on protocol messages.

In group communication, communication security measures can be applied at the replication subobject or control subobject on per-message rather than per-recipient basis. If the communication subobject manages group communication through a number of point-to-point channels, this can significantly reduce the cryptographic overhead.

The replication subobject security is also concerned with the enforcement of secure replication of objects, and prevention of malicious parties from altering the DSO state, interface or implementation.

The control subobject and the underlying operating system are responsible of secure execution of the methods of the semantics subobject. Method level access control can be provided to decide which methods can, under which constraints, be invoked in the local environment. Client behavior and method execution sequences can be monitored for intrusion and misuse detection.

3.4 Placement of Security Functionality

Consider a simple entity authentication protocol [11, p.402]. B initiates the protocol by sending a random value r_B to A. A replies with a random number r_A and a keyed hash $h_K(r_A, r_B, B)$. B then sends the value $h_K(r_B, r_A, A)$ to A allowing both parties to verify each other's authenticity through knowledge of key K shared by A and B.

$$A \leftarrow B : r_B \tag{1}$$
$$A \rightarrow B : r_A, h_K(r_A, r_B, B) \tag{2}$$
$$A \leftarrow B : h_K(r_B, r_A, A) \tag{3}$$

This (or similar) protocol is likely to be implemented at several subobjects requiring peer-object authentication. An obvious design is to separate calculation of the hash value from the protocol execution logic. Protocol implementation becomes easier as the hash function can be treated as a black box and implemented separately, possibly in hardware.

Availability of a mutually agreed upon hash-function between entities is required in steps (2) and (3). To negotiate the function and to store security state, such as cryptographic keys, a security context must be maintained by the communication parties.

The protocol logic can be implemented as a separate function or as a, say, Java object (not a Globe object, though), that is called or instantiated by objects requiring to authenticate with their peer-objects. The subobject instantiates the authentication object and defines the parameters, such as the behavior in error conditions. This complicates the object interface but provides high encapsulation of security relevant processing in a dedicated authentication protocol object.

Not all protocol errors imply an authentication failure. They may be due to network congestion, excessive workload at the peer object, or some other random condition occuring. The recovery logic must decide which action to take, whether to deal with the peer-object as un-authentic, to proceed with the service request and assume further authentication at other subobjects, to block the service request and retry authentication after a delay, or to take some other action.

Parameterization of all possible error conditions at different subobjects leads to a complex exception handler or to an increased interaction between the security subobject and conventional subobjects, This reduces functional cohesion of the security subobject and leads to weaker encapsulation.

An attempt to isolate the security functionality has, therefore, led to an increased complexity of the protocol object. Each authentication request must be related to a number of other security-relevant objects, such as security context and exception handler. Proper software engineering practice, such as thin interfaces and functional encapsulation can, however, improve the design.

Each subobject has to be given a unique identity, expressed as the identity of a local object and the particular subobject. This identity is used for authentication. The protocol execution logic can be easily separated from the subobject but the semantics of different protocol messages has to be bound to a particular subobject. This encourages implementation of the authentication protocol as part of the conventional subobject as it is mostly aware of various subobject-specific semantic conventions.

The denial of service aspects should also be kept in mind. Globe objects are typically distributed using public networks, e.g. the Internet, with limited quality of service guarantees. For example, TCP guarantees an ordered delivery of messages but not the maximum time for message transmission. Protocols that depend on a TCP connection between two hosts may cause serious performance penalties due to network congestion outside the control of any local object.

Resource allocation policies may be defined for protocol steps or stateless protocols designed to prevent denial of service. However, protocols have to be carefully designed and evaluated for optimal performance and reaction on exceptional conditions. Therefore, they should be dealt with as independent software artifacts.

Protocol implementations are also likely to require a preparedness plan in terms of exception handling to recover from situations where a critical resource, such as protocol execution time, exceeds a threshold. Recovery is very subobject specific and difficult to generalize into a common protocol implementation.

These issues complicate the decision of whether the security should be managed by a dedicated security subobject, or by each conventional subobject independently. The following section provides a detailed analysis and evaluation of the two alternatives.

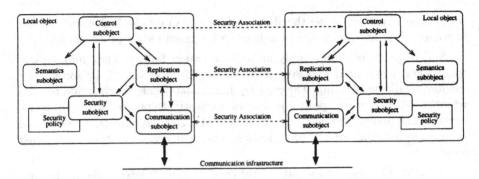

Fig. 2. A Generic DSO security design

4   The Security Design

A generic security design for a DSO, as illustrated in Fig. 2, consists of a security subobject, security policy and a number of security associations (SA). SAs describe the security state of communication channels and may be shared by multiple parties. They contain, at least, encryption and authentication keys, modes of algorithms, and other parameters such as initialization vectors, and the SA life time.

Prior to secure communication, peer objects must establish a SA through on line or off line negotiations. The initial state of the security association is downloaded during the binding. The number of security associations maintained by a local object may be different in different implementations and application environments.

The SA must be supported by a security subobject that contains the implementation of corresponding security measures. Security subobject implements a certain communication security policy. Communication security policy is fairly static but security associations may change dynamically. Additional security policies are required for access control and intrusion detection.

Replacing a communication security policy means binding to a different security subobject. Full policy-mechanism –independence, as in access control models, is hard to achieve due to the difficulties of formally expressing communication security requirements. The local object may initially contain an implementation of a number of security subobjects or download them when necessary.

There are two main alternatives for coordinating the security enforcement:

Centralized security coordination (CSC) where conventional subobjects request all security measures to be executed by the security subobject.

Distributed security coordination (DSC) where conventional subobjects execute the security measure but depend on the security subobject for critical functions, such as encryption and decryption of buffers.

In CSC, the communication security policy is followed by the security subobject that enforces the policy and executes the required security measures. In DSC, the communication security policy is followed by conventional subobjects.

In the following, the two shall be compared and evaluated against a number of security design criteria. The comparison is followed by a discussion about the security association and different security policies. A clear distinction between the two is impossible in practical systems. The implemented system is likely to be a hybrid. However, the comparison suggests that implemented systems should bear more characteristics of centralized than distributed security coordination.

4.1 Centralized Security Coordination

Communication between local objects is mediated by the security subobject. The security subobject maintains the security associations and negotiates their content with the security subobject of the peer local object. Certain security processing of messages is carried out at each passing of messages between different subobjects.

Prior to passing a protocol message to a lower or upper level in the subobject hierarchy, a subobject passes it to the security subobject. The security subobject applies the security measures to protocol messages and returns. The subobject that called the security subobject passes the security enhanced protocol message to the next subobject. The same process is repeated at each subobject and reversed at the receiving local object.

As a minimum, the security subobject only requires an interface for passing and receiving messages to and from subobjects. Each subobject interfaces with the security subobject through a common interface that may have different implementations. The calling sequence from subobjects to the security subobject can be standardized. This allows replacement of the security subobject without modifying other subobjects.

Full isolation of the security subobject is hard to achieve, mostly due to exception handling. Which action is taken if a security measure, e.g. data origin authentication, fails? The interface can be standardized to return a number of status codes the conventional subobject can use for examining the status of the security processing. For example, a standardized security exception could be thrown by the security subobject and caught by the conventional subobject. Implementations on languages such as C may return a standardized error code.

4.2 Distributed Security Coordination

The security subobject only provides basic security mechanisms to aid subobjects in the enforcement of security as part of the subobject functionality. The conventional subobject maintains the SAs and implements the security logic.

The security subobject provides basic security services, such as encryption and decryption of buffers, generation and verification of authentication codes of buffers and so on. The conventional subobject will call these measure when

Table 2. Comparison of security subobject designs

Criteria	CSC design	DSC design
Economy of mechanism	Average	Average
Duplicate functions	Good	Poor
Optimization for new hardware	Good	Good
Complete mediation	Good	Poor
Least privilege	Good	Poor
Future alterations	Good	Average
Ease of SA maintenance	Good	Poor

necessary according to the security logic. The receiving conventional subobject implements corresponding security protocols, and executes the actions of a receiving entity of the protocol.

This approach introduces a classical trade–off between assurance and diversity of security requirements. High assurance can be provided for, say, cryptographic processing on this scheme with the cost of reducing the security subobject functionality and complicating the security implementation. In CSC, equally high assurance can be achieved on cryptographic modules, but in general, higher assurance can be achieved to the general security processing.

The complexity can be reduced by standardized security libraries and proper software engineering practices. Replacing and extending the security subobject remains easy. Exception handling is logically connected to the protocol execution. Adapting to a different communication security policy remains complicated, though, alterations are required to each conventional subobject.

4.3 Comparison of Designs

Many principles of security design (e.g. [1, 14]) focus on access control models and their applications. In the following, the above design alternatives are evaluated against security design principles adapted to the DSO context. Findings are summarized in Table 2.

Economy of mechanism refers to the small size of implementation and simplicity of the design to enable appropriate testing and evaluation. It is unlikely that either design can achieve such economy of mechanisms to enable formal verification of security. Yet, this would be unnecessary in most application scenarios, due to general purpose operating systems used. There are no significant differences between the two designs.

Elimination of duplicate functionality requires that no security functionality should be implemented in multiple modules. This is a significant disadvantage of the DSC design. Many of subobject's security protocols are likely to be similar and must be implemented by each subobject even though most central security functions are implemented by the security subobject. In the CSC design, each protocol is implemented once. This improves the control of the implementation but increases the complexity of security subobject's interface.

Code optimization for new hardware is essential for performance reasons. It is likely that only cryptographic functions are implemented in hardware. In both designs, the cryptographic functions can be separated from protocols by proper software engineering or by logical separation of security protocols from security functions. Both designs provide a considerably good support for hardware implementations of different security functions.

Complete mediation requires the security subobject being consulted in each method invocation. System design should prevent subobjects from bypassing security on discretion. Complete mediation can, through design, be achieved by the CSC design. The security subobject methods are always called, even though no security measures are applied (i.e. some security functions are NULLs). With DSC, control measures and security method integrity checks have to be applied at multiple locations.

Least privilege refers to the components of the system gaining only a minimum set of accesses to sensitive data required for completing their tasks. The DSC design is problematic because of the distribution of security related processing to every subobject of the system. Each subobject must be given access to security critical components, such as security associations. The CSC design enables easier control of the privileges.

Ease of future alterations measures adaptation to different security policies. As security requires continuous maintenance, this is an important criteria. The CSC design is superior, mostly due to the single point of alterations required. In DSC design, each alteration in protocols of security functions must be implemented in each subobject. However, proper software engineering can simplify alterations.

The comparison suggests the superiority of the CSC design over the DSC design. However, the comparison only deals with the design criteria, not on performance issues. It is not clear how significant performance reduction is caused by security processing relative to, for example, network latencies when the local objects are distributed over wide area networks. Intuitively, it appears that performance penalties of different designs are not significantly different relative to the overall cost of communication. Real measurements are required to confirm the intuition.

The CSC design as has some additional advantages over the conventional subobject enforced security in, for example, **ease of SA maintenance** as discussed in the following section.

4.4   Security Association

Each local object must share at least one SA with local objects it is communicating with. In the CSC design, the security associations are maintained by the security subobject.

SA does not have any particular functionality. It is a data structure that stores the security state of communication. A significant advantage of CSC design over the DSC design is in the ease of applying different SA schemes, for example

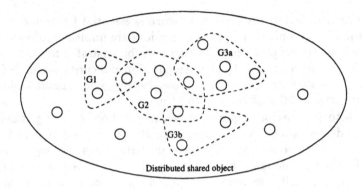

Fig. 3. A Large DSO with multiple groups of local objects

Single local object SA scheme is where a single SA is maintained by local objects and used for all security needs.

Multiple local object SAs scheme is where a number of SAs are maintained by local objects and used for different security needs but shared between all subobjects.

Single subobject SA scheme is where each subobject of a local object maintains a SA with peer subobjects and use it for all their security needs.

Multiple subobject SAs scheme is where each subobject of a local object maintains a number of SAs with peer subobjects and uses them for different security needs.

In many point-to-point applications, the single local object SA scheme is the most likely scenario. Since the subobjects of a local object are maintained in a single address space, there are no reliable means from preventing malicious local objects from violating the security of other subobjects. Therefore, multiple SAs may not be meaningful.

Different keys may be maintained for different services or security levels but shared between each subobject of a local object. If local objects operate on environments that provide separate address spaces or other tamper-proof execution environment, more complicated and fine-grained keying schemes may be relevant.

Complicated SA schemes occur also in very large DSOs (Fig. 3). Circles illustrate local objects that constitute the distributed object. Local objects are further grouped into three.

The core group ($G1$) is formulated of those local objects that are most crucial to the application, for example sites from which a WWW page can be updated. The cache group ($G2$) is a set of passive sites replicating the service, in this example the pages, without altering the content. The client groups ($G3a$ and $G3b$) contain those clients that have at a certain point of time registered with a member of the cache group to access the service. Disconnected local objects may connect to the distributed shared object in the future.

Table 3. Security subobject enforceable security policies

Policy	Purpose
Communication security policy	Static Communication security
Security Association	Dynamic communication security
Access control policies	Method invocation control
Behavior monitoring policies	Intrusion and misuse detection
Local policies	For subobject internal security

Assume that core group members deliver on line a data item, such as a newspaper, software component, or a digital media clip, for which a payment is required. In global distribution, the core group objects can not deliver the item to millions of customers. Rather a number of caching sites are established and clients access cache sites for the service.

The data item can be protected by encryption and registered clients can obtain (maybe once a payment transaction is completed) a cryptographic key to recover the item. Key distribution may depend on the level of trust of cache group members:

Untrusted cache is where the core group members do not trust members of the cache group. Caches hold encrypted data but can not decrypt it. Clients must buy the decryption key directly from core group members or a dedicated key server.

Trusted cache is where the core group members allow cache group members to have the encryption key of data elements. This means, data can be stored in plaintext on the caches and link encrypted when communicating with a client.

The level of trust of replicas depends on the application domain. Through appropriate implementations, the need for a number of SAs can be also reduced. Untrusted caches can, for example, share one SA with a core group member and another SA with a client group member. With tamper-proof hardware, data can be decrypted and reencrypted without disclosure to the caching site.

As this may be a practical impossibility, the need for flexible SA schemes remains. The above listed SA schemes can be extended by various group-security SA schemes and application specific schemes. The security subobject can independently maintain the required SA scheme without violating the security model.

4.5 Security Policy

The local object design requires a number of security policies as introduced in Table 3. System level and managerial security policies are omitted as they are beyond the scope of Globe security design.

The largely static communication security policy that describes the security measures to be applied in the communication is implicit. It is constituted by the implementation of the security subobject. Means to achieve higher policy-mechanism independence are a major area of future research.

Lower level security policy describing the ways in which the implemented security measures are executed is expressed in a more flexible manner through dynamically changing security associations. Negotiation mechanisms of security associations in fact are mechanisms for negotiating the security policy and the components of a SA used for enforcing the policy.

Access control policies describe which methods can be invoked by which clients under which circumstances. Behavior monitoring policies describe the ways in which method invocations are monitored for intrusion and misuse detection, and how deviations are handled.

Local policies must be enforced by each subobject internally. The are not concerned with the security of communication but the internal security of a subobject. For example, prevention of denial of service attacks may require local resource allocation policies at each subobject.

5   Interfaces to the Security Subobject

The DSO security has two distinctive facets: transformational security to enforce the communication security, and access control to impose restrictions on method invocations. Interfaces of the security subobject shall be discussed from both points of view. Secure operations of a DSO shall then be addressed.

5.1   Transformational Security

The security subobject must have a standardized interface and a calling sequence from other subobjects. This enables on line adoption of different communication security policies by replacing the security subobjects of relevant local objects. No alterations to other subobjects are required.

Methods of the security subobject must be paired so that each method that enforces certain security feature is associated with a method the removes or verifies the added security. For example, a method encrypting data must be paired with a method for decrypting the data. A method for calculating a MAC must be paired with a method for verifying the MAC.

Transformational security is encapsulated into two method calls per subobject. First method adds the security that the other method removes or verifies. Security-adding methods are called by the sending local object and the security-removing and verifying methods by the receiving local object. The general calling sequence (Fig. 4), is as follows:

1. A client accesses the local object through the control subobject.
2. The control subobject marshalls the arguments to the method call and prior to passing the marshalled arguments to the replication subobject, calls the SecCtrlAdd() method of the security subobject. The method does all the security related processing relevant to the control subobject and adds the required fields into the argument string.

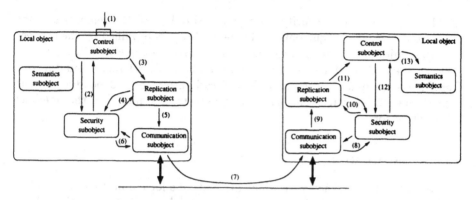

Fig. 4. Security subobject calling sequence

3. The control subobject passes the processed argument string to the replication subobject.
4. The replication subobject processes the marshalled arguments and invokes the SecReplAdd() method that implements the security relevant processing and adds the necessary data to the arguments.
5. The data is passed to the communication subobject.
6. The communication subobject processes the data and calls the SecCommAdd() method that implements the required features of communication security prior to sending the resulting data through the communication channel.
7. The data is communicated to the other local object.
8. Upon receipt of data, the receiving communication subobject calls the SecCommRemove() method of the security subobject that reverses and verifies the security measures put on place by the SecCommAdd() method.
9. The data is processed by the communication subobject and passed to the replication subobject.
10. The replication subobject calls the SecReplRemove() method that reverses and verifies the security measures put on place by the SecReplAdd() method.
11. The replication subobject processes the data and passes it to the control subobject.
12. The control subobject calls the security subobject method SecCtrlRemove() that reverses and verifies the security measures put on place by the SecCtrlAdd() method.
13. The control subobject proceeds with the method invocation. The results are passed to the remote client and the security process is repeated.

Some security methods may be NULL as there may not be security processing at each subobject. However, it is imperative that the methods are called in the above sequence to enable replacement of the security subobject without modifying other subobjects.

The interface to the security subobject is simple, and the semantics of the byte strings passed as arguments to the methods are determined by the conventional subobjects. Security processing does not need to be aware of it.

Each method must receive the arguments through reference and throw a GlobeSecurityException or return a GlobeSecurityError code, depending on the implementation language. This enables subobjects to standardize the handling of security errors.

5.2   Access Control

Access control is concerned with which methods of a local object can be invoked by which clients under which conditions. For example, access to Write() methods may be restricted to the members of the core group, and may only be granted after strong authentication and if a certain environmental condition, such as time of the day, is met. This requires a sophisticated access control scheme to mediate method invocation. This is logically placed in the control subobject.

Access control functionality can be divided into two layers: credential verification layer and access enforcement layer. As the access control decisions are made at the control subobject, there is a need to store the results of credential verification in a data structure available for the access control decision function.

The security subobject maintains a Credentials data structure that consists of a number of *(attribute, value)* pairs, where the values of different attributes are set by the security subobject once credentials are verified. The access control decision function reads the credential data structure, access control rule base, and a set of environmental variables needed in access decisions. The access control facility can be invoked by the CtrlVerifyAccess(methodID) method, prior to step (13) in Fig 4.

The access control rule base consists of a number of authorization statements describing under which conditions certain clients are allowed to invoke certain methods of the semantics subobject.

5.3   Secure Operations

Similar methods than those required for security in method invocation are required for control messages. Similar to method invocations, control messages need to be protected in transmission, and a right to invoke certain control operations may be restricted to certain local objects. Separate security associations may need to be maintained for control messages.

Previous discussion has focused on the secure invocation of methods remotely. There are also security requirements that are not related to method invocations. These requirements are mostly concerned with the dynamic aspects of DSOs, most importantly the binding of new local objects to a distributed shared objects.

Each Globe object has a unique ID. The naming service maps a symbolic name to an object ID. The location server can then be queried for the contact address of a local object to bind to. Prior to the actual binding, an object class must be retrieved from the local implementation repository, and the local object constructed from the object class.

Name servers, location servers, and implementation repositories may not belong to a single administrative domain. Therefore, clients may not equally trust all service providers. Measures are required for adequate security at all the services. Different from, say, security extensions to the Internet Domain Name Server (DNS), naming and location information may not always be public. Therefore, it is unlikely that existing standards can be directly applied in Globe.

Research is currently carried out to investigate the extent to which existing infrastructure services can be applied in Globe.

6 Conclusions

This paper has analyzed the difficulties in designing security of DSO platforms using Globe as a reference system. In particular, the objective of isolating security relevant processing from other computations constitutes a fundamental design challenge. Yet, it is essential to enable a framework where appropriate assurance of the correctness of security design and implementation can be achieved.

We have concluded that, despite certain disadvantages, it is better to centralize security enforcement into a single security subobject. The method names and calling sequences from subobjects can then be standardized to enable replacement of the security subobject without modifying other subobjects.

The interface of the security subobject consists of three types of methods. Transformational security measures are used for protecting method invocations and control messages during communication. Access control methods prevent unauthorized clients from invoking methods or unauthorized local objects from invoking control methods. Other security measures are applied for other security operations, such as secure binding.

The work is currently on progress, and certain applications have been developed in Globe and different more advanced applications scenarios are currently under research. Further research is also going on in the provision of a high level policy-mechanism independence.

As the Globe is currently implemented in Java, there are certain possibilities for replacing the implementation on-line, for example during the binding of a new client to an existing local object.

References

1. M. D. Abrams, H. J. Podell, and D. W. Gambel. Security engineering. In *Information Security, An Integrated Collection of Essays*, volume Abrams, Marshall D. and Jajodia, Sushil and Podell, Harold J., pages 330–349. IEEE Computer Society Press, Los Alamitos, CA, USA, 1995.
2. A. Anderson, D. Longley, and L. F. Kwok. Security modeling for organizations. In *Proceedings of the 2nd ACM Conference on Computer and Communications Security*, pages 241–250. ACM Press, 1994.

3. A. Bakker, M. van Steen, and A. S. Tanenbaum. From remote objects to physically distributed objects. In *Proceedings of the 7th IEEE Workshop on Future Trends of Distributed Systems*, 1999.

4. K. P. Birman. *Building Secure and Reliable Applications*. Manning Publications Corporation, Greenvich, CT, USA, 1996.

5. D. L. Brinkley and R. R. Schell. Concepts and terminology for computer security. In *Information Security, An Integrated Collection of Essays*, volume Abrams, Marshall D. and Jajodia, Sushil and Podell, Harold J., pages 40–97. IEEE Computer Society Press, Los Alamitos, CA, USA, 1995.

6. S. Castano, M. Fugini, G. Martella, and P. Samarati. *Database Security*. Addison-Wesley, Wokingham, UK, 1995.

7. International standard ISO/IEC 15408 common criteria for information technology security evaluation (parts 1-3), version 2.0, CCIB-98-026, May 1998.

8. D. Gollmann. *Computer Security*. John Wiley & Sons, Chichester, UK, 1999.

9. W. Kou. *Networking Security and Standards*. Kluwer Academic Publishers, 1997.

10. R. Kruger and J. Eloff. A common criteria framework for the evaluation of information technology security. In *Proceedings of the IFIP TC11 13th International Conference on Information Security, (SEC'97)*, pages 197–209. Chapmann & Hall, 1997.

11. A. J. Menezes, P. C. van Oorschot, and S. A. Vanstone. *Handbook of Applied Cryptography*. CRC Press, NY, USA, 1997.

12. R. Oppliger. *Authentication Systems for Secure Networks*. Artech House, Norwood, MA, USA, 1996.

13. C. Salter, O. Saydjari, B. Schneier, and J. Wallner. Toward a secure system engineering methodology. In *Proceedings of the New Security Paradigms Workshop*. ACM Press, 1998.

14. R. C. Summers. *Secure Computing: Threats and Safeguards*. McGraw-Hill, 1997.

15. M. Van Steen, P. Homburg, and A. S. Tanenbaum. Globe: A wide-area distributed system. *IEEE Concurrency*, pages 70–78, January–March 1999.

16. W. A. Wulf, C. Wang, and D. Kienzle. A new model of security for distributed systems. In *Proceedings of the ACM New Security Paradigms Workshop*, pages 34–43, Lake Arrowhead, CA, USA, 1996. ACM Press.

Self-Escrowed Public-Key Infrastructures

Pascal Paillier[1] and Moti Yung[2]

[1] Cryptography Group, Gemplus
paillier@gemplus.com
[2] CertCo New York
moti@certco.com, moti@cs.columbia.edu

Abstract. This paper introduces a cryptographic paradigm called self-escrowed encryption, a concept initiated by kleptography. In simple words, a self-escrowed public-key cryptosystem features the property that the scheme's public and private keys are connected to each other by the mean of an other cryptosystem, called the master scheme. We apply this notion to the design of auto-recoverable auto-certifiable cryptosystems, a solution to software key escrow due to Young and Yung, and provide a new cryptographic escrow system called self-escrowed public key infrastructure. In addition, we give an example of such a system based on ElGamal and Paillier encryption schemes which achieves a high level of both efficiency and security.

1 Introduction

In recent years, considerable research efforts have been invested by the cryptographic community into the quest for an efficient and fair solution to the key escrow problem. Although the widespread use of nowadays communication networks such as the Internet would require the urgent deployment of a large-scale key recovery system for law-enforcement purposes, the complexity of the problem is such that very few satisfactory proposals have appeared so far. Tamper-resistant hardware solutions, such as the Clipper and Capstone chips, arouse the users' suspicion about the (unscrutinized) cryptographic algorithms executed inside the device [12, 5, 9]; many proposed systems require the escrow authorities to get involved in interactive computations at an undesirable level; finally, other investigated constructions suffer from not resisting various kinds of attacks (*cf.* shadow-public-key non-resistance [8]) from the system users.

Young and Yung [15, 16] recently introduced the concept of auto-recoverable auto-certifiable cryptosystems (ARC), a software-based cryptographic protocol that fulfills most of identified desirable requirements. ARCs conjugate functionalities of a typical public-key infrastructure (see below for definitions) with the ability to escrow private keys of the system users. To achieve this, the certification procedure of a given public key demands the key to be submitted along with a publicly verifiable zero-knowledge proof that the escrow authorities can efficiently recover the corresponding private key. The proof forms a certificate of recoverability which has to be stored securely by the certification authority

JooSeok Song (Ed.): ICISC'99, LNCS 1787, pp. 257–268, 2000.
© Springer-Verlag Berlin Heidelberg 2000

(CA). If a key recovery procedure is authorized for some suspect user, the escrow authorities query the CA for the matching certificate which allows them to completely recover the user's private key. The same authors also proposed a particular embodiment of their concept which relies on ElGamal encryption as well as on a specific key generation technique involving an extensive (hence costly) use of double decker exponentiations [11].

In this paper, we propose a new cryptographic notion which we call *self-escrowed* encryption. We show how to employ this technique to design a cryptographic protocol that meets all specifications of an auto-recoverable cryptosystem and presents other additional advantages. In particular, it confers on escrow authorities the ability to recover private keys directly from public ones. Consequently, the storage of some certificate of recoverability is no longer required. We call such a system a *self-escrowed public key infrastructure* (or SE-PKI for short). For completeness, we provide an practical example of an SE-PKI which is based on the joint use of ElGamal [7] and Paillier [10] encryption schemes and achieves a high level of both efficiency and security.

The paper is divided as follows. The next two sections briefly recall the definitions of a public-key infrastructure and of an auto-recoverable cryptosystem. Section 2 introduces the notion of self-escrowed encryption, which is then used to define SE-PKIs in section 3. In section 4, we propose a discrete-log based self-escrowed encryption scheme and analyze the corresponding SE-PKI in terms of efficiency and security.

1.1 Public-Key Infrastructures

A public-key infrastructure (PKI) is a distributed cryptographic protocol involving system users and trusted third parties called certification authorities (CA). Let $S = \langle G, E, D \rangle$ denote an encryption scheme where $G(1^k) = (x, y)$ is a probabilistic key generator (for a certain scheme parameter k), and where $m \mapsto E_y(m)$ and $c \mapsto D_x(c)$ represent the encryption and decryption functions, respectively. A PKI based on S is a protocol that fulfills the following specifications :

1. **Setup.** CA's addresses and parameters are published and distributed.
2. **Key Generation.** Each user runs G to generate a public/private key pair (x, y) and submits y (together with an ID string including personal system attributes) to a CA.
3. **Certification Process.** The CA verifies the ID string, signs y and enters the certified key (y + signature) in the public key database.
4. **Encryption.** To send a message, a user queries the CA to obtain the public key y of the recipient and verifies the CA's signature on y. If the verification holds, the user encrypts the message m using y and sends the ciphertext $c = E_y(m)$ to the recipient.
5. **Decryption.** The recipient decrypts the ciphertext with his/her private key to recover the message $m = D_x(c)$.

1.2   Auto-Recoverable Auto-Certifiable Cryptosystems

The notion was introduced in [15]. The system is a classical public-key infrastructure to which are added escrowing mechanisms. The protocol indeed ensures that some escrow agents, called hereafter escrow authorities, are capable of recovering the private key of any user suspected to misbehave. The cryptosystem is denoted $S = \langle G, E, D, V, R \rangle$ where :

- $G(1^k) = (x, y, P)$ is a probabilistic key generator that outputs a public/ private key pair (x, y) and a publicly verifiable non-interactive zero-knowledge proof P that x is recoverable by the escrow authorities using P.
- $V(y, P) \in \{0, 1\}$ is a publicly known algorithm such that (with overwhelming probability) $V(y, P) = 1$ iff x is recoverable by the escrow authorities using P.
- R takes as inputs P and some private information and returns x, provided that (x, y, P) is a possible output of G such that $V(y, P) = 1$. Optionally (distributed key recovery), R can also be an m-tuple $(R_1, \ldots, R_m)$ such that each R_i, run on P and some private input, returns the share x_i of x w.r.t some (perfect) secret sharing scheme. Escrow authorities then collaborate to recover x. The problem of computing x given (y, P) without R is assumed to be intractable.

An ARC based on S is a protocol specified by the following :

1. **Setup.** The escrow authorities generate a set of public parameters along with the corresponding private algorithm R. The public parameters and CA's parameters are published and distributed.
2. **Key Generation.** Each user runs G to generate a public/private key pair (x, y) and a certificate of recoverability P. The user then submits the pair (y, P) (together with an ID string including personal system attributes) to a CA.
3. **Certification Process.** The CA checks the ID string and verifies that $V(y, P) = 1$. If the verification holds, the CA signs y and enters the certified key $(y + \text{signature} + \text{certificate } P)$ in the public key database.
4. **Encryption.** To send a message, a user queries the CA to obtain the public key y of the recipient and verifies the CA's signature on y. If the verification holds, the user encrypts the message m using y and sends the ciphertext $c = E_y(m)$ to the recipient.
5. **Decryption.** The recipient decrypts the ciphertext with his/her private key to recover the message $m = D_x(c)$.
6. **Key Recovery.** If key recovery is authorized for a given user, the escrow authorities query the CA for the corresponding certificate P and run R on P to recover the user's private key x.

2   Self-Escrowed Encryption Schemes

In this section, we rigorously formalize the notion of self-escrowed encryption, initiated in spirit by kleptography [13, 14].

The usual way of formally describing a public-key encryption scheme S consists in decomposing it into three distinct algorithms $S = \langle G, E, D \rangle$ where G is a key generator (e.g. a probabilistic algorithm that outputs a typical key pair (x, y) in polynomial-time), and where $m \mapsto E_y(m)$ and $c \mapsto D_x(c)$ represent the encryption and decryption algorithms parameterized by the respective keys. To provide one-wayness, the scheme is necessarily built in such a way that the public key y is derived from the secret key x by the mean of some compliant one-way function $y = F(x)$ such as integer multiplication or exponentiation in a well-chosen group .

A self-escrowed encryption scheme can be defined as an encryption scheme for which the function F, in addition to being one-way, also presents partial or total trapdoorness. When it does, F can then be expressed as some encryption function $\mathcal{E}_Y$ for some existing encryption key Y. This also means that a public/private key pair (x, y) of S such that x falls into the "trapdoorness domain" of F reaches the property that $x = \mathcal{D}_X(y)$ i.e. there exist some trapdoor information X which allows to recover x from y. This property can be captured precisely, as follows.

Definition 1. *An encryption scheme $S = \langle G, E, D \rangle$ is said to be perfectly self-escrowed when there exist an encryption scheme $\Sigma = \langle \mathcal{G}, \mathcal{E}, \mathcal{D} \rangle$ and a key pair (X, Y) of Σ such that for all key pair (x, y) of S the relation*

$$y = \mathcal{E}_Y(x) \tag{1}$$

holds. By analogy with the secret key setting, Σ is called the master encryption scheme of S, Y the master public key and X the master private key.

As we will see later in the paper, definition 1 may not be reached in a strict sense although the given scheme present some a partial access to the self-escrow property. In particular, there could exist a master key pair for which *most* of key pairs satisfy relation 1. Situations may also occur wherein the set of escrowable private keys remains of reasonable size, although being a negligible proportion of the whole private key space. Self-escrow properties can still be defined in that case by weakening the strong requirements of definition 1. This is as follows.

Definition 2. *An encryption scheme $S = \langle G, E, D \rangle$ is said to be (partially) self-escrowed when there exist an encryption scheme $\Sigma = \langle \mathcal{G}, \mathcal{E}, \mathcal{D} \rangle$, a key pair (X, Y) of Σ and a pair of probabilistic polynomial-time algorithms $\langle \mathcal{P}, \mathcal{V} \rangle$ such that for all key pair (x, y) of S satisfying relation 1, $\mathcal{P}(Y, x, y) = P$ is a non-interactive (statistical) zero-knowledge proof that relation 1 holds and P is publicly verifiable i.e. $\mathcal{V}(Y, y, P) \in \{0, 1\}$ equals 1 (with overwhelming probability) if and only if P is a valid proof for y.*

In other words, we require that any key pair fulfilling the desired property can efficiently be proven such and that the generated proof of recoverability is public and can be publicly verified by anyone : a requirement that imposes zero-knowledgeness.

Once more, it is understood that the purpose of kleptographic attacks is no different from attempting to turn a target encryption scheme into a self-escrowed cryptosystem. Kleptography is by nature closely related to key recovery techniques, they differ in spirit only.[1] In both cases, the computational dependence of public keys on other public keys (be it subliminal or publicly known) seems to be necessary.

3   Self-Escrowed Public-Key Infrastructures

A self-escrowed public-key infrastructure is a particular case of an auto-recoverable auto-certifiable cryptosystem. The major advantage of an SE-PKI resides in that the proof of recoverability generated by the user is verified by the CA and then immediately discarded, since it is of no use regarding the key recovery procedure. Consequently, this releases certification authorities from the data storage of certificates of recoverability needed in ARCs, and completely removes interaction with escrow authorities during key recovery. This novel property is achieved using self-escrowed encryption as follows.

Let $S = \langle G, E, D \rangle$ be a self-escrowed encryption scheme and $\Sigma = \langle \mathcal{G}, \mathcal{E}, \mathcal{D} \rangle$ denote its master scheme. Recall that by definition, there also exist a pair of algorithms $\langle \mathcal{P}, \mathcal{V} \rangle$ allowing to generate and verify proofs of recoverability. An SE-PKI based on S can be defined as follows.

1. **Setup.** The escrow authorities run $\mathcal{G}$ to generate a master public/private key pair (Y, X). The public parameters, including Y and CA's parameters, are published and distributed.
2. **Key Generation.** Each user runs $G(Y, 1^k)$ to generate a public/private key pair (x, y), and then runs $\mathcal{P}(Y, x, y)$ to get the proof P that $y = \mathcal{E}_Y(x)$ holds. The user then submits the pair (y, P) (together with an ID string including personal system attributes) to a CA.
3. **Certification Process.** The CA checks the ID string and runs $\mathcal{V}$ to verify that $\mathcal{V}(Y, y, P) = 1$. If the verification holds, the CA signs y and enters the certified key $(y + \text{signature})$ in the public key database.
4. **Encryption.** To send a message, a user queries the CA to obtain the public key y of the recipient and verifies the CA's signature on y. If the verification holds, the user encrypts the message m using y and sends the ciphertext $c = E_y(m)$ to the recipient.
5. **Decryption.** The recipient decrypts the ciphertext with his/her private key to recover the message $m = D_x(c)$.
6. **Key Recovery.** If key recovery is authorized for a given user, the escrow authorities recover the user's private key $x = \mathcal{D}_X(y)$ and decipher the transmitted ciphertext(s). Here again, the key recoverability may be distributed among escrow authorities using some threshold decryption scheme.

---

[1] the kleptographic adversary wishes the very *existence* of the ability to recover keys to be secret. A subliminally self-escrowed encryption scheme is called SETUP, see [13].

It is worthwhile noticing that an SE-PKI remains extremely close by construction to a regular PKI : a given PKI is self-escrowed iff the underlying encryption scheme is also self-escrowed, the two properties directly derive from each other.

4   An Efficient Self-Escrowed PKI

We now proceed to describe a practical example of a SE-PKI. As just pointed out above, this requires to set up a self-escrowed cryptosystem first. Our scheme proposal is based on the joint use of ElGamal and Paillier encryption schemes and achieves (partial) self-escrow in the sense of definition 2, as will be shown later. We begin by a brief overview of useful mathematical facts.

4.1   Self-Escrowed Discrete Log-Based Cryptosystems

Paillier Encryption. Recently, Paillier [10] introduced public-key probabilistic encryption schemes based on composite residuosity classes over $\mathbb{Z}_{n^2}^*$ where n is an RSA modulus $n = pq$. To briefly describe the trapdoor, the knowledge of the factors of n happens to allow a fast extraction of discrete logarithms modulo n^2, provided that the base $g \in \mathbb{Z}_{n^2}^*$ is of order $n\alpha$ for some α with $\gcd(n, \alpha) = 1$. In the sequel, g will be chosen of maximal order $n\lambda$ where $\lambda = \lambda(n) = \mathrm{lcm}(p-1, q-1)$ must be relatively prime to n. We define over $\mathcal{U}_n = \{u < n^2 \mid u = 1 \bmod n\}$ the integer-valued function $L(u) = (u-1)/n$ where the division takes place in $\mathbb{Z}$. The public key is then the pair (n, g) while the private key is λ or equivalently the factors p and q. Encryption of a plaintext $m < n$ is done as follows. Pick an integer r uniformly at random in $[0, 2^\ell]$ where ℓ denotes the bitlength of n, and compute the ciphertext

$$c = g^{m+n \cdot r} \bmod n^2 . \tag{2}$$

To decrypt, compute

$$m = \frac{L(c^\lambda \bmod n^2)}{L(g^\lambda \bmod n^2)} \bmod n . \tag{3}$$

The one-wayness of the scheme is known to be equivalent to the *partial* discrete logarithm problem with base g, which is thought to be intractable provided that n is hard to factor. We refer the reader to [10] for more details. In this paper, we will be considering a deterministic version of this encryption scheme, i.e. encryption of a message $m < n$ is done by a simple exponentiation $c = g^m \bmod n^2$ while decryption is carried out as in equation 3. The encryption scheme is depicted on figure 1.

From a theoretical viewpoint, making the cryptosystem deterministic somehow decreases its security level, since computing partial logarithms then reduces to computing simple discrete logarithms[2]. However, since we do not know any

---

[2] the scheme also looses semantic security.

Public Key n, g of maximal order.

Private Key $\lambda = \mathrm{lcm}(p - 1, q - 1)$.

Encryption plaintext $m < n$
ciphertext $c = g^m \bmod n^2$

Decryption ciphertext $c < n^2$

$$\text{plaintext } m = \frac{L(c^\lambda \bmod n^2)}{L(g^\lambda \bmod n^2)} \bmod n.$$

Fig. 1. Paillier's Deterministic Encryption Scheme.

way of extracting discrete logs without the secret factors, we will make the assumption that inverting the encryption function still remains an intractable problem in this context.

Self-Escrowed Diffie–Hellman. The celebrated Diffie–Hellman key exchange protocol [3] exploits the (conjectured) hardness of extracting discrete logarithms over $\mathbb{Z}_p^*$ (for some well-chosen large prime p) and the additive homomorphicity of modular exponentiation. The protocol can straightforwardly be executed using other kinds of groups over which the discrete log problem is also thought to be intractable : we focus here on the specific group $\mathbb{Z}_{n^2}^*$ where n is an RSA modulus $n = pq$ just as before. As pointed out above, the knowledge of the factors p and q is sufficient to recover the discrete logarithm of $y = g^x \bmod n^2$ provided that $x < n$. This leads to a simple self-escrowed Diffie–Hellman variant (*cf.* figure 2) in which some escrow authority, whose private key is (p, q), can easily open the session key after wiretapping data exchanged during the protocol.

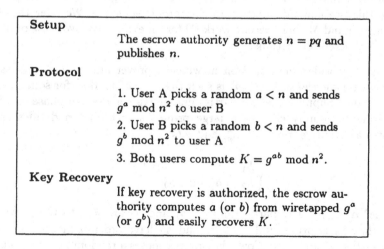

Setup	
	The escrow authority generates $n = pq$ and publishes n.
Protocol	
	1. User A picks a random $a < n$ and sends $g^a \bmod n^2$ to user B
	2. User B picks a random $b < n$ and sends $g^b \bmod n^2$ to user A
	3. Both users compute $K = g^{ab} \bmod n^2$.
Key Recovery	
	If key recovery is authorized, the escrow authority computes a (or b) from wiretapped g^a (or g^b) and easily recovers K.

Fig. 2. Self-Escrowed Diffie–Hellman Key Establishment.

Self-Escrowed ElGamal Encryption. Because ElGamal encryption is a non-interactive instance of the Diffie–Hellman protocol, the self-escrow property shown above also extends to ElGamal encryption over $\mathbb{Z}_{n^2}^*$. The resulting cryptosystem is displayed on figure 3 below. It clearly appears that the encryption (ElGamal) possesses a master scheme (Paillier).

Master Public Key	n, g of maximal order, $\ell = 2|n|$.
Master Private Key	λ
Public Key	$y = g^x \bmod n^2$ where $x <_R n$
Private Key	$x < n$
Encryption	plaintext $m < n^2$
	ciphertext $c = (my^k, g^k)$ where $k <_R 2^\ell$
Decryption	ciphertext $c = (a, b)$
	plaintext $m = a/b^x \bmod n^2$

Fig. 3. Self-Escrowed ElGamal/Paillier Encryption Scheme.

We claim :

Theorem 1. *The encryption scheme of figure 3 is self-escrowed.*

Proof (Construction of $\mathcal{P}$ and $\mathcal{V}$). To comply with definition 2, we still have to exhibit algorithms $\mathcal{P}$ and $\mathcal{V}$ with the desired properties. In our context, we clearly have $Y = (n, g)$ and $\mathcal{E}_Y(x) = g^x \bmod n^2$. Therefore the proof P to be generated must actually be a proof that $x < n$ given $y = g^x \bmod n^2$. To achieve this, we now introduce a specific proof technique. We refer the reader to Camenish and Michels' recent work [1] for an exhaustive overview of interval proofs for discrete logarithms.

We first consider the situation in which a prover interactively proves to a verifier that some element $y \in \langle g \rangle$ is such that $y = g^x \bmod n^2$ for some $x < n$ of his knowledge. Suppose the two parties first engage in a set-up phase as follows. The verifier randomly chooses a large prime p_0 such that n divides $p_0 - 1$, generates a primitive root γ of $\mathbb{Z}_{p_0}^*$ and sets

$$\begin{cases} h_1 = \gamma^{(p_0-1)/n} \bmod p_0 \\ h_2 = h_1^\alpha \bmod p_0 \end{cases}$$

for some[3] $\alpha <_R n$. The parameters p_0, h_1 and h_2 are sent to the prover who checks that p_0 is prime and that h_1 and h_2 have order n modulo p_0. These concludes the set-up phase. Now, the prover chooses a random $z <_R n$, computes

---

[3] $a <_R b$ denotes that a is chosen in the interval $[0, b-1]$ with uniform distribution.

$\bar{y} = h_1^z h_2^x \bmod p_0$ and sends $\bar{y}$ to the verifier. Finally, the two parties engage in the protocol

$$\pi = PK\{(x, z) : y = g^x \bmod n^2 \wedge \bar{y} = h_1^z h_2^x \bmod p_0\} \qquad (4)$$

which ensures the verifier that the wanted property $x < n$ is fulfilled. Note that protocol π must be carried out by the parallel execution of the two protocols

$$\begin{cases} \pi_1 = PK\{(x) : y = g^x \bmod n^2\} & \text{and} \\ \pi_2 = PK\{(x, z) : \bar{y} = h_1^z h_2^x \bmod p_0\}, \end{cases}$$

together with the same challenge. In virtue of a result due to Fujisaki and Okamoto [6], π_1 will work only if the strong-RSA assumption holds over $\mathbb{Z}_{n^2}^*$: we will therefore make this hypothesis in what follows. We refer the reader to figure 4 for an insight into the complete protocol.

Prover		Verifier
p_0 prime?		p_0 prime such that
$h_1^n \overset{?}{=} 1 \bmod p_0$		n divides $p_0 - 1$
$h_2^n \overset{?}{=} 1 \bmod p_0$	$\xleftarrow{\quad (p_0, h_1, h_2) \quad}$	$h_1, h_2 \in \mathbb{Z}_{p_0}^*$ of order n
$r_0 <_R 2^\ell$		
$z, r_1, r_2 <_R n$		
$t_0 = g^{r_0} \bmod n^2$		
$t_1 = h_1^{r_1} h_2^{r_2} \bmod p_0$	$\xrightarrow{\quad t_0, t_1, \bar{y} \quad}$	
$\bar{y} = h_1^z h_2^x \bmod p_0$		
	$\xleftarrow{\quad e \quad}$	$e \in_R \{0, 1\}^T$
$s_0 = r_0 + ex$		
$s_1 = r_1 + ez \bmod n$		
$s_2 = r_2 + ex \bmod n$	$\xrightarrow{\quad (s_0, s_1, s_2) \quad}$	$t_0 \overset{?}{=} g^{s_0}/y^e \bmod n^2$
		$t_1 \overset{?}{=} h_1^{s_1} h_2^{s_2}/\bar{y}^e \bmod p_0$

Fig. 4. An Interactive SZK Proof that $y = g^x \bmod n^2$ with $x < n$.

A part of the prover's response (s_0) being computed in $\mathbb{Z}$, we have to make sure that $|ex|$ is far smaller than $|r_0|$ e.g. T should be small before $|n|$ so that no information on x can leak out of s_0. The proof is then complete. We now see that there is no need for the set-up parameters (p_0, h_1, h_2) to be random : they can be advantageously replaced by absolute constants (depending on n only) i.e. they can be chosen once for all according to the above requirements and then included as a part of the public key Y. It is however necessary to ensure at this level that the prover chooses the random value z on his own so that the verifier cannot gain any information whatsoever about x from the knowledge of $\bar{y} = h_1^z h_2^x \bmod p_0$.

We know turn the protocol into a non-interactive zero-knowledge proof by applying the so-called Fiat-Shamir heuristic [4]. By doing so, we determine the challenge e by applying some collision-resistant hash function H to the commitment. In our case, this also determinates the two algorithms $\mathcal{P}$ and $\mathcal{V}$ that we have been looking for. These are depicted on figure 5.

Parameters	n, $\ell = 2\,|n|$, $g \in \mathbb{Z}_{n^2}^*$ of maximal order, p_0 a prime such that n divides $p_0 - 1$, h_1 and h_2 of order n in $\mathbb{Z}_{p_0}^*$
Generation $\mathcal{P}$	
Input	$x < n$ and $y = g^x \bmod n^2$.
	1. pick $z <_R n$ and compute $\bar{y} = h_1^z h_2^x \bmod p_0$
	2. pick $r_0 <_R 2^\ell$ and $r_1, r_2 <_R n$
	3. compute $e = H\left(g^{r_0} \bmod n^2, h_1^{r_1} h_2^{r_2} \bmod p_0\right)$
	4. compute $\begin{cases} s_0 = r_0 + ex \\ s_1 = r_1 + ez \ \bmod n \\ s_2 = r_2 + ex \ \bmod n \end{cases}$
Output	the proof is $P = (\bar{y}, e, s_0, s_1, s_2)$.
Verification $\mathcal{V}$	
Input	$y < n^2$ and $P = (\bar{y}, e, s_0, s_1, s_2)$.
	check whether $e \overset{?}{=} H\left(g^{s_0}/y^e \bmod n^2, h_1^{s_1} h_2^{s_2}/\bar{y}^e \bmod p_0\right)$
Output	0 or 1 according to the verification result.

Fig. 5. Generation and Verification of a Proof of Recoverability.

Note that previous works on range-bounded commitments such as [2] could also lead to a suitable pair of algorithms for $\mathcal{P}$ and $\mathcal{V}$. □
Combining our self-escrowed encryption scheme with the specifications of section 3 now gives us a self-escrowed public key infrastructure. We sum up hereafter the main features and properties that characterize our cryptosystem.

4.2 Security Aspects

One-Wayness. The problem of inverting our self-escrowed encryption function is equivalent to the Diffie–Hellman problem with base g over $\mathbb{Z}_{n^2}^*$.

Semantic Security IND-CPA. The semantic security of the scheme is equivalent to the Decision Diffie–Hellman problem with base g over $\mathbb{Z}_{n^2}^*$.

Extraction of the Private Key. The problem of computing the private key x from the public key $y = g^x \bmod n^2$ i.e. inverting the master encryption scheme is equivalent to the discrete log problem with base g over $\mathbb{Z}_{n^2}^*$.

Extraction of the Master Private Key. Since the master private key is λ or equivalently the prime factors of n, computing the master key exactly means factoring n.

4.3 Efficiency

Each step of our SE-PKI involves a few modular exponentiations at most for each party. We will provide a further analysis of computational workloads in the final version of the paper.

4.4 Other Cryptographic Features

Soundness. Key recovery can be successfully performed over any certified user key with overwhelming probability : this is ensured by definition of algorithms $\mathcal{P}$ and $\mathcal{V}$.

Traceability. We conclude on a technical remark illustrating that our cryptosystem allows user traceability under certain circumstances : suppose that some known user, say user A, is suspected to regularly encrypt illegal documents that he or she sends through a global computer system to an unknown recipient user B. Further assume that police forces' action priority is to identify user B. Of course, tracing the transmitted data packets themselves to locate the recipient is assumed to be unfeasible (mixed network f.i). Police could ask escrow authorities to recover user A's private key but this would be of no help. The contents of the transmitted information are known (this is a typical sensitive information thieving scenario). What could be done?

Suppose you catch a single ciphertext $c = (my^k, g^k)$ (by wiretapping the sender). Since m is known (or guessable), you easily compute (y^k, g^k), query an escrow authority for disclosing the exponents and get the two values $xk \bmod n$ and $k \bmod n$. You then infer $x = xk/k \bmod n$ which leads to the recipient's public key $y = g^x \bmod n^2$. Finally, querying the public-key database will give user B's ID.

4.5 Open Research

A typical research topic would be to control private key generation in such a way that the overall cryptosystem achieves shadow-public-key resistance; we believe this property can be achieved at moderate cost. Also, it would be of interest to investigate how our proposal fits the notion of escrow hierarchy introduced by Young and Yung in [16].

Acknowledgements

The authors are grateful to Marc Joye for constructive comments and suggestions. We also thank the anonymous referees for useful comments.

References

1. J. Camenisch and M. Michels. *Separability and Efficiency for Generic Group Signature Schemes.* In Advances in Cryptology, Crypto'99, LNCS 1666, pp. 413–430, Springer Verlag, 1999.

2. A. Chan, Y. Frankel and Y. Tsiounis. *Easy Come-Easy Go Divisible Cash.* In Advances in Cryptology, Eurocrypt'98, LNCS 1403, pp. 561–575, Springer Verlag, 1998.

3. W. Diffie and M. Hellman. *New Directions in Cryptography.* In IEEE Transactions on Information Theory, Vol. IT-22, No. 6, pp. 644–654, 1976.

4. A. Fiat and A. Shamir. *How to Prove Yourself: Practical Solution to Identification and Signature Problems.* In Advances in Cryptology, Crypto'86, LNCS 263, pp. 186–194, Springer Verlag, 1986.

5. Y. Frankel and M. Yung. *Escrow Encryption Systems Visited: Attacks, Analysis and Designs.* In Advances in Cryptology, Crypto'95, LNCS 963, pp. 222–235, Springer Verlag, 1995.

6. E. Fujisaki and T. Okamoto. *Statistical Zero-knowledge Protocols to Prove Modular Polynomial Relations.* In Advances in Cryptology, Crypto'97, LNCS 1294, pp. 16–30, Springer Verlag, 1997.

7. T. ElGamal *A Public Key Cryptosystem and a Signature Scheme Based on Discrete Logarithms.* In IEEE Transactions on Information Theory, Vol. IT-31, No. 4, pp. 469–472, 1985.

8. J. Killian and F. T. Leighton. *Fair Cryptosystems, Revisited - A Rigorous Approach to Key Escrow.* In Advances in Cryptology, Crypto'95, LNCS 963, pp. 208–221, Springer Verlag, 1995.

9. L. R. Knudsen, M. J. B. Robshaw and D. Wagner. *Truncated Differentials and Skipjack.* In Advances in Cryptology, Crypto'99, LNCS 1666, pp. 165–180, Springer Verlag, 1999.

10. P. Paillier. *Public-Key Cryptosystems Based on Composite-Degree Residuosity Classes.* In Advances in Cryptology, Eurocrypt '99, LNCS 1592, pp. 223–238, Springer Verlag, 1999.

11. M. Stadler. *Publicly Verifiable Secret Sharing.* In Advances in Cryptology, Eurocrypt '96, LNCS 1070, pp. 190–199, Springer Verlag, 1996.

12. A. Young and M. Yung. *The Dark Side of Black-Box Cryptography.* In Advances in Cryptology, Crypto'96, LNCS 1109, pp. 89–103, Springer Verlag, 1996.

13. A. Young and M. Yung. *Kleptography: Using Cryptography against Cryptography.* In Advances in Cryptology, Eurocrypt'97, LNCS 1233, pp. 62–74, Springer Verlag, 1997.

14. A. Young and M. Yung. *The Prevalence of Kleptographic Attacks on Discrete-Log Based Cryptosystems.* In Advances in Cryptology, Crypto'97, LNCS 1294, pp. 264–276, Springer Verlag, 1997.

15. A. Young and M. Yung. *Auto-Recoverable Auto-Certifiable Cryptosystems.* In Advances in Cryptology, Eurocrypt'98, LNCS 1403, pp. 17–31, Springer Verlag, 1998.

16. A. Young and M. Yung. *Auto-Recoverable Cryptosystems with Faster Initialization and the Escrow Hierarchy.* PKC'99, LNCS 1560, pp. 306–314, Springer Verlag, 1999.

Electronic Funds Transfer Protocol Using Domain-Verifiable Signcryption Scheme

Moonseog Seo and Kwangjo Kim

ICU, 58-4 Hwaam-dong, Yusong-gu
Taejon, 305-350, Korea
{msseo,kkj}@icu.ac.kr

Abstract. In this paper, we propose Domain-verifiable signcryption scheme, which is applied to the Electronic Funds Transfer(EFT) protocol, that only predetermined n participants within the domain of protocol participants can decrypt their own part of message and verify whole transaction. The computational cost of our scheme is as low as that of Zheng's scheme assuming that Trusted Third Party(TTP) must be used to keep partial information for participants confidential and multiverification. Our scheme does not require the role of TTP.

1 Introduction

The Electronic Funds Transfer(EFT) protocol is most widely used for transfering money between the financial institutions. The protocol requires both confidentiality and authentication services simultaneously. Efficiency is a factor that must be fulfilled in financial systems. The efficiency is achieved by applying signcryption scheme to EFT protocol. The Signcryption[9], which is first proposed by Zheng, is a new cryptographic primitive called *"catch two birds with single stone"* scheme. This simultaneously fulfills both the functions of signature and encryption in a single logical step, and reduces computational cost which is significantly lower than that required by the traditional signature-then-encryption paradigm [3, 6, 9].

In application to EFT protocol in multiple participants environment, the signcryption scheme needs modification so that only predetermined n participants within a domain can decrypt their own part of message and verify whole transaction. We call this modified signcryption scheme Domain-verifiable signcryption scheme where domain means a set of participants involved in a transaction protocol. In Zheng's signcryption scheme, the unsigncryption (decryption and signature verification) needs the recipient's private key; therefore, only the recipient can verify the signature. So, Zheng's signcryption schemes have some constraints to be used in applications where a signature needs to be validated by any others. To overcome this problem, Bao and Deng[1] modified Zheng's signcryption scheme such that verification of a signature no longer needs the recipient's private key. However, Bao and Deng's scheme is not as efficient computationally as Zheng's scheme. Also in their scheme, the message must be decrypted before it is verified by other people ending up losing confidentiality. To

JooSeok Song (Ed.): ICISC'99, LNCS 1787, pp. 269–277, 2000.
© Springer-Verlag Berlin Heidelberg 2000

maintain the confidentiality and also to be used in firewall application, Gamage, Leiwo and Zheng[4] proposed the signcryption for third-party verification. But in this scheme, whereas any verifier can verify the signature, only one person can obtain the whole plaintext message.

In EFT protocol usage, there exist many participants for one transaction. A transaction consists of secret information to be processed by each participant. Each participant requires confidentiality for his own secret information. Also all participants need authentication of that whole transaction. Signcryption schemes[1, 4, 9] proposed previously cannot be directly used in this situation.

In this paper, we propose Domain-verifiable signcryption scheme based on Gamage, Leiwo and Zheng's signcryption that can be easily applicable to the EFT and Secure Electronic Transaction(SET) protocol[7] that many participants within domain can keep their own part of message confidentially and verify the whole transaction. Also we sketch EFT protocol between two banks using the Domain-verifiable signcryption. The computational cost of our scheme is as low as Zheng's scheme with assuming that Trusted Third-Party(TTP) must be used for keeping partial information for participants confidential and multi-verification. When we use Domain-verifiable signcryption, we can construct EFT protocol without interaction of TTP.

The rest of the paper is organized as follows. The signcryption schemes proposed until now are described briefly in Section 2. The proposed scheme for domain-verification is discussed in Section 3. Section 4 provides the application of our scheme with financial EFT protocol. Finally concluding remarks are given in Section 5.

2 Related Work

We describe three signcryption schemes proposed until now. The original signcryption primitive proposed in [9] by Zheng combines the sign-then-encrypt two step process to create a secure authenticated message into a single logical step with significant savings in both computational and transmission costs. A disadvantage for some applications such as EFT protocol in which more than two participants involved is that only the intended recipient can verify the message. A modified signcryption scheme was proposed in [1] by Bao and Deng to overcome this limitation. But it has the increased computational cost while still preserving the transmission cost savings achieved by the original scheme. Two disadvantages of this modified signcryption scheme are:

- The signature verification-only mode of operation can be used only after the original recipient has recovered the plaintext message.
- The plaintext message must be forwarded to a third party for signature verification and the message confidentiality can be lost.

In [4], Camage, Leiwo and Zheng modified Bao and Deng's scheme to carry out signature verification without accessing the plaintext for preserving confidentiality of the original message without altering sign-then-encrypt paradigm. But in this scheme, whereas any verifier can verify the signature, only one person to unsigncrypt signcrypted message can obtain the whole plaintext message.

Therefore, these all schemes could not be applied directly for EFT protocol which transaction consists of partial information for each participant that requires confidentiality about his own information even against other protocol participants.

2.1 Zheng's Scheme

Task: Alice has a message to send to Bob. Alice signcrypts it so that the effect is similar to signature-then-encryption.

Public Parameters:
p: a large prime
q: a large prime factor of $p-1$
g: an element of Z_p^* of order q
$hash$: a one-way hash function
KH: a keyed one-way hash function
(E, D): the encryption and decryption algorithms of a symmetric key cipher
Alice's Key:
$x_a \in Z_q^*$: Alice's private key, $y_a = g^{x_a} \bmod p$: Alice's public key
Bob's Keys:
$x_b \in Z_q^*$: Bob's private key, $y_b = g^{x_b} \bmod p$: Bob's public key

Signcrypting: Alice randomly chooses $x \in Z_q^*$ then sets
$(k_1, k_2) = hash(y_b{}^x \bmod p)$
$c = E_{k_1}(m)$
$r = KH_{k_2}(m)$
$s = x/(r + x_a) \bmod q$.
Alice sends (c, r, s) to Bob.
Unsigncrypting: Bob computes
$(k_1, k_2) = hash((y_a g^r)^{s x_b} \bmod p)$,
$m = D_{k_1}(c)$ to recover the plaintext message, and then checks whether $KH_{k_2}(m) = r$ for signature verification. In unsigncrypting process, it is straightforward to see that x_b is involved for signature verification.

2.2 Bao and Deng's Scheme

Signcrypting: Alice randomly chooses $x \in Z_q^*$ then sets
$k_1 = hash(y_b{}^x \bmod p)$
$k = hash(g^x \bmod p)$
$c = E_{k_1}(m)$
$r = KH_k(m)$
$s = x/(r + x_a) \bmod q$.
Alice sends (c, r, s) to Bob.

Unsigncrypting: Bob computes
$t_1 = (y_a g^r)^s \mod p$
$t_2 = t_1^{x_b} \mod p$
$k_1 = hash(t_2)$
$k = hash(t_1)$,
$m = D_{k_1}(c)$ to obtain the plaintext message, then checks whether $KH_k(m) = r$
for signature verification.

Later when necessary, Bob may forward (m, r, s) to others, who can be convinced that it came originally from Alice by verifying $k = hash((y_a g^r)^s \mod p)$
and $r = KH_k(m)$.
In this signature verification, verifiers require to get the plaintext message.

2.3   Gamage, Leiwo and Zheng's Signcryption for Third-Party Verification

Signcrypting: Alice randomly chooses $x \in Z_q^*$ then sets
$k = hash(y_b{}^x \mod p)$
$y = g^x \mod p$
$c = E_k(m)$
$r = hash(y, c)$
$s = x/(r + x_a) \mod q$.
Alice sends (c, r, s) to Bob.

Unsigncrypting: Bob will compute from (c, r, s)
$y = (y_a g^r)^s \mod p$
$k = hash(y^{x_b} \mod p)$,
$m = D_k(c)$ to obtain the plaintext message.
Bob accepts signature if and only if $hash(y, c) = r$.

For partial unsigncryption with signature verification-only, any verifier will
compute from (c, r, s) and $y = (y_a g^r)^s \mod p$.
Any verifier accepts signature if and only if $hash(y, c) = r$.
This signature verification does not require access to the plaintext message.

3   Domain-Verifiable Signcryption Scheme

Within domain of protocol participants, each participant wants to be maintained
his own message included in transaction secretly even against any other participants. Also, all participants require to authenticate the transaction that consists
of participants secret partial information. We construct the Domain-verifiable
signcryption scheme that satisfys these requirements. Each participant can decrypt just his own message and all participants can verify the whole transaction.
This scheme could be applied to EFT protocol as well as any other protocols
like SET protocol that need to be kept participant's partial information secret
and to be authenticated total message by all participants simultaneously.

3.1 Scheme for Domain Verification

For consistency, we use the same notations as in Zheng's scheme except recipients' key.

Recipient B_i's Keys within Domain of n Participants ($i \in \{1, \ldots, n\}$)
$x_{b_i} \in Z_q^*$: B_i's private key
$y_{b_i} = g^{x_{b_i}} \bmod p$: B_i's public key

Signcrypting: Alice randomly chooses $x \in Z_q^*$ then sets
$k_1 = hash(y_{b_1}^x \bmod p), k_2 = hash(y_{b_2}^x \bmod p), \ldots, k_n = hash(y_{b_n}^x \bmod p)$
$k = hash(g^x \bmod p)$
$c_1 = E_{k_1}(m_1), c_2 = E_{k_2}(m_2), \ldots, c_n = E_{k_n}(m_n)$
$r_1 = KH_k(m_1||c_2|| \cdots ||c_n), r_2 = KH_k(c_1||m_2|| \cdots ||c_n), \ldots,$
$r_n = KH_k(c_1||c_2|| \cdots ||m_n)$
$s = x/(r_1 r_2 \cdots r_n + x_a) \bmod q.$
Alice sends $(c_1, c_2, \ldots, c_n, r_1, r_2, \ldots, r_n, s)$ to B_n.

Unsigncrypting: Recipient B_i computes
$t = (y_a g^{r_1 r_2 \cdots r_n})^s \bmod p$
$t_i = t^{x_{b_i}} \bmod p$
$k = hash(t)$
$k_i = hash(t_i),$
$m_i = D_{k_i}(c_i)$ to obtain B_i's own plaintext message, then checks whether
$KH_k(c_1|| \cdots ||m_i|| \cdots ||c_n) = r_i$ for signature verification.

Later when necessary, B_i may forward $(c_1, c_2, \ldots, c_n, r_1, r_2, \ldots, r_n, s)$ to any other participants, who want to decrypt his own message and can be convinced that it came originally from Alice by executing through this unsigncrypting.

3.2 Performance and Security

We should consider a situation where Domain-verifiable signcryption scheme must be used. If we use Zheng's scheme, TTP must be involved to divide message into partial messages for each participant and signcrypt the partial message for the corresponding participant[10]. But our Domain-verifiable signcryption scheme does not need TTP. While considering only exponentiation cost as the computational cost and n participants, Domain-verifiable signcryption requires $n + 1$ modulo exponentiations for signcryption and $3n$ modulo exponentiations for unsigncryption. In the general case of n participants more than 2 or 3 participants are involved, the communication bandwidth of our scheme is not lower than that of the Zheng's scheme, since the whole transaction message for n participants must be always transferred.

It can be done only within domain of protocol participants to unsigncrypt message, since a participant B_i having his own secret key x_{b_i} within a domain can obtain partial information m_i and the only person who gets m_i can try to check if $KH_k(c_1||\cdots||m_i||\cdots||c_n) = r_i$ for signature verification. Any other persons that have not secret x_{b_i} will not be able to take part in unsigncryption.

In [4], they not only provide the formal proof based on the random oracle model about the security argument about the computation of two values, $y_b{}^x \bmod p$ and $g^x \bmod p$ using the same secret x, but also show the pseudo-independence of two computed values as an adequate guarantee of security for the signature scheme. Namely, if a signer chooses the integer x uniformly and randomly, then two values are (pseudo) independent as both g and $y_b = g^{x_b} \bmod p$ are generators in Z_p^* of order q which is a prime. This ensures that the signature verification and partial recovery of bits does not leak information that can be used in an attack on breaking message confidentiality or signature forgery. We can consider to apply this method to our scheme. According to this security analysis, if a signer chooses the integer x uniformly and randomly, then $n + 1$ values such as $y_{b_1}{}^x \bmod p, \ldots, y_{b_n}{}^x \bmod p$ and $g^x \bmod p$ in Domain-verifiable signcryption are (pseudo) independent as $g, y_{b_1} = g^{x_{b_1}} \bmod p, \ldots, y_{b_n} = g^{x_{b_n}} \bmod p$ are generators Z_p^* of order q which is a prime. This guarantees that Domain-verifiable signcryption scheme has message confidentiality and signature unforgeability.

4   EFT Protocol Based on Domain-Verifiable Signcryption

EFT is considered to be any transfer of funds, other than a transaction by check, draft, or similar paper instrument, that is initiated through an electronic terminal, telephone, computer or magnetic tape for the purpose of ordering, instructing, or authorizing a financial institution to debit or credit an account. In the inter-bank EFT protocol, withdrawal accounts and deposit accounts are placed in different banks. A client should request EFT transaction to the bank that has business relations with him. The bank that receives the request draws the corresponding money from the requester's account and asks the deposit bank to deposit the same amount of money to recipient's account. The withdrawal bank that receives the result of deposit from a deposit bank informs the client who requests the EFT transaction of the final result of the funds transfer [2].

The message that clients send to the withdrawal bank will be constituted of client's information such as his own account number and PIN (Personal Identification Number), and recipient's information such as deposit bank name, deposit account number and amount of money to be transferred, etc. This message has to be encrypted and signed for privacy and integrity. In detail, client's information is encrypted for withdrawal bank and recipient's information is encrypted for deposit bank. Also the transaction for EFT protocol has to be authenticated by both withdrawal and deposit banks.

To use signcryption scheme at the inter-bank EFT protocol, we need TTP when using Zheng's scheme. But when using Domain-verifiable signcryption, we don't need TTP as shown in Fig. 1.

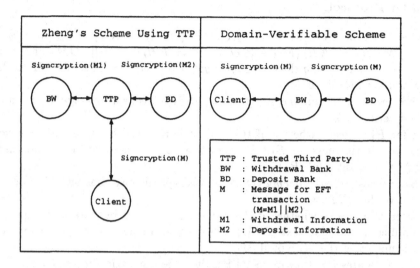

Fig. 1. EFT protocol using the signcryption schemes

4.1 Inter-Bank EFT Protocol

We use the following notations to describe this protocol.

Participants and Tools
Client: A
Withdrawal Bank: BW
Deposit Bank: BD
$Signcrypt_A(\bullet)$: Domain-verifiable signcryption by client A including signature-only mode[9]
$Unsigncrypt_A(\bullet)$: Domain-verifiable unsigncryption by client A
$Sign_A(\bullet)$: signature-only mode of signcryption by client A
$\|$: message concatenation
$hash(\bullet)$: hash algorithm

Preparation
Creation of funds transfer information: $M = M_1\|M_2\|COM$

- M_1: Client A's information such as withdrawal account number and PIN, encrypted for withdrawal bank
- M_2: Deposit information such as deposit bank and deposit account number, encrypted for deposit bank
- COM: Common data for EFT such as amount of money to be transferred, date, sequence number and recipient's name, *etc.* This data should be maintained as plaintext for the transaction processing.

Transfer Protocol

1. Client A generates $SM = (c_1, c_2, COM, r_1, r_2, s)$ where $c_1 = E_{k_1}(M_1), c_2 = E_{k_2}(M_2), r_1 = KH_k(M_1||c_2||COM), r_2 = KH_k(c_1||M_2||COM)$ and $s = x/(r_1 r_2 + x_A) \bmod q$ through $Signcrypt_A(M)$ and then A sends SM to the withdrawal Bank, BW.
2. BW processes $Unsigncrypt_A(SM)$ to decrypt his own message M_1 from c_1 and verifies SM.
3. After BW checks whether if the request is replayed by date and sequence number in the message, BW draws money from A's account in M_1 and sends SM to deposit bank BD.
4. BD processes $Unsigncrypt_A(SM)$ to decrypt his own message M_2 from c_2 and verifies SM.
5. After BD checks whether if the request is replayed by date and sequence number in the plaintext COM, BD deposits money to the corresponding account using the decrypted M_2.
6. BD generates $r = Sign_{BD}(SM||\text{Result of Deposit})$ and then sends (Result of Deposit, r) to the BW.
7. BW does the necessary job according to the result of deposit that received from BD and generates $\hat{r} = Sign_{BW}(\text{Result of Transfer})$. And then BW sends (Result of Transfer, $\hat{r}$) to client A.
8. Client A can use the received (Result of Transfer, $\hat{r}$) as receipt for counterpart of transfer.

4.2   Security Consideration

The security of the inter-bank EFT protocol based on Domain-verifiable signcryption is summarized as below.

- Confidentiality: An adversary cannot recover the message M that transferred between a client and the banks because that message is encrypted for the corresponding bank before the transmission. Specially PIN in M_1, client's secret information for making withdrawal is not compromised by any others except only withdrawal bank.
- Authentication and Integrity: To send a fund transfer message, a client must sign on that message using his own private key. The banks that received a fund transfer message can authenticate the client who sends that message using the private key for the client. Also the banks can determine the integrity of the received message, since that message is signed by the client.
- Non-repudiation: A client's signature on the fund transfer message for a transaction can be used for the evidence[5, 8] of an user's request for EFT.
- Replay attack: If an adversary tries to replay the protocol, the bank can detect the message replayed by checking whether if the date and sequence number in the message are duplicated with the message that already has received.

- Usage as receipt: The result of a transfer along with signature from the bank can be used as a receipt for the result of funds transfer to the recipient. The recipient can verify the receipt that received from requester of funds transfer using the bank's public key.

5   Concluding Remarks

We proposed Domain-verifiable signcryption scheme applicable to the situation that only predetermined n participants can decrypt and verify within a domain. This scheme is useful when each participant can decrypt his own message that is partial information of the whole transaction message and all participants can verify the whole transaction message.

As an example application, we designed inter-bank EFT protocol based on the Domain-verifiable signcyption scheme. We found that this inter-bank EFT protocol is so efficient that it can be used at the real world. The detailed designs of multi-level hierarchical key distribution or SET protocol based on our Domain-verifiable signcryption need further research.

Acknowledgments

The authors would like to express great thanks to Pil Joong Lee, Kunsoo Park and Sangjae Moon for their useful discussions on our manuscript.

References

1. F.Bao and R.H.Deng, "A signcryption scheme with signature directly verifiable by public key,"Proc. of PKC'98, LNCS, Vol. 1431, Springer-Verlag, pp. 55-59, 1998.
2. Electronic Funds Transfer Act (EFTA), 15 U.S.C. Sec. 1693.
3. T.ElGamal, "A public key cryptosystem and a signature scheme based on discrete logarithms,"IEEE Transactions on Information Theory, IT-31(4):469-472, 1985.
4. C.Gamage, J.Leiwo, and Y.Zheng, "Encrypted message authentication by firewalls,"Proc. of PKC'99, LNCS, Vol. 1560, Springer-Verlag, pp. 69-81, 1999.
5. K. Kim, S. Park, and J. Baek, "Improving fairness and privacy of Zhou-Gollmann's fair non-repudiation protocol,"Proc. of 1999 ICPP Workshops on Security(IWSEC), pp. 140-145, IEEE Computer Society, Sep. 21-22, 1999
6. C.P. Schnorr, "Efficient identification and signature for smart cards,"Advances in Cryptology- CRYPTO '89, LNCS 435, Springer-Verlag, pp. 239-251,1989.
7. Visa International and MasterCard International, Secure Electronic Transaction(SET) Specification book 1:Business Description, May 1997.
8. J. Zhou and D.Gollmann, "Observation on non-repudiation,"Advances in Cryptology- ASIACRYPT'96, LNCS 1163, Springer-Verlag, pp. 133-144, 1996.
9. Y. Zheng, "Digital signcryption or how to achieve cost(signature and encryption) << cost(signature)+cost(encryption),"Advances Cryptology-CRYPTO'97, LNCS 1294, Springer- Verlag, pp. 165-179, 1997.
10. Y. Zheng, "Signcryption and its application in efficient public key solutions,"Proc. of Information Security Workshop(ISW'97), LNCS, Vol. 1396, Springer-Verlag, pp. 291-312, 1998.

Author Index

Lecture Notes in Computer Science

For information about Vols. 1–1719
please contact your bookseller or Springer-Verlag

Vol. 1752: S. Krakowiak, S. Shrivastava (Eds.), Advances in Distributed Systems. VIII, 509 pages. 2000.

Vol. 1753: E. Pontelli, V. Santos Costa (Eds.), Practical Aspects of Declarative Languages. Proceedings, 2000. X, 327 pages. 2000.

Vol. 1754: J. Väänänen (Ed.), Generalized Quantifiers and Computation. Proceedings, 1997. VII, 139 pages. 1999.

Vol. 1755: D. Bjørner, M. Broy, A.V. Zamulin (Eds.), Perspectives of System Informatics. Proceedings, 1999. XII, 540 pages. 2000.

Vol. 1757: N.R. Jennings, Y. Lespérance (Eds.), Intelligent Agents VI. Proceedings, 1999. XII, 380 pages. 2000. (Subseries LNAI).

Vol. 1758: H. Heys, C. Adams (Eds.), Selected Areas in Cryptography. Proceedings, 1999. VIII, 243 pages. 2000.

Vol. 1759: M.J. Zaki, C.-T. Ho (Eds.), Large-Scale Parallel Data Mining. VIII, 261 pages. 2000. (Subseries LNAI).

Vol. 1760: J.-J. Ch. Meyer, P.-Y. Schobbens (Eds.), Formal Models of Agents. Poceedings. VIII, 253 pages. 1999. (Subseries LNAI).

Vol. 1761: R. Caferra, G. Salzer (Eds.), Automated Deduction in Classical and Non-Classical Logics. Proceedings. VIII, 299 pages. 2000. (Subseries LNAI).

Vol. 1762: K.-D. Schewe, B. Thalheim (Eds.), Foundations of Information and Knowledge Systems. Proceedings, 2000. X, 305 pages. 2000.

Vol. 1763: J. Akiyama, M. Kano, M. Urabe (Eds.), Discrete and Computational Geometry. Proceedings, 1998. VIII, 333 pages. 2000.

Vol. 1764: H. Ehrig, G. Engels, H.-J. Kreowski, G. Rozenberg (Eds.), Theory and Application of Graph Transformations. Proceedings, 1998. IX, 490 pages. 2000.

Vol. 1765: T. Ishida, K. Isbister (Eds.), Digital Cities. IX, 444 pages. 2000.

Vol. 1767: G. Bongiovanni, G. Gambosi, R. Petreschi (Eds.), Algorithms and Complexity. Proceedings, 2000. VIII, 317 pages. 2000.

Vol. 1768: A. Pfitzmann (Ed.), Information Hiding. Proceedings, 1999. IX, 492 pages. 2000.

Vol. 1769: G. Haring, C. Lindemann, M. Reiser (Eds.), Performance Evaluation: Origins and Directions. X, 529 pages. 2000.

Vol. 1770: H. Reichel, S. Tison (Eds.), STACS 2000. Proceedings, 2000. XIV, 662 pages. 2000.

Vol. 1771: P. Lambrix, Part-Whole Reasoning in an Object-Centered Framework. XII, 195 pages. 2000. (Subseries LNAI).

Vol. 1772: M. Beetz, Concurrent Ractive Plans. XVI, 213 pages. 2000. (Subseries LNAI).

Vol. 1773: G. Saake, K. Schwarz, C. Türker (Eds.), Transactions and Database Dynamics. Proceedings, 1999. VIII, 247 pages. 2000.

Vol. 1774: J. Delgado, G.D. Stamoulis, A. Mullery, D. Prevedourou, K. Start (Eds.), Telecommunications and IT Convergence Towards Service E-volution. Proceedings, 2000. XIII, 350 pages. 2000.

Vol. 1776: G.H. Gonnet, D. Panario, A. Viola (Eds.), LATIN 2000: Theoretical Informatics. Proceedings, 2000. XIV, 484 pages. 2000.

Vol. 1777: C. Zaniolo, P.C. Lockemann, M.H. Scholl, T. Grust (Eds.), Advances in Database Technology – EDBT 2000. Proceedings, 2000. XII, 540 pages. 2000.

Vol. 1778: S. Wermter, R. Sun (Eds.), Hybrid Neural Systems. IX, 403 pages. 2000. (Subseries LNAI).

Vol. 1780: R. Conradi (Ed.), Software Process Technology. Proceedings, 2000. IX, 249 pages. 2000.

Vol. 1781: D.A. Watt (Ed.), Compiler Construction. Proceedings, 2000. X, 295 pages. 2000.

Vol. 1782: G. Smolka (Ed.), Programming Languages and Systems. Proceedings, 2000. XIII, 429 pages. 2000.

Vol. 1783: T. Maibaum (Ed.), Fundamental Approaches to Software Engineering. Proceedings, 2000. XIII, 375 pages. 2000.

Vol. 1784: J. Tiuryn (Ed.), Foundations of Software Science and Computation Structures. Proceedings, 2000. X, 391 pages. 2000.

Vol. 1785: S. Graf, M. Schwartzbach (Eds.), Tools and Algorithms for the Construction and Analysis of Systems. Proceedings, 2000. XIV, 552 pages. 2000.

Vol. 1786: B.H. Haverkort, H.C. Bohnenkamp, C.U. Smith (Eds.), Computer Performance Evaluation. Proceedings, 2000. XIV, 383 pages. 2000.

Vol. 1787: J. Song (Ed.), Information Security and Cryptology – ICISC'99. Proceedings, 1999. XI, 279 pages. 2000.

Vol. 1790: N. Lynch, B.H. Krogh (Eds.), Hybrid Systems: Computation and Control. Proceedings, 2000. XII, 465 pages. 2000.

Vol. 1792: E. Lamma, P. Mello (Eds.), AI*IA 99: Advances in Artificial Intelligence. Proceedings, 1999. XI, 392 pages. 2000. (Subseries LNAI).

Vol. 1793: O. Cairo, L.E. Sucar, F.J. Cantu (Eds.), MICAI 2000: Advances in Artificial Intelligence. Proceedings, 2000. XIV, 750 pages. 2000. (Subseries LNAI).

Vol. 1795: J. Sventek, G. Coulson (Eds.), Middleware 2000. Proceedings, 2000. XI, 436 pages. 2000.

Vol. 1794: H. Kirchner, C. Ringeissen (Eds.), Frontiers of Combining Systems. Proceedings, 2000. X, 291 pages. 2000. (Subseries LNAI).

Vol. 1796: B. Christianson, B. Crispo, J.A. Malcolm, M. Roe (Eds.), Security Protocols. Proceedings, 1999. XII, 229 pages. 2000.

Vol. 1800: J. Rolim et al. (Eds.), Parallel and Distributed Processing. Proceedings, 2000. XXIII, 1311 pages. 2000.

Vol. 1801: J. Miller, A. Thompson, P. Thomson, T.C. Fogarty (Eds.), Evolvable Systems: From Biology to Hardware. Proceedings, 2000. X, 286 pages. 2000.

Vol. 1802: R. Poli, W. Banzhaf, W.B. Langdon, J. Miller, P. Nordin, T.C. Fogarty (Eds.), Genetic Programming. Proceedings, 2000. X, 361 pages. 2000.

Vol. 1803: S. Cagnoni et al. (Eds.), Real-World Applications and Evolutionary Computing. Proceedings, 2000. XII, 396 pages. 2000.

Vol. 1805: T. Terano, H. Liu, A.L.P. Chen (Eds.), Knowledge Discovery and Data Mining. Proceedings, 2000. XIV, 460 pages. 2000. (Subseries LNAI).